CUSHING'S SYNDROME

CUSHING'S SYNDROME

edited by

Lewis S. Blevins Jr., M.D.
Associate Professor of Medicine & Neurological Surgery
The Pituitary Center at Vanderbilt
Vanderbilt University School of Medicine
Nashville, Tennessee USA

KLUWER ACADEMIC PUBLISHERS
Boston / Dordrecht / London

Distributors for North, Central and South America:
Kluwer Academic Publishers
101 Philip Drive
Assinippi Park
Norwell, Massachusetts 02061 USA
Telephone (781) 871-6600
Fax (781) 681-9045
E-Mail: kluwer@wkap.com

Distributors for all other countries:
Kluwer Academic Publishers Group
Post Office Box 322
3300 AH Dordrecht, THE NETHERLANDS
Telephone 31 786 576 000
Fax 31 786 576 474
E-Mail: services@wkap.nl

Electronic Services <http://www.wkap.nl>

Library of Congress Cataloging-in-Publication Data

A C.I.P. Catalogue record for this book is available
from the Library of Congress. ISBN 1-4020-7131-0

Cushing's Syndrome by Lewis S. Blevins, Jr., MD

Printed on acid-free paper.

Printed in the United States of America.

TABLE OF CONTENTS

LIST OF CONTRIBUTORS

Yutaka Oki, M.D.
Hamamatsu University School of Medicine, Japan

André Lacroix, M.D., Johanne Tremblay, M.D. and Pavel Hamet, M.D.
Hôtel-Dieu du Centre hospitalier de l'Université de Montréal, Montréal, Québec, Canada

Wendell Nicholson, BS; Robert J. Workman, MD
Vanderbilt School of Medicine, Nashville, TN

Robert J. Weil, M.D. and George S. Allen, M.D., PhD.
Department of Neurosurgery, Vanderbilt University School of Medicine, Nashville, TN, USA

John Newell-Price, M.D. and Mike Besser, M.D.
Northern General Hospital, Sheffield, UK and St Bartholomew's Hospital, London, UK.

T. Joseph McKenna, M.D., FRCPI
St. Vincent's University Hospital, University College, Dublin.

Peter T. Loosen, M.D., Ph.D.
Department of Psychiatry, Vanderbilt School of Medicine, Nashville, TN, USA

A Brew Atkinson, M.D.
Royal Victoria Hospital and Queen's University, Belfast, Northern Ireland

William J. Kovacs, M.D.
Division of Endocrinology, Diabetes and Metabolism, Department of Medicine, Vanderbilt School of Medicine, Nashville, TN, USA

Tom Kason, M.D. and Lewis S. Blevins, Jr., M.D.
Division of Endocrinology, Diabetes and Metabolism, Department of Medicine, Vanderbilt School of Medicine, Nashville, TN, USA

Richard E. Goldstein, M.D., Ph.D.
Division of Surgical Oncology, Department of Surgery, Vanderbilt University School of Medicine, Nashville, TN

Michael Fowler M.D. and Lewis S. Blevins, Jr., M.D.
Division of Endocrinology, Diabetes & Metabolism, Department of Medicine, Vanderbilt School of Medicine, Nashville, TN, USA

Anthony J. Cmelak, M.D.
Department of Radiation Oncology, Vanderbilt School of Medicine, Nashville, TN

Lewis S. Blevins, Jr., M.D.
Division of Endocrinology, Diabetes and Metabolism, Department of Medicine, Vanderbilt School of Medicine, Nashville, TN, USA

Marie-Laure Raffin-Sanson, M.D. (1,3)
Yves de Keyzer, M.D. (3)
Xavier Bertagna, M.D. (2,3)
(1) Service de Médecine Interne 2, Hôpital Ambroise Paré, Boulogne/seine, France
(2) Clinique des Maladies Endocriniennes, Hôpital Cochin,
(3) Institut Cochin de Génétique Moléculaire, Faculté Cochin, Université Paris V – René Descartes, Paris, France.

Preface

I have been asked on more than one occasion about the necessity of putting together a text on Cushing's syndrome in the electronic age. There is no doubt that information about patients with Cushing's syndrome is readily available via internet-based literature searches, UpToDate®, and access to other electronic media. I do believe, however, that people still do possess, and always will, a love for books. Books are real. They can be held and savored in the office, in the home by a fire, or in the library. They do not disappear when shelved and are usually there for ready retrieval when needed unless loaned or taken by an avid interested reader. On the shelf or in our hands, books are material evidence of our current knowledge, new ideas, and possible future directions. My dear friend, James H. Christy, Professor Emeritus of Internal Medicine at Emory University in Atlanta, Georgia, deserves credit for the idea to assemble this book. Jim and I were colleagues for nearly 5 years at Emory. I will always cherish the many memorable and enlightening discussions we had pertaining to our own patients with Cushing's syndrome, relevant literature, and research ideas. It was he who arranged a meeting with the representatives of Kluwer Publishers to initiate this project and, thus, it would not have been possible without his kind intercessions. I am also indebted to and grateful for the numerous contributors who have penned chapters to transform the vision of this book into a real object that you can hold in your own hands. The contributors represent a portion of the Diaspora of Vanderbilt endocrinology. Many of them were trained by one of two giants in the field, either Dr. Grant Liddle or Dr. David Orth, and have each made seminal contributions to the physiology and pathophysiology of disorders the pituitary and adrenal glands. I know that we can expect to hear and learn more from these fine individuals in the future. Lastly, this work would not have come to fruition without the keen editorial assistance of Mr. Aaron Pannone.

It is with great admiration and respect that I dedicate this book to Dr. Gary S. Wand, Professor of Medicine and Psychiatry, at the Johns Hopkins University School of Medicine in Baltimore, Maryland. Dr. Wand received his B.A. in Biology and Chemistry from Skidmore College. He attended medical school at George Washington University in Washington, D.C. and then completed his internship and residency at that same institution. He trained in Endocrinology and Metabolism at the Johns Hopkins University School of Medicine under the supervision of the late Robert Ney and future collaborators, Richard Mains and Betty Eipper. Thereafter, he joined the faculty at Hopkins where he has risen to his current academic rank. His clinical and research interests are in the fields of pituitary disorders, signal transduction and the hypothalamic-pituitary-adrenal axis, and neuroendocrine mechanisms of addiction and tolerance. He established and currently serves as director of the Neuroendocrine Clinic at Hopkins. He also directs the operations of the Johns Hopkins Hospital Substance Abuse Initiatives and the Johns Hopkins University Drug Treatment Program Primary Care Clinics. He is a member of several societies including The Endocrine Society, The American College of Physicians, and the Research Society on Alcoholism. He has received several honors and awards for expertise in teaching, research, scholarship, and clinical medicine. A devoted husband and father, he currently resides in Baltimore, Maryland with his wife and two children.

Dr. Wand functioned as my clinical and research mentor during my endocrinology fellowship at Hopkins from 1990 to 1993. I still regard him as one of the important mentors and role models of my career. In a short period of time during my fellowship, he taught me many things about endocrinology, medicine, and life in general. I am grateful to have had the opportunity to observe and learn from his compassion towards his patients, his excellent bedside teachings, and his judicious exercise of the scientific method in the laboratory. One of the many things he has said about Cushing's syndrome that has served me well is to "...never take shortcuts in the evaluation of the patient with Cushing's syndrome ...this approach will minimize your errors in judgement."

Lewis S. Blevins, Jr., M.D.
Nashville, Tennessee
January 2002

1

PHYSIOLOGY OF THE HPA AXIS

Yutaka Oki, M.D.
Hamamatsu University School of Medicine, Japan

INTRODUCTION

Glucocorticoids are essential for life. They influencing glucose metabolism, lipid metabolism, immune function, electrolyte and fluid homeostasis, development, the central nervous system, and the adrenal medulla. Glucocorticoid secretion is regulated by hormonal interactions among the hypothalamus, pituitary, and adrenal cortex (HPA). The synthesis and secretion of cortisol, the most important glucocorticoid in humans, is regulated by adrenocorticotropin (ACTH) secreted from corticotrophs in the anterior pituitary. In response to stress, neural input from the brain stimulates the release of corticotropin-releasing hormone (CRH), arginine vasopressin (AVP), and other factors from the hypothalamus at the median eminence. These peptide hormones are carried to the anterior pituitary through hypophyseal-hypothalamic portal blood and stimulate ACTH secretion. The increase of glucocorticoid inhibits the secretion of CRH, AVP, and ACTH (negative feedback loop, Figure 1).

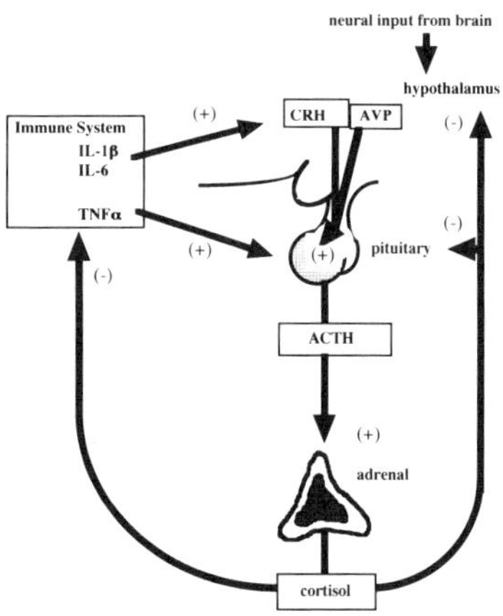

Figure 1. Immune-Hypothalamus-Pituitary-Adrenal Axis.
(+):stimulation (-):inhibition

ADRENOCORTICOTROPIN

Localization

ACTH is synthesized in corticotrophs, which constitute 10 – 20% of the population of the human anterior pituitary cells. Corticotrophs are numerous in the midsagittal region of the pituitary but are also distributed in the lateral wings of the pituitary (1). Although the majority of ACTH is synthesized in anterior pituitary corticotrophs, it is also found in the central nervous system and other tissues. In the central nervous system, ACTH and other proopiomelanocortin (POMC) peptides are expressed in the cell bodies of the infundibular nucleus of the basal hypothalamus (2-4). POMC peptides are also expressed in adrenal, testis, mononuclear cells, spleen, kidney, ovary, lung, thyroid, liver, and colon (5, 6). In these tissues, POMC peptides are not secreted because they are translated without signal peptide. POMC peptides are expressed and secreted from some neoplastic tissues, such as small cell lung cancer (ectopic ACTH secretion).

ACTH and POMC gene

ACTH is a 39-amino acid single-chain peptide (Figure 2) and is derived from a 241-amino acid precursor, POMC (Figure 3). The NH_2-terminal 24 amino acids of ACTH are identical among all species, while minor species differences are recognized in the COOH-terminal region. The first 18 amino acids are required for full biological activity. However, a longer amino acid sequence, ACTH1-24, is necessary to maintain the biological activity because of rapid degeneration of ACTH1-18 *in vivo*.

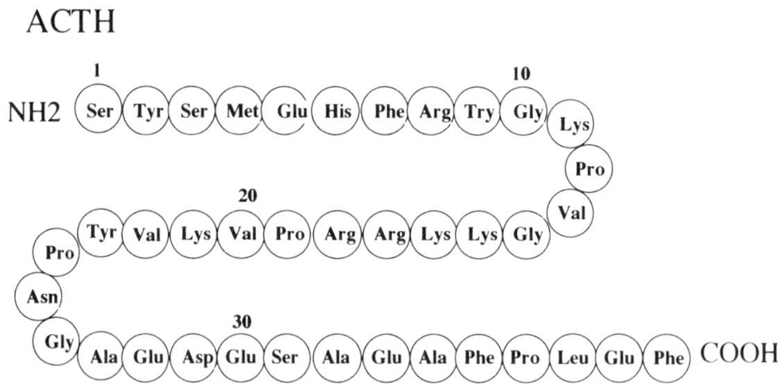

Figure 2. Amino acid sequence of human ACTH 1-39.

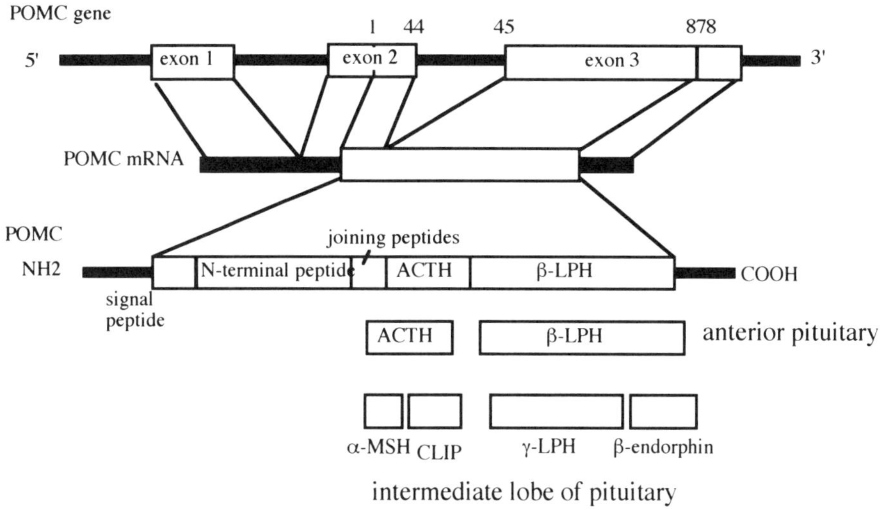

Figure 3. Synthesis of proopiomelanocortin-related peptides.
POMC: proopiomelancortin, ACTH: adrenocorticotropin.
LPH: lipotropin, MSH:melancyte-stimulating hormone.
CLIP: corticotropin-like intermediate peptide.

The human POMC gene is 8 Kb long and located on chromosome 2p23 (7). It is composed of a promoter region, three exons, and two large introns (8). The sequence of exon1 encoding 87 nucleotides is varied among species and is not translated. The second exon (152 bp) encodes untranslated region, the 26-amino acid signal peptide sequence, and the first 18-amino acid of the mature POMC. The second intron (intron B) interrupts the coding region between amino acids 18 and 19 of the mature POMC molecule. The third exon (833 bp) contains the entire coding region of the remaining POMC molecule (Figure 3) (8-10).

Regulation of POMC gene expression

In human corticotrophs, POMC mRNA levels are increased by CRH and inhibited by glucocorticoids (11). CRH binds to type I CRH receptors, which are located in the pituitary and central nervous system as described below, activates adenylate cyclase, and generates cyclic adenosine 3',5' monophosphate (AMP) (12-15). The increase in cyclic AMP is responsible for both the stimulation of POMC mRNA expression and peptide synthesis (16). In addition, cFos and AP-1 sites in the promoter region of POMC are required for the increase in the CRH-induced POMC mRNA (17).

AVP stimulates basal ACTH secretion and potentiates CRH-induced ACTH secretion synergistically (18, 19). However, AVP does not increase POMC mRNA content in the anterior pituitary (20). Glucocorticoids inhibit the expression of POMC mRNA (11). This is thought to be mediated by glucocorticoid-glucocorticoid receptor complex binding to the *cis*-acting DNA sequence of the POMC promoter (21, 22). Other mechanisms, which antagonize the transcription activators of POMC gene such as fos, jun or Nur77, may be involved in the negative regulation by glucocorticoids (23, 24). It is also postulated that non-genomic pathways, e.g. the reduction of cyclic AMP, are involved in this negative regulation of POMC mRNA (25).

POMC processing

Several biologically active peptides are derived from POMC. These are generated by posttranslational processing with prohormone convertase (PC) enzymes. The distribution of these enzymes shows the tissue-specificity. The processing of POMC occurs at Lys-Arg or Arg-Arg amino acid residues *in vivo*. This proteolytic digestion is mediated by structurally related two endopeptidases, prohormone convertase1 (PC1) and prohormone convertase2 (PC2) (26-30). These two enzymes are widely distributed in various endocrine cells and neurons (26, 28, 29). PC1 is expressed in the anterior pituitary corticotrophs and pituitary intermediate lobe cells and in the hypothalamus (26, 28, 31). PC2 is expressed in the intermediate pituitary, central nervous system, and pancreatic islet cells, but not in the anterior pituitary (26, 28, 31). Therefore, the tissue-specific distribution of these proteolytic enzymes reflects the tissue-specific differences in POMC processing, with the result that PC1 produces POMC-derived peptides including ACTH and that PC2 produces α-melanocyte stimulating hormone and β-endorphin in the intermediate pituitary which humans do not have, and brain (Figure 3).

Effects of ACTH on the adrenal cortex

ACTH primarily stimulates cortisol synthesis and secretion by binding to a seven-transmembrane cell surface receptor, the melanocortin receptor 2 (MC2-R, Table 1) (32). ACTH up-regulates MC-R2, and enhances ACTH-induced cortisol secretion (33). ACTH binds to MC-R2 and activates adenylate cyclase resulting in increase in intracellular cyclic AMP concentration, which subsequently activates cyclic AMP-dependent protein kinase A. Extracellular Ca^{2+}, which enters through T-type Ca^{2+} channels, is required for ACTH-induced steroidogenesis (34). The effect of ACTH on the adrenal cortex is divided into two time courses (35, 36). The acute effect of

ACTH occurs within minutes. It activates cholesterol delivery to mitochondria, and stimulates the conversion of cholesterol to pregnenolone by CYP11A1, which is the initial step of steroidogenesis (37). ACTH chronically increases the synthesis of most steroidogenic enzymes and adrenal growth (38). Therefore, long-term ACTH exposure enlarges the adrenals as seen in Cushing's disease. When plasma ACTH levels are decreased, as in hypopituitarism or long-time glucocorticoid administration, all of the steroidogenic CYP enzymes and protein synthesis decline and then the adrenal glands atrophy (36).

Table 1. Melanocortin receptor family.

subtype	amino acids	distribution	ligand specificity
MC1-R	317	melanocyte	αMSH>ACTH>βMSH>γMSH
MC2-R	297	adrenal cortex	ACTH
MC3-R	361	brain placenta	αMSH = βMSH= γMSH=ACTH
MC4-R	333	brain	αMSH=ACTH= βMSH>> γMSH
MC5-R	325	brain	αMSH>ACTH> βMSH> γMSH

Regulation of ACTH secretion

ACTH secretion from the anterior pituitary is regulated by several factors as shown in Figure 1. CRH and AVP, which are synthesized in hypothalamus, are the major intrinsic ACTH secretagogues, and glucocorticoids negatively regulate ACTH secretion. These ligands bind to their specific receptors and ACTH secretion is mediated by the signal transduction pathways including cyclic AMP-protein kinase A, Ca^{2+}/phopholipid-dependent protein kinase C, or glucocorticoid. These signal transduction pathways change the phosphorylation of specific cellular substrate and intracellular calcium concentration regulating ACTH secretion.

CRH

CRH is the most potent ACTH secretagogue (19). CRH was initially characterized in sheep and was found to have 41 amino acids (39). Human CRH is identical to rat CRH, and differs by seven amino acids from ovine

CRH. The human CRH gene is located in chromosome 8 and synthesizes a larger precursor of CRH, preproCRH (191 amino acids, Figure 4) (40-43). CRH is processed from preproCRH by cleaving at flanking basic amino acid pairs.

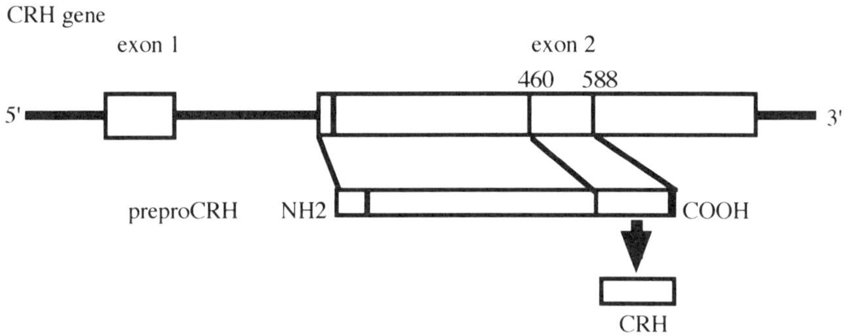

Figure 4. Synthesis of corticotropin-releasing hormone.

CRH is synthesized by neurons in the parvocellular region of the hypothalamic paraventricular nucleus and is co-localized with AVP (44-48). These neurons project to the median eminence and release CRH into the hypophyseal portal blood, in which immunoreactive CRH can be detected by radioimmunoassay (49). CRH is also widely expressed in the central nervous system, spinal cord, adrenal medulla, gastrointestinal tract, pancreas, testis, and placenta (50-55). CRH binds to its specific receptors. CRH receptors (CRH-R) have at least three subtypes, CRH-R1, CRH-R2α, and CRH-R2β (Table 2) (15, 56-59). CRH-R1 is expressed in the anterior pituitary and central nervous system whereas CRH-R2s are expressed in the central nervous system and peripheral tissues. CRH-R1 is involved in ACTH release and is down-regulated by activation of cyclic AMP-dependent protein kinase A (CRH), activation of Ca^{2+}/phospholipid-dependent protein kinase C (AVP), or glucocorticoids (60, 61). CRH binds to CRH-R1, stimulates the adenylate cyclase through G-protein, generates cyclic AMP, and subsequently activates protein kinase A (62). The CRH-induced ACTH release requires the entry of the extracellular Ca^{2+} into the cells through L-type voltage-dependent Ca^{2+} channels (63). Non-selective sodium channels not voltage-dependent sodium channels may be involved in CRH-induced Ca^{2+} influx and ACTH secretion (64). In perifused rat anterior pituitary cells, CRH stimulates ACTH release as long as the cells are exposed to CRH (mono-phasic pattern) (65).

Table 2. Distribution of CRH receptors.

CRH-R1	pituitary cerebellum cerebral cortex olfactory bulb
CRH-R2α	hypothalamus paraventricular n. ventromedial n. supraoptic n. lateral septum olfactory bulb
CRH-R2β	heart skeletal muscle lung intestine choroid plexus hypothalamus arteriole

AVP

AVP is a weaker ACTH secretagogue than CRH, but has important roles in regulating ACTH release in conjunction with CRH (19). AVP is produced in the paraventricular nucleus and carried to anterior pituitaryin portal blood (66). The gene encoding AVP is located on chromosome 20 and produces the precursor to AVP. The precursor also gives rise to neurophysin II that is the AVP-binding protein. A single subclass of AVP receptors, V3 (V1b), is found in corticotrophs. Most corticotrophs have AVP receptors and bind AVP. These receptors also bind oxytocin. The AVP receptors on corticotrophs are distinct from the V1a hepatic/vasopressor receptors and the V2 renal receptors. The V1 receptors activate phospholipase C (PLC), whereas the V2 receptors activate adenylate cyclase (67, 68). Therefore, desmopressin, a V2 agonist, has a negligible effect on ACTH release. The V3 receptors are down-regulated by the bilateral adrenalectomy, which elevates AVP concentrations in hypophyseal portal blood, and this down-regulation is reversed by glucocorticoid administration (69, 70). AVP binds to V3 receptors, and activates PLC via G-protein (67). The activation of PLC hydrolyzes phosphatidylinositol 4,5-bisphosphate to inositol 1,4,5-trisphosphate, which mobilizes intracellular Ca^{2+} from intracellular Ca^{2+} stores, and diacylglycerol, which activates Ca^{2+}/phospholipid-dependent protein kinase C (71-73). In perifused rat anterior pituitary cells, the mobilization of intracellular Ca^{2+} by IP3 induces the rapid and spike-shaped release of ACTH, and the activation of PKC, which requires extracellular Ca^{2+}, produces the sustained plateau release of ACTH (bi-phasic) (65, 74-76). AVP alone does not increase the intracellular cyclic AMP, but augments CRH-induced cyclic AMP levels by inhibiting phosphodiesterase activity, resulting in the synergistic ACTH release (77). Unlike CRH, AVP alone does not increase POMC transcription (20).

OTHER FACTORS

CRH and AVP are physiologically important ACTH secretagogues, but other factors, such as oxytocin, angiotensin II, catecholamines, vasoactive intestinal polypeptide (VIP), peptide histidine isoleucine (PHI), gastrin-releasing peptide (GRP), and atrial natriuretic peptide (ANP) may be involved in regulating ACTH secretion. Cytokines, such as interleukin-1 (IL-1), IL-6, and tumor necrosis factor α (TNFα), also have important roles in HPA axis regulation.

Oxytocin, Angiotensin II, Atrial Natriuretic Peptide

Oxytocin stimulates ACTH release, probably through the AVP receptors and signal transduction pathway (78). Angiotensin II binds to type 1 angiotensin II receptors and activates phosphatidyl inositol pathways resulting in stimulation of ACTH release (79, 80). Angiotensin II has a direct stimulatory effect on ACTH release from the anterior pituitary *in vitro* but may act mainly by central mechanisms to stimulate CRH release from the hypothalamus (39, 78, 81). ANP inhibits CRH-induced ACTH release by binding to type B natriuretic peptide receptors and activating guanylyl cyclase (82).

Catecholamines

Catecholamines stimulate ACTH release apparently by secretion of CRH and/or AVP into hypophyseal portal blood. The parvocellular cells in the PVN are innervated by catecholamine containing neurons. Activation of central catecholamine stimulates ACTH release, which appears to be mediated via α1-adrenergic receptors (83). It is unlikely that peripheral catecholamines have a direct effect on the anterior pituitary.

Cytokines

The immune system, such as cytokines, and the HPA axis interact closely with each other. TNFα released by macrophages stimulates ACTH release from cultured anterior pituitary cells, suggesting that TNFα has a direct stimulatory effect on ACTH release by the anterior pituitary (84). Bacterial endotoxin stimulates ACTH, cortisol, and AVP through TNFα. Prostaglandins may be involved in the endotoxin-induced response, because cycloxygenase inhibitors block the endotoxin-induced increase in ACTH, cortisol, and TNF (85-87). IL-1 stimulates ACTH release by acting on the hypothalamus to stimulate secretion of CRH and possibly AVP (88-90). IL-1β is more potent than IL-1α (91). IL-1 activates adrenergic neurons in the brain stem and hypothalamus, which may stimulate CRH secretion from the hypothalamus (92, 93). Prostaglandins may be involved in IL-1-induced CRH release, because indomethacin blocks IL-1-induced ACTH release *in vivo* (91). IL-6 also induces ACTH secretion by stimulating CRH secretion from the hypothalamus via a prostaglandin-dependent mechanism (94). IL-2 has a stimulatory effect on CRH secretion from the hypothalamus or on ACTH secretion from the anterior pituitary (95-97).

Choline

The hypothalamus contains hypothalamic cholinergic innervations and cholinergic stimulation increases ACTH release (98). Atropin stimulates ACTH release via central CRH secretion *in vivo* and acetylcholine stimulates CRH release *in vitro* (99).

Serotonin

In humans, serotonin stimulates stress-induced ACTH release by stimulating CRH release. In animals, serotonin increases ACTH secretion also by stimulating CRH secretion from the hypothalamus (100, 101). Serotonin appears to have no direct effect on the anterior pituitary.

GABA

The secretion of ACTH is inhibited by γ-aminobutyric acid (GABA). Valproate, a GABA-transaminase inhibitor, decreases ACTH secretion in humans (102). In rats, GABA acts at the hypothalamus to increase dopamine and norepinephrine in the median eminence, and to inhibit serotonin-induced

CRH release (103).

Opiate

Opiate peptides inhibit ACTH secretion in humans by binding to δ- or κ-opiate receptors (104). In rats, however, opiate agonists stimulate ACTH secretion by binding to central μ- or κ-receptors (105, 106). The tonic inhibition by opiates is blocked by naloxone resulting in the stimulation of ACTH release (107, 108). This effect of opiates occurs at a level above the pituitary, because morphine has no effect on ACTH release by cultured anterior pituitary cells (104, 109).

Urocortin

Urocortin, a newly discovered CRH-related peptide, has the ability to bind to all subtypes of CRH receptors and to stimulate ACTH secretion from the rat anterior pituitary *in vivo* and *in vitro* more potently than CRH (110-112). However, endogenous urocortin may not be involved in regulating ACTH release, because immunoreactive urocortin has not been detected in the median eminence and anti-urocortin serum does not inhibit the adrenalectomy-induced ACTH release in rats (113).

GLUCOCORTICOID FEEDBACK

Glucocorticoids are the negative regulators of ACTH secretion. This negative regulation occurs at both the pituitary to inhibit ACTH release and the hypothalamus to inhibit CRH and AVP release and probably at higher centers.

Type II glucocorticoid receptors are distributed in the brain including PVN where CRH and AVP are synthesized (114). The PVN is thought to be a site of glucocorticoid negative feedback, because dexamethasone decreases CRH release (115). The absence of corticosteroid-binding globulin in the brain may be involved in the sensitivity of the brain to glucocorticoids (116). In the PVN, dexamethasone decreases mRNA expression of CRH, and inhibits the rise in CRH and AVP contents (117, 118).

In the anterior pituitary, glucocorticoid feedback on ACTH secretion is mediated mainly via type II glucocorticoid receptors (119). Glucocorticoid negative feedback regulation on ACTH release from the anterior pituitary cells has been considered to be divided into at least two major time domains

(120). Fast (rapid) feedback occurs in seconds to minutes and delayed feedback occurs in hours to days. It is possible that, at the individual time domains, glucocorticoid inhibits ACTH release by different mechanisms respectively. Fast feedback is dependent on the rate of increase in glucocorticoid concentration rather than absolute glucocorticoid levels, and inhibits CRH-induced cyclic AMP levels. However, the precise mechanism of fast feedback is still unclear. Delayed feedback is proportional to the glucocorticoid dose, potency, and duration of administration. There is general agreement that, at the slow domain, the suppression of POMC gene transcription leads to the decrease in the hormone precursor and the subsequent reduction in the amount of releasable ACTH (121).

PHYSIOLOGICAL ACTH AND CORTISOL SECRETION

Pulsatile secretion

In vivo, ACTH and cortisol are secreted in episodic pulsatile bursts with slow declines of cortisol as a result of its shorter clearance from plasma. The frequency of ACTH pulsatile secretion is 12 – 40 bursts in 24 hours. The normal circadian rhythm is produced by changing the amplitudes but not frequency of ACTH secretion (122, 123).

Circadian rhythm

An endogenous circadian rhythm, which may be generated in the supraoptic nucleus, produces the unique pattern of ACTH and cortisol secretion by modulating CRH release from PVN. Plasma ACTH and cortisol levels reach the highest at about the time of waking in the morning between 6:00 and 9:00 a.m., decline through the day to intermediate levels around 4:00 p.m., and are the lowest an hour or two after beginning sleep (122). All of the POMC-related peptides are secreted in a circadian rhythm.

Hypoglycemia

Many physical and psychological stresses activate the HPA axis and increase plasma ACTH and cortisol levels. Insulin-induced hypoglycemia causes 5- to 6-fold increase in plasma ACTH within 60 minutes after the insulin injection. Cortisol levels increase over 2-fold at 60 – 90 minutes. Insulin-induced hypoglycemia causes a greater ACTH response than CRH alone or the combination CRH plus AVP. Therefore, multiple factors may be involved in hypoglycemia-induced ACTH secretion. Hypoglycemia increases plasma CRH, AVP, angiotensin II, and catecholamine levels (124, 125). Although peripheral CRH levels may not accurately reflect plasma CRH levels in the hypophyseal portal flow, CRH may be the major secretagogue in response to insulin-induced hypoglycemia (126). AVP may also be involved in hypoglycemia-induced ACTH secretion, and may increase ACTH secretion synergistically with CRH (127, 128). Hypoglycemia increases catecholamines, which may act at the hypothalamus to stimulate CRH release. Epinephrine may play more important roles in hypoglycemia-induced ACTH secretion than norepinephrine, because epinephrine is increased more rapidly and to a greater degree than norepinephrine by insulin-induced hypoglycemia (125). The central administration of α-adrenergic agent stimulates hypothalamic CRH, AVP, and ACTH release in response to hypoglycemia (100).

Surgery

Surgery is one of the most potent stresses in activating the HPA axis. Plasma ACTH levels increase at the incision and during the surgery and the greatest response is observed during reversal of anesthesia, extubation, and the early post-operative recovery period (129).

Exercise

Acute or intense exercise increases ACTH, β-endorphin, and cortisol levels (130). The regulation of ACTH may be under tonic inhibition by opiates. Naloxone enhances exercise-induced ACTH secretion. Extremely well conditioned athletes sometimes experience less mental stress partly due to analgesia induced by the exercise (131).

Hypovolemia

Hypovolemia such as hemorrhage increases ACTH secretion (132). This may be mediated by CRH, AVP, and catecholamine (133, 134). Stretch receptors are involved in the hemorrhage-induced ACTH secretion via vagal and glosopharyngeal nerves through the brain stem to the hypothalamus (135).

Immune stress

In response to immune stress, monocytes, macrophages, and lymphocytes produce cytokines, which activate the HPA axis. Bacterial endotoxin stimulates the release of IL-1, IL-6, and TNFα, which act on the levels of hypothalamus, pituitary and adrenal. These cytokines are negatively regulated by glucocorticoid (negative feedback loop: interaction between endocrine and immune systems, Figure 1).

ADRENAL CORTEX

Anatomy

The adrenals are extraperitoneal at the upper pole of the kidneys and lie on each side of vertebral column, lateral to the eleventh thoracic (right) and the first lumbar vertebrae (left). Each adult adrenal is 2 to 3 cm wide and 4 to 6 cm long, and the weighs about 4 g, regardless of age, sex, or body weight, but can be enlarged by pathological ACTH stimulation such as Cushing's disease. The adrenal is encapsulated with a thin layer of areola connective tissue and a thick fibrous capsule. The right adrenal is usually higher and further laterally than the left. The adrenal blood is supplied by several short arteries that are branches of the inferior phrenic artery, the superior, middle, and inferior adrenal arteries. Arterial blood enters into the sinusoidal circulation of the cortex, flows into the adrenal medulla, and finally drains toward the central vein so that medulla chromaffin cells are exposed to very high concentrations of adrenal steroid hormones. The adrenal vein empties directly into vena cava on the right and into the left renal vein on the left. In adults, the adrenal cortex is about 90% of the adrenal and surrounds the medulla. The cortex consists of three histologically and functionally distinct zones: the zona glomerulosa, the zona fasciculata, and the zona reticularis. The zona glomerulosa constitutes about 15% of the adrenal cortex and is located under the capsule. It produces aldosterone but not cortisol. The glomerulosa cells are small and have low cytoplasmic volume because of their low lipid content. The zona fasciculata produces cortisol but not aldosterone. It is located between the zona glomerulosa and the zona reticularis and constitutes about 75% of the adrenal cortex. The fasciculata cells cannot be clearly demarcated from the glomerulosa cells. The fasciculata cells contain rich cholesterol and cholesterol-esters. The zona reticularis produces the major adrenal androgens: dehydroepiandrosterone, dehydroepiandrosterone sulfate, etc. The reticularis cells also produce cortisol although at a much lower rate than the fasciculata cells. The thin layer between the zona glomerulosa and the zona fasciculata, the transitional zone, consists of cells which contain neither aldosterone synthetase nor 11β-hydroxylase. Therefore, those cells cannot produce either aldosterone or cortisol. These cells may be progenitor of the glomerulosa cells and the fasciculata cells.

ACTH has a remarkable effect on the adrenal cortex, especially the zona fasciculata and reticularis. ACTH increases adrenal blood flow and cortisol secretion within minutes after its administration, and adrenal weight within hours. The clear cells located at the junction of the zona fasciculata and reticularis lose lipid and become compact. Chronic ACTH stimulation changes the phenotype of the glomerulosa cells to that of the fasciculata cells, and also converts the phenotype of the inner fasciculata cells to that of the reticularis cells. This change extends outwards until the reticularis may replace the zona fasciculata, causing hypertrophy and hyperplasia. The NH2-terminal fragment of POMC, epidermal growth factor, fibroblast growth factor, and insulin-like growth factor-I are also involved in the regulation of adrenal weight in adults (136-140).

Glucocorticoid Biosynthesis

The steroid hormones synthesized by the adrenal cortex are derived from the basic steroid ring structure, which consists of three cyclohexane rings, and one cyclopentane ring (Figure 5). All human steroid hormones are synthesized from cholesterol. About 80% of cholesterol for steroid hormone synthesis are derived from circulating plasma low-density lipoprotein (LDL). Adrenal cortex utilizes LDL cholesterol by a specific receptor mediated mechanism. LDL receptors located on microvilli bind circulating LDL, internalize it, and become concentrated in "coated pits" at the base of the villi.

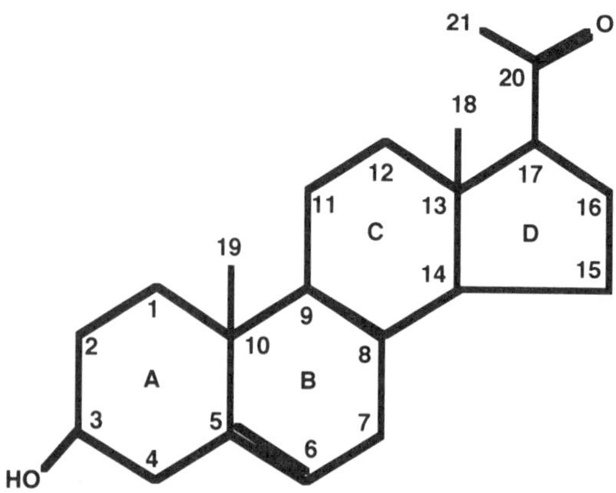

Figure 5. Steroid ring structure.

The coated vesicles fuse with lysosomes and lysosomal acid lipase cleaves cholesterol esters to free cholesterol. The rate limiting process in steroidogenesis is the transport of free cholesterol to inner mitochondrial membrane where CYP11A1 (P450scc) is located. This transport is regulated by a protein, StAR (steroidogenic acute regulatory protein), which is rapidly induced by cyclic AMP.

Adrenal cortex synthesizes the steroid hormones using four CYP enzymes (Figure 6). The steroidogenic CYP enzymes modify ring carbons of cholesterol. CYP11A1 cleaves the side chain from C21 of cholesterol in adrenal mitochondria, and produces pregnenolone. Pregnenolone is transported to endoplasmic reticulum and converted to progesterone by 3β-hydroxysteroid dehydrogenase (3β-HSD). Progesterone is converted to 17α-hydroxyprogesterone by CYP17 (P450c17), which has 17α-hydroxylation and 17,20-lase activities. Both progesterone and 17α-hydroxyprogesterone undergo 21-hydroxylation by CYP21A2 (P450c21), and 17α-hydroxyprogesterone is converted to 11-deoxycortisol. Finally, a mitochondrial enzyme, CYP11B1 (P450c11β), converts 11-deoxycortisol to cortisol. The expression of all steroidogenic P450 enzymes requires a transcription activator, steroidogenic factor-1 (141, 142). Cortisol is the adrenal steroid produced in the greatest quantity, and its secretion rate is estimated to be 8 – 25 mg/day (143-145).

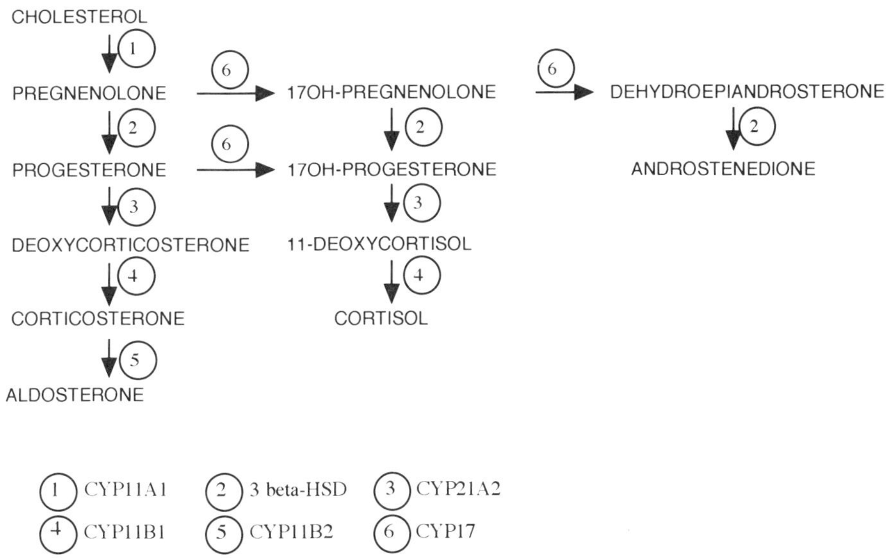

Figure 6. Steroid biosynthesis in the adrenal cortex.

Glucocorticoid metabolism

Cortisol is cleared with a half-time of 80 - 120 min. Catabolism of cortisol to inactive derivatives takes place in the liver and kidney. In the hepatic pathway, cortisol is in reversible equilibrium with cortisone. These steroids are initially inactivated by the reduction of the Δ^4 double bond. This is followed by the reduction of 3-keto groups by 3β-HSD to yield tetrahydrocortisol and tetrahydrocortisone. These tetrahydro-derivatives are reduced further by 20-HSD to yield cortol and cortolone. Finally, these are conjugated with gluclonic acid to be water-soluble. Most of cortisol metabolites are excreted in the urine as 17-hydroxycorticosteroid (17-OHCS). Small amount of cortisol is metabolized by 6β-hydroxylase in the liver. 6β-hydroxylation is a minor pathway, but is activated during pregnancy, or following the administration of estrogens, rifampcin, phenytoin, phenobarbital or mitotane.

In the kidney, 11β-HSD type II converts cortisol to cortisone to inhibit the mineralcorticoid activity of cortisol because cortisol has the same affinity to mineralcorticoid receptors as aldosterone (146, 147).

Circulating cortisol

The cortisol released by the adrenal cortex is free, but approximately 90 to 97% of the circulating cortisol is bound by plasma protein such as corticosteroid-binding globulin (CBG) and albumin (148). The major function of these proteins is as a reservoir of steroids. The steroids are inactive when bound by plasma protein. CBG is primarily produced in the liver and is a 383-amino-acid glycoprotein with a molecular weight of 59 kd (149). About 90% of bound-form cortisol is associated with CBG which binds cortisol with high affinity (Kd: 3×10^{-8} M). CBG can bind up to 25 μg/dl of cortisol (150). Therefore, the free cortisol levels in plasma are highly correlated with the total cortisol levels at physiological concentrations. However, when the total cortisol levels rise more than 25 μg/dl, the proportion of free cortisol is increased. A number of endogenous steroids can also bind to CBG but most synthetic steroids except prednisolone do not bind to CBG.

ACTIONS OF GLUCOCORTICOIDS

Steroids including glucocorticoids exert their effects by interacting with cytosolic receptor proteins. Free glucocorticoids enter the cells by passive diffusion fashion, and then bind to the glucocorticoid receptor (GR), of which molecular weight is about 94 kd. Two highly conserved "zinc fingers" in the central part of the receptor constitute the DNA-binding domain, which is involved in receptor dimerization, nuclear translocation, and transactivation (151, 152). The glucocorticoid receptor is activated by binding to glucocorticoid. The inactive GR is part of a multiprotein complex that consists of the receptor, two molecules of hsp90, and one molecule of hsp56 (153-156). Glucocorticoid binding induces a conformational change in the GR protein and the glucocorticoid-GR complex dissociates from the hsp complex. Then the glucocorticoid-GR complex is translocated from the cytoplasm to the nucleus (153, 154). A homodimer of the glucocorticoid-GR complex binds to a short, palindromically arranged DNA sequence termed glucocorticoid-responsible element (GRE) in the promoter region of the glucocorticoid-responsive genes, and affects the protein synthesis (157, 158). As an alternative mechanism of the glucocorticoid effect, the glucocorticoid-GR complex interacts with other transcription factors such as fos and jun families, which activate the AP-1 site, and suppresses the activities of those transcription factors (Figure 7) (159-162). This is observed in the negative transcriptional regulation of immune genes.

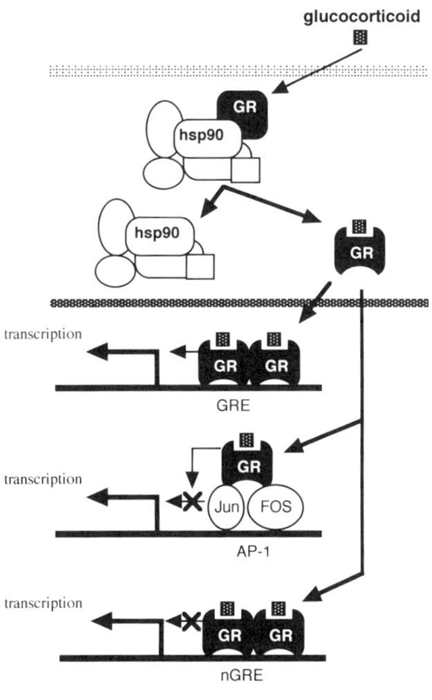

Figure 7. Molecular mechanics of glucocorticoid action.
GR: glucocorticoid receptor
GRE: glucocorticoid-responsible element
nGRE: negative GRE

Effects on metabolism

It is well known that glucocorticoids increase the glucose-supply. Glucocorticoids are required for the maintenance of hepatic glycogen stores by activating glycogen synthase and by inactivating glycogen-phosphorylase.

Glucocorticoids activate hepatic gluconeogenesis. Glucocorticoids increase the availability of substrates such as glucogenic amino acids from peripheral tissues, e.g. skeletal muscle. Glucocorticoids also activate the induction of the enzymes required for gluconeogenesis such as glucose-6-phosphate and phosphoenolpyruvate carboxykinase. The action of other gluconeogenic hormones such as glucagon and catecholamines requires the permissive effect of glucocorticoids.

In addition to the activation of gluconeogenesis, glucocorticoids also inhibit the transportation of glucose into the cells by repression of the glucose-transporters(163).

Effects on immune system

Glucocorticoid excess suppresses immunological and inflammatory responses and synthetic glucocorticoids are employed in the treatment of inflammatory disease, autoimmune disease, and allergy. They are also used to suppress the host immune system and prevent rejection during organ transplantation.

Glucocorticoids alter the movement of immune cells. Glucocorticoids decrease the number of lymphocytes in blood with greater depletion of T cells than B cells by redistributing lymphocytes from the circulation to the spleen, lymph nodes, thoracic duct, and bone marrow (164, 165). In addition, glucocorticoids accelerate the apoptosis of the lymphocytes (166). Glucocorticoids also induce the release of neutrophils from the bone marrow resulting in an increase in the number of neutrophils in the circulation.

Although glucocorticoids alter the redistribution of lymphocytes, this is not sufficient to explain the immunosuppressive effect of glucocorticoids. It is also well known that glucocorticoids suppress the expression of inflammatory cytokines, which are key mediators of the immune response. Most inflammatory genes such as cytokines that are suppressed by glucocorticoids have no responsive elements to glucocorticoid-GR complex in their promoters. This suggests that the GR antagonizes the positive transcription factors such as nuclear factor κ-B and AP-1 in the inflammatory genes (167).

Effects on fluid and electrolyte homeostasis

Patients with glucocorticoid excess have hypertension. Cortisol can bind to mineralcorticoid receptors, but is converted to cortisone by 11β-HSD in the kidney (146, 147). Therefore, the physiological concentration of cortisol does not show the mineralcorticoid activity, but patients with cortisol excess such as Cushing's syndrome often present hypertension and hypokalemia. However, patients receiving some synthetic glucocorticoids such as dexamethasone, which has no ability to bind to mineralcorticoid receptors, also have hypertension. This is not due to the mineralcorticoid activity, but the precise mechanism of glucocorticoid-induced hypertension is unclear. In patients with glucocorticoid deficiency, free water clearance is decreased and plasma AVP concentrations are elevated resulting in hyponatremia (168). Glucocorticoids negatively regulate AVP mRNA levels in the parvocellular division of the PVN, and glucocorticoid deficiency increases AVP mRNA levels and plasma AVP concentrations (169). Therefore, the replacement therapy with glucocorticoids decreases plasma AVP concentrations and increases free water clearance, and thus normalizes the water balance.

Effects on bone

The chronic administration of glucocorticoids in excess causes osteoporosis by inhibiting osteoblast function (170). Glucocorticoids also inhibit intestinal calcium absorption and renal calcium excretion (171).

Effects on adrenal medulla

The adrenal medulla receives the venous effluent directly from the adrenal cortex and thus is exposed to higher concentrations of cortisol than any other tissues. Glucocorticoids affect the enzymes mediating adrenal catecholamine biosynthesis such as phenylethanolamine N-methyltransferase (PNMT), tyrosine hydroxylase, and dopamine β-hydroxylase. The high concentrations of glucocorticoids are required especially for PNMT, which catalyzes the formation of epinephrine from norepinephrine (172). Therefore, the principal catecholamine is epinephrine in the adrenal medulla. In contrast, the principal product in other tissues producing catecholamine such as sympathetic nerve ganglion is norepinephrine.

Effects on emotion

In glucocorticoid excessive and deficient states, mood changes are commonly observed. Patients with Cushing's syndrome or those receiving high doses of glucocorticoids are usually depressed (173). However, patients receiving glucocorticoid therapy at physiological or slightly higher doses usually have a feel of well-being. Patients with Addison's disease are also depressed, irritable, seclusive , apathetic, and anorexic.

In depressive syndrome, glucocorticoid production is increased and the circadian rhythm of the HPA axis is blunted. In this state, glucocorticoids suppress the HPA axis,

REFERENCES

1. Doniach I 1985 Histopathology of the pituitary. Clin Endocrinol Metab 14:765-89.
2. Bugnon C, Bloch B, Lenys D, Fellmann D 1979 Infundibular neurons of the human hypothalamus simultaneously reactive with antisera against endorphins, ACTH, MSH and beta-LPH. Cell Tissue Res 199:177-96.
3. Desy L, Pelletier G 1978 Immunohistochemical localization of alpha-melanocyte stimulating hormone (alpha-MSH) in the human hypothalamus. Brain Res 154:377-81.
4. Parker CR Jr., Barnea A, Tilders FJ, Porter JC 1981 Characterization of immunoreactive alpha-melanocyte stimulating hormone (alpha-MSHi) in human brain tissue. Brain Res Bull 6:275-80.
5. Buzzetti R, McLoughlin L, Lavender PM, et al. 1989 Expression of pro-opiomelanocortin gene and quantification of adrenocorticotropic hormone-like immunoreactivity in human normal peripheral mononuclear cells and lymphoid and myeloid malignancies. J Clin Invest 83:733-7.
6. DeBold CR, Nicholson WE, Orth DN 1988 Immunoreactive proopiomelanocortin (POMC) peptides and POMC-like messenger ribonucleic acid are present in many rat nonpituitary tissues. Endocrinology 122:2648-57.
7. Owerbach D, Rutter WJ, Roberts JL, et al. 1981 The proopiocortin (adrenocorticotropin/beta-lipoprotein) gene is located on chromosome 2 in humans. Somatic Cell Genet 7:359-69.
8. Whitfeld PL, Seeburg PH, Shine J 1982 The human pro-opiomelanocortin gene: organization, sequence, and interspersion with repetitive DNA. DNA 1:133-43.
9. Chang AC, Cochet M, Cohen SN 1980 Structural organization of human genomic DNA encoding the pro-opiomelanocortin peptide. Proc Natl Acad Sci U S A 77:4890-4.
10. Takahashi H, Teranishi Y, Nakanishi S, Numa S 1981 Isolation and structural organization of the human corticotropin--beta-lipotropin precursor gene. FEBS Lett 135:97-102.
11. Suda T, Tozawa F, Yamada M, et al. 1988 Effects of corticotropin-releasing hormone and dexamethasone on proopiomelanocortin messenger RNA level in human corticotroph adenoma cells in vitro. J Clin Invest 82:110-4.
12. Aguilera G, Harwood JP, Wilson JX, et al. 1983 Mechanisms of action of corticotropin-releasing factor and other regulators of corticotropin release in rat pituitary cells. J Biol Chem 258:8039-45.
13. Chen R, Lewis KA, Perrin MH, Vale WW 1993 Expression cloning of a human corticotropin-releasing-factor receptor. Proc Natl Acad Sci U S A 90:8967-71.

14. Labrie F, Veilleux R, Lefevre G, et al. 1982 Corticotropin-releasing factor stimulates accumulation of adenosine 3', 5'-monophosphate in rat pituitary corticotrophs. Science 216:1007-8.
15. Perrin MH, Donaldson CJ, Chen R, et al. 1993 Cloning and functional expression of a rat brain corticotropin releasing factor (CRF) receptor. Endocrinology 133:3058-61.
16. Reisine T, Rougon G, Barbet J, Affolter HU 1985 Corticotropin-releasing factor-induced adrenocorticotropin hormone release and synthesis is blocked by incorporation of the inhibitor of cyclic AMP-dependent protein kinase into anterior pituitary tumor cells by liposomes. Proc Natl Acad Sci U S A 82:8261-5.
17. Autelitano DJ 1998 Stress-induced stimulation of pituitary POMC gene expression is associated with activation of transcription factor AP-1 in hypothalamus and pituitary. Brain Res Bull 45:75-82.
18. Rivier CL, Plotsky PM 1986 Mediation by corticotropin releasing factor (CRF) of adenohypophysial hormone secretion. Annu Rev Physiol 48:475-94.
19. Vale W, Vaughan J, Smith M, et al. 1983 Effects of synthetic ovine corticotropin-releasing factor, glucocorticoids, catecholamines, neurohypophysial peptides, and other substances on cultured corticotropic cells. Endocrinology 113:1121-31.
20. Levin N, Blum M, Roberts JL 1989 Modulation of basal and corticotropin-releasing factor-stimulated proopiomelanocortin gene expression by vasopressin in rat anterior pituitary. Endocrinology 125:2957-66.
21. Nakai Y, Usui T, Tsukada T, et al. 1991 Molecular mechanisms of glucocorticoid inhibition of human proopiomelanocortin gene transcription. J Steroid Biochem Mol Biol 40:301-6.
22. Therrien M, Drouin J 1991 Pituitary pro-opiomelanocortin gene expression requires synergistic interactions of several regulatory elements. Mol Cell Biol 11:3492-503.
23. Autelitano DJ 1994 Glucocorticoid regulation of c-fos, c-jun and transcription factor AP-1 in the AtT-20 corticotrope cell. J Neuroendocrinol 6:627-37.
24. Philips A, Maira M, Mullick A, et al. 1997 Antagonism between Nur77 and glucocorticoid receptor for control of transcription. Mol Cell Biol 17:5952-9.
25. Iwasaki Y, Aoki Y, Katahira M, et al. 1997 Non-genomic mechanisms of glucocorticoid inhibition of adrenocorticotropin secretion: possible involvement of GTP-binding protein. Biochem Biophys Res Commun 235:295-9.
26. Korner J, Chun J, Harter D, Axel R 1991 Isolation and functional expression of a mammalian prohormone processing enzyme, murine prohormone convertase 1. Proc Natl Acad Sci U S A 88:6834-8.
27. Seidah NG, Gaspar L, Mion P, et al. 1990 cDNA sequence of two distinct pituitary proteins homologous to Kex2 and furin gene products: tissue-specific mRNAs encoding candidates for pro-hormone processing proteinases [published erratum appears in DNA Cell Biol 1990 Dec;9(10):789]. DNA Cell Biol 9:415-24.
28. Seidah NG, Marcinkiewicz M, Benjannet S, et al. 1991 Cloning and primary sequence of a mouse candidate prohormone convertase PC1 homologous to PC2, Furin, and Kex2: distinct chromosomal localization and messenger RNA distribution in brain and pituitary compared to PC2. Mol Endocrinol 5:111-22.
29. Smeekens SP, Avruch AS, LaMendola J, et al. 1991 Identification of a cDNA encoding a second putative prohormone convertase related to PC2 in AtT20 cells and islets of Langerhans. Proc Natl Acad Sci U S A 88:340-4.
30. Smeekens SP, Steiner DF 1990 Identification of a human insulinoma cDNA encoding a novel mammalian protein structurally related to the yeast dibasic processing protease Kex2. J Biol Chem 265:2997-3000.
31. Lindberg I 1991 The new eukaryotic precursor processing proteinases. Mol Endocrinol 5:1361-5.
32. Mountjoy KG, Robbins LS, Mortrud MT, Cone RD 1992 The cloning of a family of genes that encode the melanocortin receptors. Science 257:1248-51.
33. Mountjoy KG, Bird IM, Rainey WE, Cone RD 1994 ACTH induces up-regulation of

ACTH receptor mRNA in mouse and human adrenocortical cell lines. Mol Cell Endocrinol 99.
34. Barbara JG, Takeda K 1995 Voltage-dependent currents and modulation of calcium channel expression in zona fasciculata cells from rat adrenal gland. J Physiol (Lond) .
35. Simpson ER, Waterman MR 1983 Regulation by ACTH of steroid hormone biosynthesis in the adrenal cortex. Can J Biochem Cell Biol 61:692-707.
36. Simpson ER, Waterman MR 1988 Regulation of the synthesis of steroidogenic enzymes in adrenal cortical cells by ACTH. Annu Rev Physiol 50:427-40.
37. Lin D, Sugawara T, Strauss JF, 3rd, et al. 1995 Role of steroidogenic acute regulatory protein in adrenal and gonadal steroidogenesis [see comments]. Science 267:1828-31.
38. Hall PF 1985 Trophic stimulation of steroidogenesis: in search of the elusive trigger. Recent Prog Horm Res 41:1-39.
39. Vale W, Spiess J, Rivier C, Rivier J 1981 Characterization of a 41-residue ovine hypothalamic peptide that stimulates secretion of corticotropin and beta-endorphin. Science 213:1394-7.
40. Arbiser JL, Morton CC, Bruns GA, Majzoub JA 1988 Human corticotropin releasing hormone gene is located on the long arm of chromosome 8. Cytogenet Cell Genet 47:113-6.
41. Jingami H, Mizuno N, Takahashi H, et al. 1985 Cloning and sequence analysis of cDNA for rat corticotropin-releasing factor precursor. FEBS Lett 191:63-6.
42. Shibahara S, Morimoto Y, Furutani Y, et al. 1983 Isolation and sequence analysis of the human corticotropin-releasing factor precursor gene. Embo J 2:775-9.
43. Thompson RC, Seasholtz AF, Herbert E 1987 Rat corticotropin-releasing hormone gene: sequence and tissue-specific expression. Mol Endocrinol 1:363-70.
44. Bloom FE, Battenberg EL, Rivier J, Vale W 1982 Corticotropin releasing factor (CRF): immunoreactive neurones and fibers in rat hypothalamus. Regul Pept 4:43-8.
45. Kiss JZ, Mezey E, Skirboll L 1984 Corticotropin-releasing factor-immunoreactive neurons of the paraventricular nucleus become vasopressin positive after adrenalectomy. Proc Natl Acad Sci U S A 81:1854-8.
46. Olschowka JA, O'Donohue TL, Mueller GP, Jacobowitz DM 1982 Hypothalamic and extrahypothalamic distribution of CRF-like immunoreactive neurons in the rat brain. Neuroendocrinology 35:305-8.
47. Sawchenko PE, Swanson LW, Vale WW 1984 Co-expression of corticotropin-releasing factor and vasopressin immunoreactivity in parvocellular neurosecretory neurons of the adrenalectomized rat. Proc Natl Acad Sci U S A 81:1883-7.
48. Swanson LW, Sawchenko PE, Rivier J, Vale WW 1983 Organization of ovine corticotropin-releasing factor immunoreactive cells and fibers in the rat brain: an immunohistochemical study. Neuroendocrinology 36:165-86.
49. Plotsky PM, Vale W 1984 Hemorrhage-induced secretion of corticotropin-releasing factor-like immunoreactivity into the rat hypophysial portal circulation and its inhibition by glucocorticoids. Endocrinology 114:164-9.
50. Grino M, Chrousos GP, Margioris AN 1987 The corticotropin releasing hormone gene is expressed in human placenta. Biochem Biophys Res Commun 148:1208-14.
51. Nicholson WE, DeCherney GS, Jackson RV, Orth DN 1987 Pituitary and hypothalamic hormones in normal and neoplastic adrenal medullae: biologically active corticotropin-releasing hormone and corticotropin. Regul Pept 18:173-88.
52. Nieuwenhuijzen Kruseman AC, Linton EA, Lowry PJ, et al. 1982 Corticotropin-releasing factor immunoreactivity in human gastrointestinal tract. Lancet 2:1245-6.
53. Petrusz P, Merchenthaler I, Maderdrut JL, et al. 1983 Corticotropin-releasing factor (CRF)-like immunoreactivity in the vertebrate endocrine pancreas. Proc Natl Acad Sci U S A 80:1721-5.
54. Petrusz P, Merchenthaler I, Ordronneau P, et al. 1984 Corticotropin-releasing factor (CRF)-like immunoreactivity in the gastro-entero-pancreatic endocrine system. Peptides 5:71-8.

55. Sasaki A, Liotta AS, Luckey MM, et al. 1984 Immunoreactive corticotropin-releasing factor is present in human maternal plasma during the third trimester of pregnancy. J Clin Endocrinol Metab 59:812-4.
56. Kishimoto T, Pearse Rn, Lin CR, Rosenfeld MG 1995 A sauvagine/corticotropin-releasing factor receptor expressed in heart and skeletal muscle [published erratum appears in Proc Natl Acad Sci U S A 1995 Apr 25;92(9):4074]. Proceedings of the National Academy of Sciences of the United States of America 92:1108-12.
57. Liaw CW, Lovenberg TW, Barry G, et al. 1996 Cloning and characterization of the human corticotropin-releasing factor-2 receptor complementary deoxyribonucleic acid. Endocrinology 137:72-7.
58. Lovenberg TW, Chalmers DT, Liu C, De SE 1995 CRF2 alpha and CRF2 beta receptor mRNAs are differentially distributed between the rat central nervous system and peripheral tissues. Endocrinology 136:4139-42.
59. Lovenberg TW, Liaw CW, Grigoriadis DE, et al. 1995 Cloning and characterization of a functionally distinct corticotropin-releasing factor receptor subtype from rat brain. Proc Natl Acad Sci U S A 92:836-40.
60. Luo X, Kiss A, Rabadan DC, Aguilera G 1995 Regulation of hypothalamic and pituitary corticotropin-releasing hormone receptor messenger ribonucleic acid by adrenalectomy and glucocorticoids. Endocrinology 136:3877-83.
61. Pozzoli G, Bilezikjian LM, Perrin MH, et al. 1996 Corticotropin-releasing factor (CRF) and glucocorticoids modulate the expression of type 1 CRF receptor messenger ribonucleic acid in rat anterior pituitary cell cultures. Endocrinology 137:65-71.
62. Giguere V, Labrie F, Cote J, et al. 1982 Stimulation of cyclic AMP accumulation and corticotropin release by synthetic ovine corticotropin-releasing factor in rat anterior pituitary cells: site of glucocorticoid action. Proc Natl Acad Sci U S A 79:3466-9.
63. Won JG, Orth DN 1990 Roles of intracellular and extracellular calcium in the kinetic profile of adrenocorticotropin secretion by perifused rat anterior pituitary cells. I. Corticotropin-releasing factor stimulation. Endocrinology 126:849-57.
64. Takano K, Yasufuku TJ, Teramoto A, Fujita T 1996 Corticotropin-releasing hormone excites adrenocorticotropin-secreting human pituitary adenoma cells by activating a nonselective cation current. J Clin Invest 98:2033-41.
65. Watanabe T, Orth DN 1987 Detailed kinetic analysis of adrenocorticotropin secretion by dispersed rat anterior pituitary cells in a microperifusion system: effects of ovine corticotropin-releasing factor and arginine vasopressin. Endocrinology 121:1133-45.
66. Whitnall MH 1993 Regulation of the hypothalamic corticotropin-releasing hormone neurosecretory system. Prog Neurobiol 40:573-629.
67. Thibonnier M, Berti-Mattera LN, Dulin N, et al. 1998 Signal transduction pathways of the human V1-vascular, V2-renal, V3-pituitary vasopressin and oxytocin receptors. Prog Brain Res 119:147-61.
68. Kjaer A 1993 Vasopressin as a neuroendocrine regulator of anterior pituitary hormone secretion. Acta Endocrinol (Copenh) 129:489-96.
69. Antoni FA, Holmes MC, Kiss JZ 1985 Pituitary binding of vasopressin is altered by experimental manipulations of the hypothalamo-pituitary-adrenocortical axis in normal as well as homozygous (di/di) Brattleboro rats. Endocrinology 117:1293-9.
70. Koch B, Lutz BB 1985 Specific receptors for vasopressin in the pituitary gland: evidence for down-regulation and desensitization to adrenocorticotropin-releasing factors. Endocrinology 116:671-6.
71. Berridge MJ, Irvine RF 1984 Inositol trisphosphate, a novel second messenger in cellular signal transduction. Nature 312:315-21.
72. Nishizuka Y 1984 The role of protein kinase C in cell surface signal transduction and tumour promotion. Nature 308:693-8.
73. Nishizuka Y 1988 The molecular heterogeneity of protein kinase C and its implications for cellular regulation. Nature 334:661-5.
74. Oki Y, Nicholson WE, Orth DN 1990 Role of protein kinase-C in the adrenocorticotropin

secretory response to arginine vasopressin (AVP) and the synergistic response to AVP and corticotropin-releasing factor by perifused rat anterior pituitary cells. Endocrinology 127:350-7.
75. Oki Y, Peatman TW, Qu ZC, Orth DN 1991 Effects of intracellular Ca2+ depletion and glucocorticoid on stimulated adrenocorticotropin release by rat anterior pituitary cells in a microperifusion system. Endocrinology 128:1589-96.
76. Won JG, Oki Y, Orth DN 1990 Roles of intracellular and extracellular calcium in the kinetic profile of adrenocorticotropin secretion by perifused rat anterior pituitary cells. II. Arginine vasopressin, oxytocin, and angiotensin-II stimulation. Endocrinology 126:858-68.
77. Abou-Samra AB, Harwood JP, Manganiello VC, et al. 1987 Phorbol 12-myristate 13-acetate and vasopressin potentiate the effect of corticotropin-releasing factor on cyclic AMP production in rat anterior pituitary cells. Mechanisms of action. J Biol Chem 262:1129-36.
78. Watanabe T, Oki Y, Orth DN 1989 Kinetic actions and interactions of arginine vasopressin, angiotensin-II, and oxytocin on adrenocorticotropin secretion by rat anterior pituitary cells in the microperifusion system. Endocrinology 125:1921-31.
79. Gasc JM, Shanmugam S, Sibony M, Corvol P 1994 Tissue-specific expression of type 1 angiotensin II receptor subtypes. An in situ hybridization study. Hypertension 24:531-7
80. Saavedra JM 1992 Brain and pituitary angiotensin. Endocr Rev 13:329-80.
81. Rivier C, Vale W 1983 Effect of angiotensin II on ACTH release in vivo: role of corticotropin-releasing factor. Regul Pept 7:253-8 alpha-MSH. We suggest that desacetyl alpha-MSH, rather than alpha-MSH, is the predominant alpha-MSHi in adult human brain tissue.
82. Gilkes AF, MacKay KB, Cramb G, Guild SB 1992 Atrial natriuretic peptide effects in AtT-20 pituitary tumour cells. Mol Cell Endocrinol 89:39-45.
83. Itoi K, Suda T, Tozawa F, et al. 1994 Microinjection of norepinephrine into the paraventricular nucleus of the hypothalamus stimulates corticotropin-releasing factor gene expression in conscious rats. Endocrinology 135:2177-82.
84. Milenkovic L, Rettori V, Snyder GD, et al. 1989 Cachectin alters anterior pituitary hormone release by a direct action in vitro. Proc Natl Acad Sci U S A 86:2418-22.
85. Rettori V, Milenkovic L, Beutler BA, McCann SM 1989 Hypothalamic action of cachectin to alter pituitary hormone release. Brain Res Bull 23:471-5.
86. Sharp BM, Matta SG 1993 Prostaglandins mediate the adrenocorticotropin response to tumor necrosis factor in rats. Endocrinology 132:269-74.
87. Terao A, Oikawa M, Saito M 1993 Cytokine-induced change in hypothalamic norepinephrine turnover: involvement of corticotropin-releasing hormone and prostaglandins. Brain Res 622:257-61.
88. Kavelaars A, Ballieux RE, Heijnen CJ 1989 The role of IL-1 in the corticotropin-releasing factor and arginine- vasopressin-induced secretion of immunoreactive beta-endorphin by human peripheral blood mononuclear cells. J Immunol 142:2338-42.
89. Sapolsky R, Rivier C, Yamamoto G, et al. 1987 Interleukin-1 stimulates the secretion of hypothalamic corticotropin-releasing factor. Science 238:522-4.
90. Uehara A, Gottschall PE, Dahl RR, Arimura A 1987 Interleukin-1 stimulates ACTH release by an indirect action which requires endogenous corticotropin releasing factor. Endocrinology 121:1580-2.
91. Katsuura G, Gottschall PE, Dahl RR, Arimura A 1988 Adrenocorticotropin release induced by intracerebroventricular injection of recombinant human interleukin-1 in rats: possible involvement of prostaglandin. Endocrinology 122:1773-9.
92. Rivier C, Vale W, Brown M 1989 In the rat, interleukin-1 alpha and -beta stimulate adrenocorticotropin and catecholamine release. Endocrinology 125:3096-102.
93. Weidenfeld J, Abramsky O, Ovadia H 1989 Evidence for the involvement of the central adrenergic system in interleukin 1-induced adrenocortical response. Neuropharmacology 28:1411-4.

94. Mastorakos G, Chrousos GP, Weber JS 1993 Recombinant interleukin-6 activates the hypothalamic-pituitary-adrenal axis in humans. J Clin Endocrinol Metab 77:1690-4.
95. Denicoff KD, Durkin TM, Lotze MT, et al. 1989 The neuroendocrine effects of interleukin-2 treatment. J Clin Endocrinol Metab 69:402-10.
96. Fukata J, Usui T, Naitoh Y, et al. 1989 Effects of recombinant human interleukin-1 alpha, -1 beta, 2 and 6 on ACTH synthesis and release in the mouse pituitary tumour cell line AtT-20. J Endocrinol 122:33-9.
97. Smith LR, Brown SL, Blalock JE 1989 Interleukin-2 induction of ACTH secretion: presence of an interleukin-2 receptor alpha-chain-like molecule on pituitary cells. J Neuroimmunol 21:249-54.
98. Makara GB, Stark E 1976 The effects of cholinomimetic drugs and atropine on ACTH release. Neuroendocrinology 21:31-41.
99. Suda T, Yajima F, Tomori N, et al. 1987 Stimulatory effect of acetylcholine on immunoreactive corticotropin-releasing factor release from the rat hypothalamus in vitro. Life Sci 40:673-7.
100. Fish HR, Chernow B, O'Brian JT 1986 Endocrine and neurophysiologic responses of the pituitary to insulin-induced hypoglycemia: a review. Metabolism 35:763-80.
101. Petraglia F, Facchinetti F, Martignoni E, et al. 1984 Serotoninergic agonists increase plasma levels of beta-endorphin and beta-lipotropin in humans. J Clin Endocrinol Metab 59:1138-42.
102. Jones MT, Gillham B, Beckford U, et al. 1981 Effect of treatment with sodium valproate and diazepam on plasma corticotropin in Nelson's syndrome. Lancet 1:1179-8.
103. McCann SM, Ono N, Khorram O, et al. 1987 The role of brain peptides in neuroimmunomodulation. Ann N Y Acad Sci 496:173-81.
104. Bochicchio D, Ambrosi B, Faglia G 1988 Loperamide, an opiate analog, differently modifies the adrenocorticotropin responses to corticotropin-releasing hormone and lysine vasopressin in patients with Addison's disease. Neuroendocrinology 48:611-4.
105. Pfeiffer A, Herz A, Loriaux DL, Pfeiffer DG 1985 Central kappa- and mu-opiate receptors mediate ACTH-release in rats. Endocrinology 116:2688-90.
106. Iyengar S, Kim HS, Wood PL 1986 Kappa opiate agonists modulate the hypothalamic-pituitary-adrenocortical axis in the rat. J Pharmacol Exp Ther 238:429-36.
107. Grossman A, Gaillard RC, McCartney P, et al. 1982 Opiate modulation of the pituitary-adrenal axis: effects of stress and circadian rhythm. Clin Endocrinol (Oxf) 17:279-86.
108. Volavka J, Cho D, Mallya A, Bauman J 1979 Naloxone increases ACTH and cortisol levels in man [letter]. N Engl J Med 300:1056-7.
109. Rittmaster RS, Cutler GB, Jr., Sobel DO, et al. 1985 Morphine inhibits the pituitary-adrenal response to ovine corticotropin-releasing hormone in normal subjects. J Clin Endocrinol Metab 60:891-5.
110. Donaldson CJ, Sutton SW, Perrin MH, et al. 1996 Cloning and characterization of human urocortin. Endocrinology 137:2167-70.
111. Ozawa M, Oki Y, Watanabe F, et al. 1998 Effect of urocortin and its interaction with adrenocorticotropin (ACTH) secretagogues on ACTH release. Peptides 19:513-8.
112. Vaughan J, Donaldson C, Bittencourt J, et al. 1995 Urocortin, a mammalian neuropeptide related to fish urotensin I and to corticotropin-releasing factor [see comments]. Nature 378:287-92.
113. Masuzawa M, Oki Y, Ozawa M, et al. 1999 Corticotropin-releasing factor but not urocortin is involved in adrenalectomy-induced adrenocorticotropin release. J Neuroendocrinol 11:71-4.
114. Reul JM, de KE 1985 Two receptor systems for corticosterone in rat brain: microdistribution and differential occupation. Endocrinology 117:2505-11.
115. Plotsky PM, Otto S, Sapolsky RM 1986 Inhibition of immunoreactive corticotropin-releasing factor secretion into the hypophysial-portal circulation by delayed glucocorticoid feedback. Endocrinology 119:1126-30.
116. Dallman MF, Akana SF, Cascio CS, et al. 1987 Regulation of ACTH secretion: variations

on a theme of B. Recent Prog Horm Res 43:113-73.
117. Kovacs K, Kiss JZ, Makara GB 1986 Glucocorticoid implants around the hypothalamic paraventricular nucleus prevent the increase of corticotropin-releasing factor and arginine vasopressin immunostaining induced by adrenalectomy. Neuroendocrinology 44:229-34.
118. Kovacs KJ, Mezey E 1987 Dexamethasone inhibits corticotropin-releasing factor gene expression in the rat paraventricular nucleus. Neuroendocrinology 46:365-8.
119. Dayanithi G, Antoni FA 1989 Rapid as well as delayed inhibitory effects of glucocorticoid hormones on pituitary adrenocorticotropic hormone release are mediated by type II glucocorticoid receptors and require newly synthesized messenger ribonucleic acid as well as protein. Endocrinology 125:308-13.
120. Keller-Wood M, Dallman MF 1984 Corticosteroid inhibition of ACTH secretion. Endocr Rev 5:1-24.
121. Eberwine JH, Jonassen JA, Evinger MJ, Roberts JL 1987 Complex transcriptional regulation by glucocorticoids and corticotropin-releasing hormone of proopiomelanocortin gene expression in rat pituitary cultures. Dna 6:483-92.
122. Veldhuis JD, Iranmanesh A, Johnson ML, Lizarralde G 1990 Twenty-four-hour rhythms in plasma concentrations of adenohypophyseal hormones are generated by distinct amplitude and/or frequency modulation of underlying pituitary secretory bursts. J Clin Endocrinol Metab 71:1616-23.
123. Veldhuis JD, Iranmanesh A, Johnson ML, Lizarralde G 1990 Amplitude, but not frequency, modulation of adrenocorticotropin secretory bursts gives rise to the nyctohemeral rhythm of the corticotropic axis in man. J Clin Endocrinol Metab 71:452-63.
124. DeCherney GS, DeBold CR, Jackson RV, et al. 1987 Effect of ovine corticotropin-releasing hormone administered during insulin-induced hypoglycemia on plasma adrenocorticotropin and cortisol. J Clin Endocrinol Metab 64:1211-8.
125. Watabe T, Tanaka K, Kumagae M, et al. 1987 Hormonal responses to insulin-induced hypoglycemia in man. J Clin Endocrinol Metab 65:1187-91.
126. Plotsky PM 1988 Hypophysiotropic regulation of stress-induced ACTH secretion. Adv Exp Med Biol 245:65-81.
127. Ellis MJ, Schmidli RS, Donald RA, et al. 1990 Plasma corticotrophin-releasing factor and vasopressin responses to hypoglycaemia in normal man. Clin Endocrinol (Oxf) 32:93-100.
128. Baylis PH, Zerbe RL, Robertson GL 1981 Arginine vasopressin response to insulin-induced hypoglycemia in man. J Clin Endocrinol Metab 53:935-40.
129. Udelsman R, Norton JA, Jelenich SE, et al. 1987 Responses of the hypothalamic-pituitary-adrenal and renin-angiotensin axes and the sympathetic system during controlled surgical and anesthetic stress. J Clin Endocrinol Metab 64:986-94.
130. Bramnert M, Hokfelt B 1987 The influence of naloxone on exercise-induced increase in plasma pituitary hormones and the subjectively experienced level of exhaustion in healthy males. Acta Endocrinol (Copenh) 115:125-30.
131. Pestell RG, Hurley DM, Vandongen R 1989 Biochemical and hormonal changes during a 1000 km ultramarathon. Clin Exp Pharmacol Physiol 16:353-61.
132. Pitts AF, Preston Md, Jaeckle RS, et al. 1990 Simulated acute hemorrhage through lower body negative pressure as an activator of the hypothalamic-pituitary-adrenal axis. Horm Metab Res 22:436-43.
133. Plotsky PM, Bruhn TO, Vale W 1985 Hypophysiotropic regulation of adrenocorticotropin secretion in response to insulin-induced hypoglycemia. Endocrinology 117:323-9.
134. Plotsky PM, Bruhn TO, Vale W 1985 Evidence for multifactor regulation of the adrenocorticotropin secretory response to hemodynamic stimuli. Endocrinology 116:633-9.
135. Gann DS 1966 Carotid vascular receptors and control of adrenal corticosteroid secretion. Am J Physiol 211:193-7.

136. Mesiano S, Mellon SH, Jaffe RB 1993 Mitogenic action, regulation, and localization of insulin-like growth factors in the human fetal adrenal gland. J Clin Endocrinol Metab 76:968-76.
137. Mesiano S, Mellon SH, Gospodarowicz D, et al. 1991 Basic fibroblast growth factor expression is regulated by corticotropin in the human fetal adrenal: a model for adrenal growth regulation. Proc Natl Acad Sci U S A 88:5428-32.
138. Townsend SF, Dallman MF, Miller WL 1990 Rat insulin-like growth factor-I and -II mRNAs are unchanged during compensatory adrenal growth but decrease during ACTH-induced adrenal growth. J Biol Chem 265:22117-22.
139. Voutilainen R, Miller WL 1987 Coordinate tropic hormone regulation of mRNAs for insulin-like growth factor II and the cholesterol side-chain-cleavage enzyme, P450scc, in human steroidogenic tissues. Proc Natl Acad Sci U S A 84:1590-4.
140. Dallman MF 1984 Control of adrenocortical growth in vivo. Endocr Res 10:213-42.
141. Lala DS, Rice DA, Parker KL 1992 Steroidogenic factor I, a key regulator of steroidogenic enzyme expression, is the mouse homolog of fushi tarazu-factor I. Mol Endocrinol 6:1249-58.
142. Morohashi K, Honda S, Inomata Y, et al. 1992 A common trans-acting factor, Ad4-binding protein, to the promoters of steroidogenic P-450s. J Biol Chem 267:17913-9
143. Schoneshofer M, Wagner GG 1977 Sex differences in corticosteroids in man. J Clin Endocrinol Metab 45:814-7.
144. Zumoff B, Fukushima DK, Weitzman ED, et al. 1974 The sex difference in plasma cortisol concentration in man. J Clin Endocrinol Metab 39:805-8.
145. Zumoff B, Fukushima DK, Hellman L 1974 Intercomparison of four methods for measuring cortisol production. J Clin Endocrinol Metab 38:169-75.
146. Albiston AL, Smith RE, Obeyesekere VR, Krozowski ZS 1995 Cloning of the 11 beta HSD type II enzyme from human kidney. Endocr Res 21:399-409.
147. Stewart PM, Mason JI 1995 Cortisol to cortisone: glucocorticoid to mineralocorticoid. Steroids 60:143-6.
148. Dunn JF, Nisula BC, Rodbard D 1981 Transport of steroid hormones: binding of 21 endogenous steroids to both testosterone-binding globulin and corticosteroid-binding globulin in human plasma. J Clin Endocrinol Metab 53:58-68.
149. Hammond GL, Smith CL, Goping IS, et al. 1987 Primary structure of human corticosteroid binding globulin, deduced from hepatic and pulmonary cDNAs, exhibits homology with serine protease inhibitors. Proc Natl Acad Sci U S A 84:5153-7.
150. Brien TG 1981 Human corticosteroid binding globulin. Clin Endocrinol (Oxf) 14:193-212.
151. Hollenberg SM, Weinberger C, Ong ES, et al. 1985 Primary structure and expression of a functional human glucocorticoid receptor cDNA. Nature 318:635-41.
152. Weinberger C, Hollenberg SM, Ong ES, et al. 1985 Identification of human glucocorticoid receptor complementary DNA clones by epitope selection. Science 228:740-2.
153. Smith DF 1993 Dynamics of heat shock protein 90-progesterone receptor binding and the disactivation loop model for steroid receptor complexes. Mol Endocrinol 7:1418-29.
154. Pratt WB 1993 The role of heat shock proteins in regulating the function, folding, and trafficking of the glucocorticoid receptor. J Biol Chem 268:21455-8.
155. Smith DF, Toft DO 1993 Steroid receptors and their associated proteins. Mol Endocrinol 7:4-11.
156. Pratt WB 1992 Control of steroid receptor function and cytoplasmic-nuclear transport by heat shock proteins. Bioessays 14:841-8.
157. Strahle U, Klock G, Schutz G 1987 A DNA sequence of 15 base pairs is sufficient to mediate both glucocorticoid and progesterone induction of gene expression. Proc Natl Acad Sci U S A 84:7871-5.
158. Scheidereit C, Geisse S, Westphal HM, Beato M 1983 The glucocorticoid receptor binds to defined nucleotide sequences near the promoter of mouse mammary tumour virus.

Nature 304:749-52.
159. Konig H, Ponta H, Rahmsdorf HJ, Herrlich P 1992 Interference between pathway-specific transcription factors: glucocorticoids antagonize phorbol ester-induced AP-1 activity without altering AP-1 site occupation in vivo. Embo J 11:2241-6.
160. Schule R, Rangarajan P, Kliewer S, et al. 1990 Functional antagonism between oncoprotein c-Jun and the glucocorticoid receptor. Cell 62:1217-26.
161. Yang-Yen HF, Chambard JC, Sun YL, et al. 1990 Transcriptional interference between c-Jun and the glucocorticoid receptor: mutual inhibition of DNA binding due to direct protein-protein interaction. Cell 62:1205-15.
162. Jonat C, Rahmsdorf HJ, Park KK, et al. 1990 Antitumor promotion and antiinflammation: down-modulation of AP-1 (Fos/Jun) activity by glucocorticoid hormone. Cell 62:1189-204.
163. Weinstein SP, Paquin T, Pritsker A, Haber RS 1995 Glucocorticoid-induced insulin resistance: dexamethasone inhibits the activation of glucose transport in rat skeletal muscle by both insulin- and non-insulin-related stimuli. Diabetes 44:441-5.
164. Fauci AS 1975 Mechanisms of corticosteroid action on lymphocyte subpopulations. I. Redistribution of circulating T and b lymphocytes to the bone marrow. Immunology 28:669-80.
165. Fauci AS, Dale DC 1974 The effect of in vivo hydrocortisone on subpopulations of human lymphocytes. J Clin Invest 53:240-6.
166. Blewitt RW, Abbott AC, Bird CC 1983 Mode of cell death induced in human lymphoid cells by high and low doses of glucocorticoid. Br J Cancer 47:477-86.
167. McKay LI, Cidlowski JA 1999 Molecular control of immune/inflammatory responses: interactions between nuclear factor-kappa B and steroid receptor-signaling pathways. Endocr Rev 20:435-59.
168. Raff H 1987 Glucocorticoid inhibition of neurohypophysial vasopressin secretion. Am J Physiol 252:R635-44.
169. Davis LG, Arentzen R, Reid JM, et al. 1986 Glucocorticoid sensitivity of vasopressin mRNA levels in the paraventricular nucleus of the rat. Proc Natl Acad Sci U S A 83:1145-9.
170. Hahn TJ, Halstead LR, Teitelbaum SL, Hahn BH 1979 Altered mineral metabolism in glucocorticoid-induced osteopenia. Effect of 25-hydroxyvitamin D administration. J Clin Invest 64:655-65.
171. Klein RG, Arnaud SB, Gallagher JC, et al. 1977 Intestinal calcium absorption in exogenous hypercortisonism. Role of 25-hydroxyvitamin D and corticosteroid dose. J Clin Invest 60:253-9.
172. Hersey RM, DiStefano V 1979 Control of phenylethanolamine N-methyltransferase by glucocorticoids in cultured bovine adrenal medullary cells. J Pharmacol Exp Ther 209:147-52.
173. Loosen PT, Chambliss B, DeBold CR, et al. 1992 Psychiatric phenomenology in Cushing's disease. Pharmacopsychiatry 25:192-8.

2
CLINICAL MANIFESTATIONS AND PRESENTATION OF CUSHING'S SYNDROME

A Brew Atkinson, M.D.
Royal Victoria Hospital and Queen's University, Belfast, Northern Ireland

CLINICAL HISTORY

Cushing's syndrome occurs as a result of prolonged glucocorticoid excess (1-5). This chapter reviews the salient clinical features of the syndrome which, regardless of aetiology, is a constellation of clinical and biochemical abnormalities resulting from chronic excess of cortisol. While there are classical features of Cushing's syndrome, the diagnosis is increasingly being made earlier, before major complications have occurred. An early diagnosis is of advantage to patients as mortality from the untreated syndrome is high (see below).

Many of the clinical findings are common to the various types of Cushing's syndrome. The classical appearance of plethoric facies, centripetal obesity, ecchymoses, fragile skin due to poor quality collagen, livid striae and hirsutism is still seen. However, in my experience it is by no means as frequent as previously observed as many cases are being diagnosed at an earlier and more subtle stage. Perhaps the commonest differential diagnosis is simple obesity, particularly in patients with glucose intolerance and/or hirsutism.

It is nearly 70 years since Harvey Cushing's original and classic description of the syndrome (1) but the presentations seen in the original patients remain astonishing in their disparity and comprehensiveness. His description of Minnie G (Figure 1) still bears repetition.

MINNIE G

"One of a numerous and healthy family, though slight and undersized, she was well until 16 years of age, having escaped the customary children's ailments. Her menses which started at the age of 14 were regular for two years and then suddenly ceased. She began to grow stout and in the two years prior to admission her weight had increased from 112 to 137 pounds. She suffered greatly from headaches, nausea and vomiting sometimes accompanying the more severe attacks. She complained also of aching pains in the eyes which latterly had become prominent, and there had been occasional periods of seeing double. Other noteworthy symptoms were insomnia, tinnitus, extreme dryness of the skin, frequent sore throats, shortness of breath, palpitation, purpuric outbreaks, recurring nose bleeds, and marked constipation accompanied by bleeding piles. A definite growth of hair had appeared on the face with thinning of hair on the scalp. She had become increasingly round-shouldered. Muscular weakness had become extreme and there was constant complaint of backache and epigastric pains.

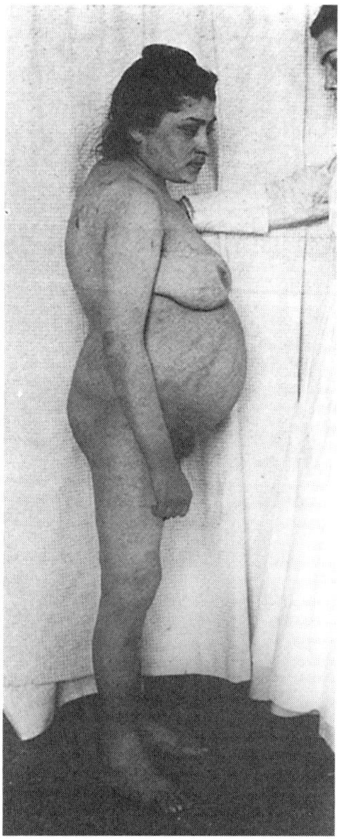

Figure 1 – Minnie G. (1)

Physical examination showed an undersized, kyphotic young woman 4 feet 9 inches in height (145 cm), of most extraordinary appearance. Her round face was dusky and cyanosed and there was an abnormal growth of hair, particular noticeable on the sides of the forehead, upper lip and chin. The mucous membranes were of bright colour despite her history of frequent bleedings. Her abdominous body had the appearance of a full-term pregnancy. The breasts were hypertrophic and pendulous and there were pads of fat over the supraclavicular and posterior cervical regions. The cyanotic appearance of the skin was particularly apparent over the body and lower extremities which were spotted by subcutaneous ecchymoses. Numerous purplish striae were present over the stretched skin of the lower abdomen and also over shoulders, breasts and hips; and a fine hirsuties was present over the back, hips and around the umbilicus. The skin which everywhere was rough and dry, showed considerable pigmentation, particularly around the eyelids, groins, pubes and areolae of the breasts. The peculiar tense and painful adiposity affecting face, neck and trunk was in marked contrast to her comparatively spare extremities.

From a neurological aspect nothing was notable other than what at the time were taken to be signs of intracranial pressure: namely, headaches, slight exophthalmos, diplopia, puffiness of the eyelids and congestion of the optic discs (due, as would now appear, to deposition of intra-orbital fat). The cranial X-ray showed what for the day was regarded as a normal sella turcica. The epiphyseal lines (radial and phalangeal) were still roentgenologically visible. Not only did the skin bruise easily but spontaneous ecchymoses frequently appeared. The systolic blood pressure was consistently high".

Cushing then states that in the intervening years six other examples of the same or a highly similar disorder had been carefully studied at the Brigham Hospital. The patients were all comparatively young women who, in association with a more or less abrupt amenorrhoea, had become rapidly obese with a peculiar tense and more or less painful adiposity chiefly affecting head, neck and trunk. They were all plethoric in appearance, all had become abnormally hirsute, all but one showed purplish cutaneous striae. Vascular hypertension with a high red blood count and haemoglobin percentage was usually present, and all complained of aches and pains and general enfeeblement.

In addition to these 7 cases Cushing reviewed 12 other cases considered to be further examples of the syndrome. Among these cases he also described childhood presentations, the finding of a basophil adenoma of the pituitary gland in one patient, of adrenal carcinoma with metastases in another and the successful use of irradiation therapy for presumed pituitary disease (patient 11). His collation of symptoms is given in Table 1 and is still highly relevant. The prevalence of the various symptoms and signs have been well reviewed subsequently (6-10). However, many of these symptoms are not uncommon and so the study of Ross and Linch (10) is of particular importance. They attempted to provide discriminant indices for use in the diagnosis of Cushing's syndrome, which as can be clearly seen from the above often includes conditions commonly seen in obese females e.g. disturbed menstruation, hirsutism and hypertension. To determine the discriminatory value of particular symptoms that can be detected at the initial examination they compared the prevalence of each feature in a collected series of 711 patients from the literature with Cushing's syndrome with a series of 159 mostly obese patients reported by Nugent et al (7) in whom the diagnosis of Cushing's syndrome had been suspected but not biochemically substantiated. In the latter series there was a high prevalence of glucose intolerance and menstrual abnormality reflecting the reason for referral to an endocrine unit.

Table 1. Clinical Features of the Syndrome as collated by Cushing (1)

Consistent

- Adiposity confined to face, neck and trunk
- Round-shoulderedness (kyphosis)
- Sexual dystrophy - early amenorrhoea in females
  - impotence in males
- Hypertrichosis of face and trunk in females
- Dusky or plethoric skin with purplish lineae atrophicae
- Hypertension
- Erythraemia
- Backaches, abdominal pain, fatigability and weakness
- Susceptible to infections

Less consistent

- Acrocyanosis
- Ecchymoses (often spontaneous)
- Aching eyes (associated with exophthalmos)
- Dry skin
- Polyphagia, polydipsia and polyuria
- Oedema of lower limbs
- Insomnia
- Polymorphonuclear leucocytosis

Table 2 lists the common features of the syndrome while Table 3 shows the discriminant indices. It can be seen that the presence of bruising, muscle weakness and hypertension are the most discriminating features of the disease that can be detected at the initial examination. Impaired glucose tolerance, headaches and menstrual irregularity are common in Cushing's syndrome but they are not uncommon in patients with simple obesity and are therefore of limited discriminatory value. In contrast to Cushing's own original description the distribution of the obesity was not a reliable indicator of Cushing's syndrome since less than half the patients in the series of Ross and Linch had truncal obesity.

Table 2. Clinical Features of Cushing's Syndrome (adapted from Ref 10)

	% in series			% in series	
Symptoms	Present	Collected	Signs	Present	Collected
Weight gain	91	..	Obesity:	97	90
Menstrual irregularity	84	80	Truncal	46	..
Hirsutism in women	81	79	Generalised	55	..
Lethargy/depression	62	..	Plethora	94	80
Headache	47	41	Facial rounding	88	88
Thirst/frequency	44	30	Bruising	62	63
Backache	43	50	Striae	56	68
Muscle weakness	29	..	Muscle weakness	56	65
Shortness of breath	26	..	Buffalo hump	54	..
Recurrent infections	25	..	Ankle oedema	50	56
Abdominal pain	21	..	Acne	21	..
Fractures	19	36	Pigmentation	4	..
Loss of scalp hair	13	..			

	% in series	
Other findings	Present	Collected
Hypertension	74	87
Diabetes:	50	55
Overt	13	19
Abnormal glucose tolerance	37	36
Osteoporosis	50	64
Renal calculi	15	19

Present series = Ross and Linch (10) n = 70
Collected series, n = 711

It remains essential to make the diagnosis early. As long ago as 1982 Ross and Linch (10) commented that the disease should not now be seen in the fully developed form because the diagnosis should have been made long before all the advanced features appear. This is generally true although onset can be very rapid and florid in some cases of the ectopic ACTH syndrome and in adrenal carcinoma. The reason for wanting to make a diagnosis as early as possible is that there is a very poor outlook if the disease is overlooked. Cushing himself (1) reported that the average duration of the disease from onset to death was slightly over five years, the extremes in his published series being three and seven years. Plotz, Knowlton and Ragan (11) discussed the natural history of Cushing's syndrome in a study encompassing 33 of their own patients and another 189 cases from the literature. Seventeen of their 33 cases died within five years of the known onset of disease while in their review of 114 cases where an autopsy had been performed, an extremely high prevalence of arteriosclerosis and of osteoporosis was noted. The major causes of death were infection, cardiovascular complications and neoplastic disease. In the light of this extremely poor prognosis, it seems wise to remember the original and classic descriptions of the syndrome but also to be ready to investigate any patient in whom there appears to be a reasonable possibility of the disease being present.

Table 3: Discriminant Indices of Clinical Features in Cushing's Syndrome (from reference 10)

	Discriminant index in series	
Clinical feature	Present	Collected
Bruising	10.3	10.5
Myopathy	8.0	7.1
Hypertension	4.4	5.1
Plethora	3.0	3.6
Oedema	2.9	3.3
Hirsutism	2.8	2.7
Red striae	2.5	3.1
Menstrual irregularity	1.6	1.6
Truncal obesity	1.6	..
Headaches	1.3	1.1
Acne	0.9	..
Generalised obesity	0.8	..
Impaired glucose tolerance	0.7	0.7

INFECTIONS IN CUSHING'S SYNDROME

Cushing described a susceptibility to pulmonary infection in five of the reported series. Plotz, Knowlton and Reagen (11) also reported an increased incidence of infections. In 1984 Graham and Tucker (12) reported a Nashville series of six patients with Cushing's syndrome and opportunistic infections and they compared these patients with another 17 reported elsewhere in the literature. In their cases the hypercortisolism was extreme. Importantly, signs and symptoms of infection were often masked by the hypercortisolism. Fever was an inconsistent and unreliable sign of infection. In the absence of fever signs of pneumonia were attributed to tumour. In addition changes in mental state were blamed on steroid psychosis when metabolic encephalopathy may have been more significant. Even bacterial and fungal meningitis were initially not recognised. Mortality was high (14 of 23) and in those who survived a key prognostic feature was early reduction of cortisol production to near normal. Four infectious processes were preponderant, namely cryptococcosis, aspergillosis, nocardiosis and pneumocystis carinii. These opportunistic infections are most likely to occur in patients with ectopic ACTH syndrome or adrenal carcinoma where the cortisol levels are rather more likely to be extremely high. The clinician obviously must be very alert to this potentially life-threatening complication of the syndrome.

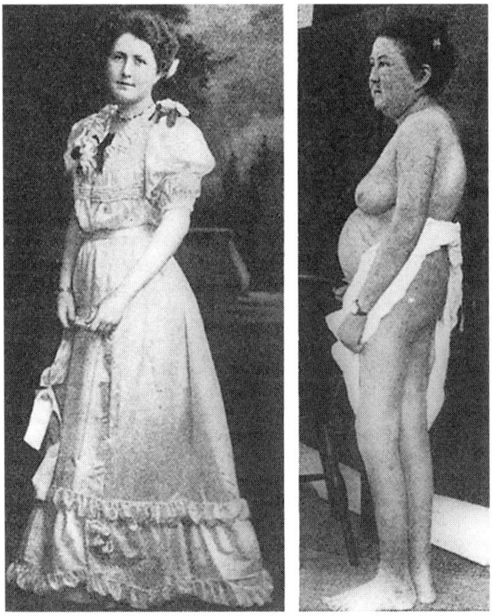

Figure 2. Before and 5 years later with development of hypercortisolism (from reference 1)

VASCULAR COMPLICATIONS

As already discussed, vascular complications contribute to the increased mortality rate of patients with Cushing's syndrome. In 1976 Sjöberg, Blombäck and Granberg reported that seven of 19 consecutive patients with Cushing's syndrome had thromboembolic manifestations before or after adrenalectomy (13). They also had increased levels of coagulation factors V and VIII as well as of prothrombin. These results and those of Dal Bo Zanon and colleagues (14) found incidences of thromboembolic complications of 37 and 27 per cent respectively. By contrast Small et al (15) suggested in a series of 43 patients a lower incidence and suggested that it was no more common than expected in the general population. Nevertheless they noted a 9% incidence of arterial events and an 11% occurrence of deep venous thrombosis or pulmonary embolism. They noted that all of the patients with arterial events were smokers as were four of six with venous thrombosis. Clinicians involved in management should be aware of this complication as a possibility. Prophylactic anticoagulation is indicated for all patients undergoing abdominal surgery. Smoking should be actively discouraged. In our experience (unpublished observations) venous thrombosis after pituitary surgery is unusual.

It is agreed that insulin resistance is increased in the syndrome (16,17) as are plasma levels of very low density lipoprotein (VLDL) while high density lipoprotein is lowered (18).

SEXUAL DYSFUNCTION

Disturbance of menses, hirsutism etc., in women have been discussed above. In men plasma testosterone levels are often suppressed, leading to complaints of loss of libido and decreased sexual potency (19).

PSYCHIATRIC MANIFESTATIONS

These are discussed in detail elsewhere in this book but, in brief, it has been shown that pathological anxiety, depression, mania and psychotic illness can all be associated. Of note is Kelly's finding (20) that when Cushing's syndrome was diagnosed, significant psychiatric illness, usually depression, was present or had been a feature of the syndrome in 120/209 (57%) of patients.

ORTHOPAEDIC MANIFESTATIONS

The major manifestation is osteoporosis and its complications and these have been discussed above. In addition, two recent cases of spontaneous rupture of the Achilles tendon have been reported as missed presentations of the disease. Spontaneous rupture is unusual but is seen in rheumatoid arthritis and systemic lupus erythematosus and has also been well described in patients receiving long term steroid therapy. In the two cases with endogenous hypercortisolism the diagnosis was made one and seven months later (21).

EXOPHTHALMOS

As already discussed exophthalmos was noted by Cushing himself and has subsequently been reported in other series. Kelly (22) measured exophthalmos in 31 patients with active Cushing's syndrome and found that it exceeded 16 mm (>2 SD above normal mean) in 45 per cent of cases of active Cushing's syndrome. Fortunately this rarely causes symptoms and diminishes when cortisol concentrations become normal. It is therefore best regarded as an alerting sign to the possibility of hypercortisolism being present.

SYMPTOMS ASSOCIATED WITH LINKED CONDITIONS

The diseases causing or associated with Cushing's syndrome have their own diagnostic features. These are briefly discussed below.

Carney's complex

Cushing's syndrome (on the basis of bilateral nodular adrenal hyperplasia) can occur as part of this familial complex. Other components are mesenchymal tumours (myxomas), pigmented skin lesions, spotty facial pigmentation, other evidence of functional endocrine tumours (pituitary and testicular) and peripheral nerve tumours (see references 23-25).

Ectopic ACTH and ectopic CRH syndrome

The former is more common and in the case of small carcinoids of the lung and pancreas may be very indolent without specific diagnostic features and therefore can be extremely difficult to distinguish from pituitary-dependent Cushing's syndrome (4,26,27). However, in oat cell carcinoma of the lung and in other cancers (28,29) the carcinomatous features, e.g., weight loss, cachexia and haemoptysis may be the major presenting features. The common causes of ectopic ACTH syndrome and ectopic CRH syndrome are shown in Table 4 (see also reference 30).

Table 4. Causes of ectopic ACTH and CRH syndromes

Ectopic ACTH		
	Lung carcinoma	
	Undetermined carcinoma	
	Carcinoid	thyroid
		pancreas
		bronchial
		g.i. tract
	Pancreatic carcinoma	
	Phaeochromocytoma	
	Medullary carcinoma of thyroid	
Ectopic CRH		
	Pancreatic carcinoma	
	Carcinoid	bronchial
		Thymus
	Medullary carcinoma of thyroid	
	Prostate carcinoma	

CHILDHOOD CUSHING'S SYNDROME

This condition is unusual but it is relevant that the presenting symptoms differ from those seen in the adult (31,32). Growth failure and weight gain are principal features while signs of virilism and hypertension are common. Psychiatric manifestations are less common than in the adult group. The findings in 59 children and adolescents studied at the N.I.H. are shown in Table 5. In most cases the onset is insidious over months to years. The majority have a pituitary aetiology but adrenal hyperplasia and tumours including carcinoma as well as the ectopic ACTH syndrome have all been reported.

Table 5. Symptoms and signs in 59 children and adolescents (31)

	No. affected	Per Cent
Weight gain	53/59	90
Growth retardation	35/42	83
Menstrual irregularities	14/18	78
Hirsutism	29/37	78
Obesity	44/59	75
Violaceous striae	36/59	61
Acne	28/59	47
Hypertension	28/59	47
Fatigue or weakness	26/59	44
Early secondary sexual development	8/21	38
Bruising	15/59	25
Mental changes	11/59	19
Hyperpigmentation	8/59	14
Muscle weakness	7/59	12
Acanthosis nigricans	7/59	12
Bone age delayed (> 2SD)	4/37	11
Bone age accelerated (> 2SD)	3/37	8
Sleep disturbances	5/59	8

All symptoms and signs seen in at least 8 per cent of cases are listed.
From Magiakou et al (31)

INTERMITTENT, PERIODIC AND CYCLICAL CUSHING'S SYNDROME

Cushing's first patient (MG) had a prolonged remission of symptoms. Since then this has also been reported by others. It may well be more common than previously reported. In our unit's own experience we have described levels of 24 hour urinary free cortisols which fluctuate markedly and in some patients the symptoms also fluctuate. Of 14 successive patients with Cushing's syndrome reported by our group, nine collected sequential daily urine samples for the estimation of cortisol to creatinine ratio in their first early morning urine. Five were shown to have cyclical Cushing's syndrome (33). Varying cycle lengths of between 12 hours and 85 days have been reported. In some cases the troughs are within the normal range. We have also described this situation postoperatively in patients who have been slow to lose their clinical features of the syndrome yet have some normal cortisol values postoperatively (34). Recently Liebowitz and colleagues (35) reported a patient whose cortisol fluctuated spontaneously between severe hypercortisolism associated with features of Cushing's syndrome to recurrent symptomatic hypocortisolism requiring glucocorticoid therapy. In addition Wallace and colleagues (36) described a patient who developed clinical and biochemical Cushing's syndrome in three pregnancies. After each delivery there was prompt resolution of the syndrome.

The implications of cyclical and intermittent forms of the disease are that a high index of suspicion of the syndrome is required in patients with symptoms and signs of Cushing's syndrome but with normal cortisol values when first evaluated.

A clinical decision must always be made for each patient as to when diagnostic testing has been sufficient to rule out the syndrome in that particular clinical case. This can be extremely difficult even for the most experienced endocrinologist.

ALTERNATIVE DIAGNOSES

Although the diagnosis is usually made on the basis of biochemical testing the clinical features described in detail above are helpful. The most common diagnostic dilemmas concern cases of obesity, depression and alcoholism while occasionally fictitious hypercortisolism and familial cortisol resistance have to be considered.

Weight increase is usually greater in patients with simple obesity than in those with hypercortisolism where the obesity may be more apparent than the rest. However, one has to remember that generalised obesity does occur accounting for 50 per cent of cases in one series. The diagnosis then depends on the presence or absence of other clinical features, the use of discriminant indices (see above) and the use of laboratory tests (see later chapters).

Table 6. Alternative diagnoses to endogenous Cushing's syndrome

Obesity
Depression
Alcohol-induced pseudo-Cushing's syndrome
Primary cortisol resistance
Fictitious Cushing's syndrome

Patients with Cushing's syndrome frequently have marked changes which can vary from frank psychosis to mild behavioural problems. Patients with florid depression share many of the biochemical abnormalities seen in patients with Cushing's syndrome but do not usually have the clinical stigmata of hypercortisolism unless obesity with its associated problems is present. The differential diagnosis is not usually too great a problem for the endocrinologist experienced in management of true Cushing's syndrome.

There have been a wide variety of reports of alcohol-induced pseudo-Cushing's syndrome, where patients present with a Cushingoid appearance and characteristic clinical features of obesity, plethora and striae (37,38). In many of such reported patients, once the history of excessive alcohol intake had been established, the biochemical hypercortisolism and clinical manifestations disappeared with abstention. The precise aetiology of the condition has not been elucidated.

Primary cortisol resistance is a rare disorder, described in only a few families (39-41). Most cases have not had the clinical features of Cushing's syndrome, but such cortisol resistance is of some importance in the context of hypertension, as raised blood pressure and hypokalaemic alkalosis can be the presenting signs. One patient described had hirsutism, male pattern baldness and menstrual irregularities. Patients have raised cortisol excretion and impaired suppression of cortisol after dexamethasone administration. However, it has been shown that they have normal cortisol and growth hormone responses to hypoglycaemia. The diagnosis should be considered in patients who appear not to have the classical glucocorticoid features of Cushing's syndrome, particularly if hypertension or hypokalaemia are present.

When it is not possible to distinguish between mild true hypercortisolism, and/or depression Yanovski and colleagues (1993) showed that a combined dexamethasone and CRH stimulation test differentiated well before the conditions (42).

Factitious hypercortisolism is quite rare (43,44) and the diagnosis is often difficult. In all cases reported to date the history provided important clues. There was often access to the medical profession by means of employment or family members. Many had undergone had multiple prior surgeries for unrelated illnesses. Rapid alterations in steroid values are seen making the condition difficult to distinguish from cyclical disease though Cizza et al (44) point out correctly that in fictitious cases it is common for very high levels on one day to be followed on the next by very low values. Valuable pointers were the presence of atrophic adrenal glands on scanning and discrepant plasma and urine glucocorticoids. Urinary HPLC should be performed when the diagnosis is being actively considered as synthetic glucocorticoids can be detected by this methodology. Management of this group of patients is difficult and no extended follow-up results are available.

SUMMARY

Cushing's syndrome is a relatively unusual but very unpleasant condition with a poor prognosis if left untreated. When any of the symptoms and signs discussed above are present it should always be considered thoughtfully. When the clinical diagnosis has been made or is seriously entertained detailed assessment should follow because of the potential pitfalls in elucidating the cause and because of the wide variety of treatment options. At this stage patients should be referred to endocrinology centres with special interest and expertise in the disease.

REFERENCES

1. Cushing H. The basophil adenomas of the pituitary body and their clinical manifestations. Bull Johns Hopkins Hospital 1932; 50:137-95.
2. Liddle GW. Cushing's syndrome. Ann NY Acad Sci 1977; 297:594-602.
3. Besser GM, Edwards CRW. Cushing's syndrome. Clin Endocrinol Metab 1972; 1:451-90.
4. Orth DN. The old and the new in Cushing's syndrome. N Engl J Med 1984; 310:649-51.
5. Atkinson AB. "Cushing's syndrome." In Handbook of Hypertension 15, JIS Robertson, ed. Elsevier, Amsterdam, 1992, pp390-419.
6. Soffer LJ, Iannaccone A, Gabrilove JL. Cushing's syndrome. A study of 50 patients. Am J Med 1961; 30:129-46.
7. Nugent CA, Warner HR, Dunn JT, Tyler FH. Probability theory in the diagnosis of Cushing's syndrome. J Clin Endocrinol Metab 1964; 24:621-30.
8. Ross EJ, Marshall-Jones P, Friedman M. Cushing's syndrome: diagnostic criteria. Quart J Med 1966; 35:149-53.
9. Crapo L. Cushing's syndrome: a review of diagnostic tests. Metabolism 1979; 28:955-7
10. Ross EJ, Linch DC. Cushing's syndrome – killing disease: discriminatory value of signs and symptoms aiding early diagnosis. Lancet 1982; ii:646-9.
11. Plotz CM, Knowlton AI, Ragan C. The natural history of Cushing's syndrome. Am J Med 1952; 13:547-614.
12. Graham BS, Tucker WS. Opportunistic infections in endogenous Cushing's syndrome. Ann Intern Med 1984; 101:334-8.
13. Sjöberg HE, Blombäck M, Granberg PO. Thromboembolic complications, heparin treatment and increase in coagulation factors in Cushing's syndrome. Acta Med Scand 1976; 199:95-8.
14. Dal Bo Zanon R, Fornasiero L, Boscaro M, Cappellatog Fabris F, Girolami M. Increased factor VIII associated activities in Cushing's syndrome: a probable hypercoagulable state. Thromb Haemo 1982; 47:116-7.
15. Small M, Lowe GDO, Forbes CD, Thomson JA. Thromboembolic complications in Cushing's syndrome. Clin Endocrinol 1983; 19:503-11.

16. Nosadini R, Del Prato S, Tiengo A, Valerio A, Muggeo M, Opocher G, Mantero F, Duner E, Marescotti C, Mollo F, Belloni F. Insulin resistance in Cushing's syndrome. J Clin Endocrinol Metab 1983; 57:529-36.
17. Heaney AP, Harper R, Ennis CN, Rooney DP, Sheridan B, Atkinson AB, Bell PM. Insulin action and hepatic glucose cycling in Cushing's syndrome. Clin Endocrinol 1997; 46:735-44.
18. Taskinen M, Nikkila EO, Peckonen R, Sane T. Plasma lipoproteins, lipolytic enzymes, and very low density lipoprotein triglyceride turnover in Cushing's syndrome. J Clin Endocrinol Metab 1983; 57:619-26.
19. McKenna TJ, Lorber D, Lacroix A, Rabin D. Testicular activity in Cushing's disease. Acta Endocrinologica 1979; 91:501-10.
20. Kelly WF. Psychiatric aspects of Cushing's syndrome. Quart J Med 1996; 89:543-51.
21. Mousa A, Jones S, Toft A, Perros P. Spontaneous rupture of Achilles tendon: missed presentation of Cushing's syndrome. Br Med J 1999; 319:560-1.
22. Kelly W. Exophthalmos in Cushing's syndrome. Clin Endocrinol 1996; 45:167-70.
23. Carson DJ, Sloan JM, Cleland J, Russell CF, Atkinson AB, Sheridan B. Cyclical Cushing's syndrome presenting as short stature in a boy with recurrent atrial myxomas and freckled skin pigmentation. Clin Endocrinol 1988; 28:173-80.
24. Young WF, Carney JA, Musa BU, Wulffraat NM, Lens JW, Drexhage H. Familial Cushing's syndrome due to primary pigmented nodular adrenocortical disease. Reinvestigation 50 years later. N Engl J Med 1989; 321:1659-64.
25. Carney JA. The Carney complex (myxomas, spotty pigmentation, endocrine overactivity, and schwannomas). Dermatol Clin 1995; 13:19-26.
26. Trainer PJ, Grossman A. The diagnosis and differential diagnosis of Cushing's syndrome. Clin Endocrinol 1991; 34:317-30.
27. Atkinson AB. The role of bilateral inferior petrosal sinus sampling in the investigation of ACTH-dependent hypercortisolism. Clin Endocrinol 1999; 51:149-50.
28. Meador CK, Liddle GW, Island DP, Nicholson WE, Lucas CP, Nuckton JG, Luetscher JA. Cause of Cushing's syndrome in patients with tumors arising from nonendocrine tissue. J Clin Endocrinol Metab 1962; 22:693-703.
29. Liddle GW, Nicholson WE, Island DP, Abe K, Lowder SC. Clinical and laboratory studies of ectopic humoral syndromes. Rec Prog Horm Research 1969; 25:283-305.
30. Wajchenberg BL, Medonca BB, Liberman B, Pereira MAA, Carneiro PC, Wakamatsu A, Kirschner MA. Ectopic adrenocorticotropic hormone syndrome. Endocrine Reviews 1994; 15:752-87.
31. Magiakou MA, Mastorakos G, Oldfield EH, Gomez MT, Doppman JL, Cutler GB, Nieman LK, Chrousos GP. Cushing's syndrome in children and adolescents. Presentation, diagnosis and therapy. N Engl J Med 1994; 331:629-36.
32. Weber A, Trainer PJ, Grossman AB, Afshar F, Medbak S, Perry LA, Plowan PN, Rees LH, Besser GM, Savage MO. Investigation, management and therapeutic outcome in 12 cases of childhood and adolescent Cushings syndrome. Clin Endocrinol 1995; 43:19-28.
33. Atkinson AB, Kennedy AL, Carson DJ, Hadden DR, Weaver JA, Sheridan B. Five cases of cyclical Cushing's syndrome. Br Med J 1985; 291:1453-57.
34. Atkinson AB, McCance DR, Kennedy L, Sheridan B. Cyclical Cushing's syndrome first diagnosed after pituitary surgery: a trap for the unwary. Clin Endocrinol 1992; 36:297-300.
35. Liebowitz G, White A, Hadani M, Gross DJ. Fluctuating hyper-hypocortisolaemia: a variant of Cushing's syndrome. Clin Endocrinol 1997; 46:759-63.
36. Wallace C, Toth EL, Lewanczuk RZ, Siminoski K. Pregnancy-induced Cushing's syndrome in multiple pregnancies. J Clin Endocrinol Metab 1996; 81:15-21.
37. Rees LH, Besser GM, Jeffcoate WJ, Goldie DJ, Marks V. Alcohol-induced pseudo-Cushing's syndrome. Lancet 1977; i:726-8.
38. Smals AHG, Njo KT, Knoben JM, Ruland CM, Kloppenborg PWC. Alcohol-induced Cushingoid syndrome. J Roy Coll Phys 1977; 12:36-41.

39. Chrousos GD, Vingerhoeds ACM, Brandon D, Eil C, Pugeat M, De Vroede M, Loriaux DL, Lipsett MB. Primary cortisol resistance in man. A glucocorticoid receptor-mediated disease. J Clin Invest 1982; 69:1261-9.
40. Iida S, Gomi M, Moriwaki K, Itoh Y, Hirobe K, Matsuzawa Y, Katagiri S, Yonezawa T, Turui S. Primary cortisol resistance accompanied by a reduction in glucocorticoid receptors in two members of the same family. J Clin Endocrinol Metab 1985; 60:967-71.
41. Lamberts SWJ, Poldermans D, Zweens M, de Jong FH. Familial cortisol resistance: differential diagnostic and therapeutic aspects. J Clin Endocrinol Metab 1986; 63:1328-33.
42. Yanovski JA, Cutler GB Jr, Chrousos GP, Nieman LK. Corticotropin-releasing hormone stimulation following low-dose dexamethasone administration. J Am Med Ass 1993; 269:2232-8.
43. O'Hare JP, Vale JA, Wood S, Corrall RJM. Factitious Cushing's syndrome. Acta Endocrinol 1986; 111:165-7.
44. Cizza G, Nieman LK, Doppman JL, Passaro MD, Czerwiec FS, Chrousos GP, Cutler GB Jr. Factitious Cushing syndrome. J Clin Endocrinol Metab 1996; 81:3573-7.

3
PSYCHIATRIC MANIFESTATIONS OF CUSHING'S SYNDROME

Peter T. Loosen, M.D., Ph.D.
Department of Psychiatry, Vanderbilt School of Medicine, Nashville, TN, USA

INTRODUCTION

The constellation of clinical signs and symptoms that we now call "Cushing's syndrome" was first described by Cushing in 1932 (1). Cushing's syndrome results from chronic exposure to excessive concentrations of glucocorticoids. The physical features of Cushing's syndrome include weight gain, central obesity (*i.e.,* accumulation of adipose tissue in the facial, nuchal, truncal and girdle areas), facial plethora, hirsutism, cutaneous striae, muscle wasting, weakness, and osteoporosis. Other common manifestations of hypercortisolemia include hypertension, glucose intolerance or diabetes mellitus, impotence and testicular atrophy, amenorrhea, mild erythrocytosis, lymphopenia, eosinopenia and frequent fungal infections (2-4). Cushing's syndrome can be divided into two major categories, each of which has two major causes. ACTH-dependent Cushing's syndrome can be caused by excessive ACTH secretion by a pituitary microadenoma (referred to as "Cushing's disease") or by ACTH secretion by a non-pituitary tumor ("ectopic ACTH syndrome"), usually a small cell carcinoma of the lung or a bronchial, thymic or pancreatic carcinoid tumor, which together account for 75% of all cases. ACTH-independent Cushing's syndrome is caused by adrenocortical tumors that secrete cortisol autonomously or by administration of pharmacologic doses of synthetic glucocorticoids (2-5).

Cushing's syndrome has the makings of a true psychoneuroendocrine disorder. Its major laboratory findings (*i.e.,* hypercortisolemia, loss of the normal diurnal cortisol rhythm, increased renal excretion of cortisol and cortisol metabolites and abnormal resistance to dexamethasone suppression) are common to major depression (6, 7). Conversely, regardless of the etiology of glucocorticoid excess, patients with Cushing's syndrome often report psychiatric symptoms, most often changes in mood and cognition (8-15). However, in both Cushing's syndrome and major depression the psychopathological significance of the glucocorticoid excess (*i.e.,* its association with behavior) has remained elusive, although recent work with corticotropin-releasing hormone (CRH) has shown that the pathophysiology of hypercortisolemia in depression and Cushing's disease seems to reflect abnormalities in different loci (*e.g.,* central in depression and pituitary in Cushing's disease)(6).

ADRENAL HORMONE CHANGES IN DEPRESSION

Diagnosis of Depression

Mood disorders are diagnostically separated into two major categories according to the Diagnostic and Statistical Manual of Mental Disorders (DSM-IV)(16): major depression (*i. e.,* patients having depressive episodes only) and bipolar disorder (*i. e.,* patients having depressive and manic episodes). It is estimated that from 13% - 20% of the population have some depressive symptom at any given time. However, the lifetime prevalence rate of major depression in the United States was found to range between 2% and 4% for men and between 5% to 9% for women; depression is thus more than two times as common in women than in men. The lifetime prevalence for bipolar disorder was found to range from 0.8% to 1.1% for men and from 0.5% to 1.3% for women. Risk factors differ for major depression and bipolar disorder. For major depression they are: being female, particularly in the age 35 to 45 years; having a family history of depression or alcoholism; having had childhood experiences in a disruptive, hostile, and generally negative environment; having had recent negative life events; lacking intimate confiding relationships; and having had a baby in the preceding 6 months. Risk factors for bipolar disorder are: being female; having a family history of bipolar disorder; and coming from an upper socioeconomic class (17).

When describing hypothalamic-pituitary-adrenal (HPA) axis changes in depression, one is often unsure whether these changes are causes, perpetuating influences, or secondary effects. In addition, psychiatric diagnoses -- often based on descriptive phenomenology -- rarely describe distinct entities and there are frequently overlapping symptoms between two or more diagnoses. Some psychiatric disorders, particularly depression, also tend to grade into normalcy rather than separate from it. The need to reach consensus about clinical definitions of more severe depressions has been recognized by the development of specific diagnostic criteria (*e.g.,* DSM-IV). Recent studies of Cushing's syndrome have utilized these diagnostic tools, focusing especially on major depression. The diagnostic criteria for major depression as defined by DSM-IV are summarized in Table 1. It is noteworthy that item D needs to be disregarded when one makes the diagnosis of major depression during Cushing's syndrome.

Table 1. Diagnostic Criteria for Major Depressive Episode According to DSM-IVH

A. Five (or more) of the following symptoms have been present during the same 2-week period and represent a change from previous functioning; at least one of the symptoms is either (1) depressed mood, or (2) loss of interest or pleasure.
(1) depressed mood most of the day, nearly every day, as indicated either by subjective report (e.g., feels sad or empty) or observation made by others (appear tearful).
(2) markedly diminished interest or pleasure in all, or almost all, activities most of the day, nearly every day
(3) significant weight loss when not dieting or weight gain, or decrease or increase in appetite nearly every day
(4) insomnia or hypersomnia nearly every day
(5) psychomotor agitation or retardation nearly every day.
(6) fatigue or loss of energy nearly every day
(7) feelings of worthlessness or excessive or inappropriate guilt (which may be delusional) nearly every day
(8) diminished ability to think or concentrate, or indecisiveness, nearly every day
(9) recurrent thought of death, recurrent suicidal ideation without a specific plan, or a suicide attempt or a specific plan for committing suicide

B. The symptoms do not meet criteria for a Mixed Episode

C. The symptoms cause clinically significant distress or impairment in social, occupational, or other important areas of functioning.

D. The symptoms are not due to the direct physiological effects of a substance (e.g., a drug of abuse, a medication) or a general medical condition (e.g., hypothyroidism).

E. The symptoms are not better accounted for by Bereavement, i.e., after the loss of a loved one, the symptoms persist for longer than 2 months ore are characterized by marked functional impairment, premorbid preoccupation with worthlessness, suicidal ideation, psychotic symptoms, or psychomotor retardation.

The Diagnostic and Statistical Manual of Mental Disorders , 4th edition (16)

HPA Axis Activity

About half of all patients with major depression exhibit evidence of cortisol hypersecretion that returns to normal once the depression is cured (6, 7, 18). The evidence of cortisol hypersecretion includes increased ACTH secretion, increased plasma concentration and urinary excretion of cortisol and its metabolites, alterations in the normal circadian rhythm of cortisol secretion, loss of glucocorticoid fast feedback, relative resistance to glucocorticoid negative feedback inhibition of ACTH secretion, and blunted ACTH response to CRH (6, 7).

Although endocrinology and psychiatry use different methodologies for the DST, both define cortisol non-suppression after dexamethasone as a positive test outcome. In psychiatric patients one 1 mg (only a few studies have employed 2 mg) of dexamethasone are administered orally at approximately 23:00 h, followed by two plasma samples taken the following afternoon and evening (usually at 16:00 h and 23:00 h). A plasma cortisol level of 5 μg/dL (50 ng/ml or 138 nmole/L) or greater in one or both of these two samples is then used to define non-suppression or a positive test outcome. Results based on more than two thousand depressed (and other psychiatric) patients indicate an overall sensitivity of 44% among all patients with major depression given 1 mg dexamethasone (31% in a smaller group of patients who received 2 mg of dexamethasone); sensitivity was significantly higher (65%) in elderly depressed patients. Within the depression spectrum, there was a striking increase in the rate of DST non-suppression from grief reactions (10%) and dysthymic disorders (23%), to major depressive disorders (44%), melancholia (50%), psychotic affective disorders (69%), and depression with serious suicidality (78%)(19). Preliminary evidence suggests that the DST may have utility in predicting or monitoring treatment outcome in major depression (19). An hyperactive HPA axis has also been noted in stroke patients with major depression, in pain patients with major depression, and in patients with anorexia nervosa, alcoholism, obsessive compulsive disorder, and anxiety disorders but not in schizophrenic patients (6, 7, 18, 19).

It appears that DST abnormalities correlate mostly with initial insomnia, weight loss, loss of sexual interest, ruminative thinking, retardation, and agitation (20). HPA axis dysregulation may also contribute to cognitive dysfunction, although it is not known whether hypercortisolemia and cognitive impairment in depression are indirectly associated or causally related (19).

A blunted ACTH response after CRH administration is another commonly observed HPA axis abnormality in major depression (6, 7). The CRH stimulation test has helped to resolve an old endocrinological dilemma, namely, whether the hypercortisolism of depression and Cushing's disease share a common pathophysiological basis (they do not, the data point to a hypothalamic locus in depression and to a pituitary locus in Cushing's disease); it also has proved clinically useful in distinguishing pseudo-Cushing's disease from true Cushing's disease (6).

The hypercortisolism of major depression is specific for the depressive episode, *i. e.,* it returns to normal once the depression is cured (6, 7, 18, 19). It is therefore possible that the hypercortisolism is not simply a result or concomitant of depression but contributes importantly to the clinical manifestation of depression. This hypothesis has been explored in studies which used the steroid synthesis inhibitors ketokonazole, aminoglutethimide, or metyrapone (often in combination) to manipulate plasma glucocorticoid concentrations in hypercortisolemic depressed patients. Early results of these studies are promising in showing reduction (and in some cases complete remission) of depressive symptoms with the sole use of antiglucocorticoid agents (18, 21-25).

Pathophysiologic Considerations

The most consistent biological finding in major depression is that some patients show hypercortisolism, at times of a magnitude resembling that seen in Cushing's disease. The following lines of evidence are in support of the hypothesis that the hypercortisolism in depression represents a defect at or above the hypothalamus, resulting in CRH hypersecretion (6, 26): [a] plasma cortisol levels often show "early escape" after dexamethasone administration, suggesting that increased CRH stimulation may "override" the negative feed-back effects of dexamethasone on the corticotroph cell; [b] plasma ACTH responses to oCRH are attenuated in major depression, indicating down-regulation of corticotroph cell receptors; [c] normal controls given a continuous infusion of CRH have the pattern and magnitude of hypercortisolism seen in major depression; [d] there is preliminary evidence of increased CRH concentrations in the cerebrospinal fluid (CSF) of some patients with depression (27) (this finding, though, was not confirmed by studies utilizing serial CSF sampling (28, 29)); and [e] the intracerebroventricular administration of CRH in laboratory animals produces symptoms classically associated with depression (*i.e.,* decreased food consumption, disturbed sleep patterns, decreased sexual behavior, and significant changes in activity). It is unclear at this time whether hyperfunctioning CRH systems are a result of alterations in monoaminergic systems in depression or whether changes in monoaminergic circuits are secondary to CRH hypersecretion (26). It is noteworthy that the hypercortisolism of major depression is not associated with the physical signs (*e.g.,* weight gain, central obesity, facial plethora, hirsutism, cutaneous striae, and muscle wasting) or the biochemical abnormalities (*e.g.,* electrolyte changes, hypertension, diabetes, osteoporosis) of Cushing's syndrome. This is perhaps the result of the greater severity and chronicity of the hypercortisolism in Cushing's syndrome.

BEHAVIORAL DISTURBANCES IN CUSHING'S SYNDROME

Psychiatric Diagnoses and Symptoms

Although behavioral symptoms are rather common in the acute Cushing's disease (8-15), their precise prevalence is difficult to ascertain from the literature. Only a few studies included consecutive, unselected patients; used standardized measures of psychiatric morbidity; or included control groups of patients with other illnesses. The latter criticism is especially important as it pertains to depression, for depression is frequently observed in various medical disorders (30-32). When a patient with Cushing's disease is described as "depressed" it is often unclear whether s/he is suffering from demoralization about his/her medical condition, or is actually suffering from a syndrome similar in symptom profile to major depression.

Table 2 compares the frequency of psychiatric diagnoses in Cushing's disease with that of an illness of comparable disease morbidity (*i.e.,* rheumatoid arthritis). Table 2 shows that 50% of patients with acute Cushing's disease suffered from major depression; that 18% had an anxiety disorder; that only 7% had bipolar disorder; that only 4% suffered from substance abuse; and that the diagnosis of schizophrenia was absent. Sonino et al. (14) reported that depression in acute Cushing's syndrome was associated with older age, female sex, higher pretreatment urinary cortisol levels, relatively more severe clinical condition, and absence of pituitary adenoma. Table 2 reveals further that the diagnosis of major depression is much more common in Cushing's disease (50%) than in rheumatoid arthritis or Graves disease (both 23%).

Table 2. Frequency of Psychiatric Diagnoses in Patients with Cushing's Syndrome (CS) and Patients with Rheumatoid Arthritis (RA)

	CS		RA	
	N	%	N	%
Major Depression	274[a]	50	16	23[f, **]
Unipolar	57[b]	58	0	0
Bipolar	21[c]	7	0	0
Anxiety Disorders	44[d]	18	0	0
Panic Disorder	14[e]	21	0	0
Substance Abuse	3[e]	4	1	7
Schizophrenia	0	0	0	0

[a]Based on 550 patients (8-15). [b]Based on 98 patients (8;9;11;12). [c]Based on 307 patients (8;9;11;12;15). [d]Based on 244 patients (9;12;15). [e]Based on 68 patients (9;11;12). [f]Based on 74 patients (9;33). **Another study directly compared 66 patients with CS and 60 patients with Graves=s disease. Depression was significantly more common in CS than in Graves= disease (62% vs 23%, respectively; $P<0.001$)(10). Between 22% and 27% of medically ill patients are depressed(30-32); as can be expected, depression is more common in severe than in mild medical illness (30;32).

It is notable that we (12) found a higher incidence of anxiety (79%) and panic-related symptoms in our patients with Cushing's disease than did other authors (9, 15). Starkman and Schteingart (34) noted that anxiety was the most frequent (63%) neuropsychiatric symptom in their patients with acute Cushing's disease, but they did not report on the incidence of a formal anxiety disorder diagnosis. Pathophysiologically, one might expect symptoms of anxiety in patients with Cushing's disease. The locus coeruleus is thought to be involved in both anxiety and panic; it receives rich CRH innervation, contains a moderate density of CRH receptors, and is activated following iontophoretic or intracerebroventricular administration of CRH (the latter produces effects in laboratory animals that can be construed as generating anxiety). Furthermore, the concentrations of CRH are selectively increased in the locus coeruleus following application of acute or chronic stress (35).

Although many patients with acute Cushing's disease meet DSM-IV criteria for major depression (except for the exclusion of an Organic Mental Disorder), the depression seen in Cushing's disease has, as Starkman *et al.* (36)

have so clearly put it "certain unique characteristics. Irritability is a prominent and consistent feature. Depressive affect is intermittent, with episodes of 1-3 days duration, recurring very frequently at irregular intervals. Because of this, patients do not experience the relentless depressed mood and unremitting persistent anhedonia so characteristic of most patients with primary endogenous depression. Psychomotor retardation, although present in many patients, is usually not so pronounced as to be clinically obvious and is usually apparent only in retrospect after improvement with treatment. The majority of these patients are not withdrawn, apathetic, monosyllabic, unspontaneous, or hopeless. Significant cognitive impairment including disorder of memory is a consistent and prominent clinical feature of patients with Cushing's disease." Our experience was similar; we (12), too, observed that the time course of the mood disturbance was intermittent rather than sustained. Compared to patients with major depression, patients with Cushing's disease appeared to be more dysphoric and irritable, occasionally losing their temper. Their self-rating of depression reflected more disease severity than did the depression scores of the physician's rating. This discrepancy may have been the result of physicians attributing such somatic components as loss of libido, insomnia, and loss of energy to Cushing's disease rather than to depression. An alternative explanation is that patients with Cushing's disease often minimize their psychological symptoms, particularly when they are seriously ill (8). Table 3 confirms and extends these observations; it also shows that one cannot assume that symptoms such as loss of energy and change in appetite are simply the results of the many endocrine and metabolic changes that accompany Cushing's disease. Although psychiatric disorders and symptoms are very common in patients with acute Cushing's disease, only a few patients are actually seen by a psychiatrist during the course of their illness. It is our experience that most patients with Cushing's disease welcome the opportunity to discuss their behavioral problems in greater detail. Several of our patients wept spontaneously during the initial interview, emphasizing how much they had suffered from erratic and severe mood swings, from irritability they could not easily control, from occasional suicidal impulses they did not understand and were frightened by and, above all, from the anxiety of interpreting these symptoms as sure signs of "losing their minds."

Table 3. Frequency of Psychiatric Symptoms (%) in Patients with Cushing's Syndrome (CS) and Patients with Major Depressive Disorder (MDD)

	CS[a]		MDD[b]	
	N	%	N	%
Decreased energy	34	97	71	75
Irritability	30	86	NR	NR
Impaired memory	29	83	NR	NR
Depressed mood	27	77	88	93
Decreased libido	24	69	42	50*
Middle insomnia	24	69	64	67
Anxiety	23	66	74	78
Impaired concentration	23	66	71	76*
Crying	23	66	NR	NR
Restlessness	21	60	44	46
Late insomnia	20	57	55	58
Hopelessness	15	40	38	40
Guilt	13	37	29	31
Early insomnia	10	29	61	64
Paranoid thoughts	3	9	9	9

[a]Based on 35 patients (34). [b]Based on 95 patients (20). NR: Not recorded.
***Symptoms not recorded in all patients.**

The neuropsychiatric symptoms of Cushing' disease are usually of gradual onset. Patients may experience weight gain, loss of libido, irritability, depression, emotional turmoil, and insomnia for months and sometimes for years before the disease is diagnosed (8-10, 12, 15, 34, 37). In view of the insidious onset of Cushing's disease, it is often impossible to determine with certainty whether the early behavioral symptoms truly preceded the onset of the disease or whether they occurred during its prodromal phase. Kelly (15) reported that psychiatric illness, usually depression, preceded the onset of all symptoms and signs of Cushing's disease in 12% of patients. Sonino et al. (10) found depression to appear in the prodromal phase of Cushing's disease in 27% of patients. Hudson et al. (9) reported that the onset of depression preceded the onset of the medical signs and symptoms of Cushing's disease by at least 1 year in two (15%) patients, occurred within the same year as the medical signs and symptoms in three (23%), and followed the onset of the onset of medical signs and symptoms by at least 1 year in eight (62%). We observed that major depression was the most common (65%) early neuropsychiatric symptom in Cushing's disease, followed by panic disorder (41%), irritability (18%), decreased libido (18%), and insomnia (12%) (12).

The neuropsychiatric symptoms of Cushing' disease appear state related, *i.e.,* they occur during the active illness and are likely to improve when plasma cortisol concentrations normalize postoperatively (8, 10, 13, 15, 34, 36-40). However, we (12) and others (10, 13, 15, 37, 39, 41, 42) observed that the improvement of neuropsychiatric symptoms does not always parallel the resolution of endocrine symptoms. Sonino et al. (10) and Kelly (15) reported that psychiatric illness, usually depression, was still present in about one third of patients after the Cushing's disease had been cured. Dorn et al. (13) demonstrated that 67% had significant psychopathology before cure; after cure, overall psychopathology decreased to 54% at three months, 36% at six months, and 24% at 12 months. It is possible that the Cushing's disease triggers the onset of a depressive episode in some individuals who are at risk for depression. In this case, one would not expect a quick resolution of neuropsychiatric symptoms after endocrine cure. Here it is noteworthy that both Cushing's disease and major depression are more than twice as common in women than in men, that both are most often initially diagnosed between the ages of 25 and 40 years, and that both share similar psychiatric family histories (to be discussed below).

The quality of neuropsychiatric symptoms in Cushing's disease argues against a non-specific response to severe physical illness (36). As discussed above, certain neuropsychiatric symptoms can occur even before patients are aware that they have physical problems. Further, depression is not experienced simply as demoralization common to patients with medical illness; rather, there is iiritability as well as episodic sadness and crying often in the absence of depressive thought (36). Finally, patients with Cushing's disease seem to experience depression significantly more often than do patients with general medical conditions, or other medical disorders of comparable severity, namely, rheumatoid arthritis and Graves disease (Table 2).

Comorbid depression connotes poor prognosis and outcome in patients with Cushing's disease, as it does in other medical conditions. Patients with Cushing's disease and depression suffer from a more severe form of illness, both in terms of cortisol production and clinical presentation (14), they are also more likely to relapse after successful treatment (41). Among patients in general medical settings, physically ill patients with depression have increased pain and physical illness and decreased social and role functioning. Moreover, the combination of advanced coronary artery disease and depressive symptoms is associated with roughly twice the reduction in social functioning as with either condition alone, suggesting that the effects of depressive symptoms and chronic medical conditions on functioning can be additive. Depression and its associated symptoms have also been shown to be a major risk factor for both the development of cardiovascular disease and death after an index myocardial infarction. Treatment of depression in patients with cardiovascular disease appears to improve their dysphoria and other signs and symptoms of depression, improves quality of life, and perhaps even increases longevity (43).

Psychiatric Family Histories

The familial prevalence rate of major depression in patients with Cushing's disease varies widely in the literature, ranging between 3% (8), 27% (37), and 56% (12). This variance is likely to be the result of methodological differences. All studies used the family history method rather than the more accurate family interview method (where information is obtained from both patients and several family members); in some the interviewers were not blind to the medical diagnosis of the subject; in others data were not collected in a prospective fashion or active controls were not included.

In addition to a 56% familial prevalence rate of depression in Cushing's disease, we noted that in Cushing's disease and major depression more female than male relatives suffered from major depression, whereas more male than female relatives suffered from substance abuse. We concluded that the diagnosis of depression during active Cushing's disease, the high prevalence of depression in relatives of patients with Cushing's disease, and the specific gender differences in relatives of patients with Cushing's disease all support the notion that Cushing's disease, major depression, and substance abuse are related, though the understanding of the nature of this association is complicated by the etiological heterogeneity of each of these disorders (12).

Predisposing Factors

Are there specific factors which predispose certain individuals to either Cushing's disease or its associated neuropsychiatric changes? The occurrence of major depression early during the course of Cushing's disease makes it possible that, in its early stages, Cushing's disease is associated with increased activity of endogenous CRH; hyperactivity of endogenous CRH has been postulated to be involved in both mental stress and major depression (6, 18). Indeed, sustained hypersecretion of CRH during depression may represent a kind of chronic stress syndrome (6). Similarly, the etiology of Cushing's disease has been linked to mental stress as early as 1913 when Cushing, in an address at the opening of the Phipp's Clinic of the Johns Hopkins Hospital, suggested that "we are not blind to the fact that there may lurk in the background particularly in the case of glandular hyperplasias .... some primary derangement of the nervous system (44)." Later work proposed an etiological link between major depression and Cushing's disease (12, 37, 41, 45-47).

The hypothesis that Cushing's disease may have a psychosomatic component is based on the observations that factors known to predispose to major depression, namely a family history of depression or suicide or a history of early bereavement, are present in a large percentage of patients (37, 45); that the occurrence of depression can sometimes precede, or coincide with, the earliest signs and symptoms of Cushing's disease (8, 10, 12, 15, 34, 37); that there is a high familial prevalence rate of depression in Cushing's disease (12, 37); and that psychiatric illness may be more common in the early phase of (pituitary) Cushing's disease than in the early phase of adrenal adenoma (15) .

There is no evidence that a consistent pattern of premorbid functioning or a specific personality structure predispose an individual to Cushing's disease or its neuropsychiatric sequelae. Thus, neuropsychiatric changes in Cushing's disease can not be viewed as merely an exacerbation of previous neuropsychiatric changes, nor are they manifested only in individuals with a genetic or constitutional vulnerability to psychiatric disorders (8).

Pathophysiological Considerations

Although neuropsychiatric symptoms have been noted early and consistently in Cushing's disease, we know little about their underlying pathophysiology. The many associations between hypercortisolism and behavior have been addressed along five main avenues (18, 48): [a] it is unclear whether the neuropsychiatric changes described should be attributed to the hypercortisolemia, or to increased levels of ACTH, β-endorphin, and other proopio-melanocortin (POMC)-derived peptides (18). Some authors have suggested that patients with an adrenal adenoma (who have high cortisol but low ACTH and β-endorphin levels) are less psychiatrically disturbed than are patients with a pituitary adenoma (who have high cortisol, ACTH, and β-endorphin levels)(34, 37, 49). However, this was not uniformly confirmed (8, 10, 15, 39); [b] the ACTH/β-endorphin hypersecretion in Cushing's disease could arise from hypothalamic dysfunction (44). In support of this hypothesis it has been noted that emotional stress can cause ACTH or cortisol hypersecretion (6, 18), that patients with major depression have cortisol hypersecretion which is poorly suppressible by dexamethasone, just as in patients with Cushing's disease, and that cyclic forms of CS are responsive to treatment with valproic acid (5); [c] those areas in brain with the highest affinity for glucocorticoids (*i.e.,* the hippocampal formation [HF], septum, and hypothalamus) are also ideally situated to regulate the neurotransmitter systems that affect attention, perception, and mood (18). Glucocorticoids directly affect catecholamine, indoleamine, and GABA synthesis, release, and receptor sensitivity (48, 50). ACTH has been found to have similar effects. Interactions of both glucocorticoids and ACTH with other hormonal systems may also be relevant for the understanding of behavioral effects (48); [d] evidence of specific effects of ACTH and CRH on attention, affect, and social behavior suggest that direct effects (*i.e.,* not mediated by other neurotransmitters) of these peptides on mental function also must be considered (6, 26, 48); [e] glucocorticoid-induced structural brain changes may be responsible for the behavioral symptoms noted. Some depressed hypercortisolemic patients (51, 52) and some patients with exogenous (53, 54) and endogenous (55) Cushing's syndrome can have cerebral atrophy on CAT scan. In laboratory animals high doses of glucocorticoids are known to damage the HF which serves an important role in both new learning/intermediate memory (56, 57) and glucocorticoid inhibition of HPA axis activity (50, 58). Damage to the HF due to lesions, aging, or glucocorticoid excess can lead to cortisol hypersecretion during moderate psychological stress, impaired shut-off of cortisol secretion after such stress, increased CRH and vasopressin mRNA and peptide concentrations in the paraventricular nuclei of the hypothalamus, and decreased sensitivity to dexamethasone suppression of HPA function (50, 58). These abnormalities in HPA axis function are not dissimilar from those observed in about half of all patients with major depression (18), raising the question of whether chronic, untreated major depression may not in part be a self-perpetuating disorder. The hypercortisolism in Cushing's disease and major depression also may, by damaging the HF, be directly involved in the cognitive decline commonly observed in these conditions (18). Starkman *et al.* (59), using Magnetic Resonance Imaging (MRI) in 12 patients with Cushing's disease, demonstrated that about 30% had a reduced HF volume, that there were significant and specific correlations between HF volume and scores for verbal paired associate learning and verbal recall, and that HF volume was negatively correlated with plasma cortisol levels. Axelson *et al.* (60) reported no differences in HF volume between depressed patients and controls, and between DST positive and DST negative depressed patients. However, they observed a relationship between HF volume and 2300 h plasma cortisol concentrations after covariance adjustment for age and sex. They also observed a significant negative correlation between HF volume and both age of depressive onset and number of hospitalization. Small reductions in HF volume have also been described in other stress-sensitive disorders, namely depression (61) and post-traumatic stress disorder (62).

EFFECTS OF ADRENOCORTICAL HORMONES ON BEHAVIOR

Physicians who treat their patients with high doses of glucocorticoids (*e.g.*, rheumatologists, allergists, and transplant specialists) are aware that behavioral side effects are rather common during glucocorticoid administration. ACTH, cortisone, and prednisone are most commonly associated with mental disturbances, but it is likely that all glucocorticoids have the potential to induce such disturbances (18, 63). Depending on study design and population selection criteria, neuropsychiatric side effects may range between 3% (64), 5% (65), 46% (66), and 57% (63). For example, the Boston Collaborative Drug Surveillance Program (64) reported a frequency of 3% among 718 patients receiving prednisone. However, the study included only patients who were emotionally stable and free from mental disorders before treatment. A review of 29 studies of the clinical efficacy of steroids in various medical illnesses reported that severe psychiatric reactions occurred B usually early in the course of treatment B in approximately 5% of patients who suffered mainly from affective and/or psychotic symptoms (65). In contrast, other investigators reported a 57% incidence of mental disturbances in patients with systemic lupus erythematosus who were treated with prednisone (63). This high incidence is probably exaggerated as systemic lupus erythematosus is known to exhibit mental symptoms even without treatment.

Can we define the exact nature of glucocorticoid-induced neuropsychiatric disturbances? Like in spontaneous Cushing's disease, affective symptoms are the most common reactions in iatrogenic Cushing's syndrome. However, mood elation appears to be more common in iatrogenic Cushing's syndrome than in spontaneous Cushing's disease (63, 67). In 12 medically healthy normal volunteers given prednisone (80 mg qd for five days), Wolkowitz et al. (67) observed mild hypomania in 67% and mild depression in 8% of cases. Goolker and Schein (66) treated 80 patients with ACTH or cortisone; 43 (54%) showed no psychiatric reaction, or one consistent with the relief of their somatic symptoms. In the remaining 37 patients, the authors noted 45 affective reactions; 70% belonged in the category of cheerfulness or alternating mood, only 13% belonged in the category of apathy or depression.

The euphoric effects vary in degree and appropriateness. Some patients may show only cheerfulness and well-being. Others may become inappropriately elated, showing hypomania or even frank mania. The noted American writer Reynold Price (68) gave a moving personal account of the behavioral side effects of dexamethasone administration: " ... the mutilations of Cushing's syndrome ... included a swollen and sodden moon face, the start of a buffalo hump of fat on the upper back, a roaring appetite for food and - worst - exhausting and unpredictable swings of mood (p 68) ... I would be unrealistically cheerful one moment, with bursts of popular songs from my youth, or despairing the next (p 78 f)." Other common symptoms of iatrogenic Cushing's syndrome are emotional lability, anxiety, psychosis, and insomnia (64, 67, 69, 70).

Women appear to be more likely than men to develop glucocorticoid-induced neuropsychiatric symptoms (63, 65). However, the data do not allow to discern whether females are generally more vulnerable to such mental disturbances, or whether more females than males require glucocorticoid treatment. Moreover, there is evidence that glucocorticoid dosage is important (63-65). Among 718 patients treated with prednisone in the Boston Collaborative study (64), neuropsychiatric symptoms were recorded in 1% of patients receiving 40mg/day or less; in 5% of those receiving 41 to 80mg/day; and in 18% of those receiving 80mg/day or more. Thus, increasing doses may increase the risk of mental disturbances; to avoid such complications one should give the lowest therapeutic dose. However, the finding of Goolker and Schein (66) that it was difficult to predict by dosage alone which patients may develop neuropsychiatric symptoms suggests that other factors may also be involved. Psychiatric premorbidity does not seem to be among those factors. Glucocorticoid-induced neuropsychiatric problems are as likely to occur in patients without psychiatric history as they are in patients with psychiatric history; they also seem to develop independently of pre-existing personality traits (63, 70). In summary, glucocorticoid-induced neuropsychiatric symptoms are more likely to develop in patients receiving daily doses of steroids in excess of 40 mg/day of prednisone or its equivalent (*i.e.,* five to eight times the maintenance dosage in adrenally deficient patients). However, neither dosage nor duration of treatment appear to have any effect on the time of onset, duration, type, or severity of the disturbance.

Glucocorticoid-induced psychiatric symptoms usually abate when the steroid is discontinued or its dosage reduced (63-65, 69, 70). When dose reduction or discontinuation are undesirable, phenothiazines are the drug of choice (65). Tricyclic antidepressants have not proven useful in this condition (65, 70). However, lithium may be useful as a preventive agent; a short-term lithium prophylaxis may thus be considered in those patients who have a prior history of developing behavioral disturbances during steroid therapy (63). The frequency of psychiatric reactions during steroid therapy and their impact on the general well-being of the patient make it necessary to fully inform the patient about these side effects before steroid therapy is initiated. A recent retrospective study of seven patients who had received steroids intermittently found that psychiatric reactions were common and that seven of the eight patients were not warned by their physician of the possible psychiatric side effects (71).

ACKNOWLEDGMENTS

Supported in part by Veterans Administration Medical Research Funds, NIH Grant RR-00095 (to the General Clinical Research Center of Vanderbilt University), NIAAA Grant AA-07732, and NIMH Grant MH-45173. I am grateful to the patients on the General Clinical Research Center who taught me much about Cushing's syndrome.

REFERENCES

1. Cushing H. The basophil adenomas of the pituitary body and their clinical manifestation (pituitary basophilism). Bull Johns Hopkins Hosp 1932; 50:137-195.
2. Orth DN, Kovacs WJ, deBold CR. The adrenal cortex. In: Wilson JD, Foster DW, editors. Williams Textbook of Endocrinology. Philadelphia: W.B. Saunders, 1992: 489-619.
3. Yanovski JA, Cutler GBJ. Glucocorticoid action and the clinical features of Cushing's syndrome. Endocrinol Metab Clin North Am 1994; 23(3):487-509.
4. Krieger DT. Physiopathology of Cushing's disease. Endocr Rev 1983; 4:22-43.
5. Orth DN. Medical progress: Cushing's syndrome. N Engl J Med 1995; 332:791-803.
6. Gold PW, Chrousos G, Kellner C et al. Psychiatric implications of basic and clinical studies with corticotropin-releasing factor. Am J Psychiatry 1984; 141:619-627.
7. Nathan KI, Musselman DL, Nemeroff CB. Biology of mood disorders. In: Schatzberg AF, Nemeroff CB, editors. Textbook of Psychopharmacology. Washington: American Psychiatric Press, 1995.
8. Haskett RF. Diagnostic categorization of psychiatric disturbance in Cushing's syndrome. Am J Psychiatry 1985; 142:911-916.
9. Hudson JI, Hudson MS, Griffing GT, Melby JC, Pope HGJ. Phenomenology and family history of affective disorder in Cushing's disease. Am J Psychiatry 1987; 144:951-953.
10. Sonino N, Fava GA, Belluardo P, Girelli ME, Boscaro M. Course of depression in Cushing's syndrome: response to treatment and comparison with Graves' disease. Hormone Research 1993; 39(5-6):202-206.
11. Dorn LD, Burgess ES, Dubbert B et al. Psychopathology in patients with endogenous Cushing's syndrome: 'Atypical' or melancholic features. Clinical Endocrinology 1995; 43(4):433-442.
12. Loosen PT, Chambliss B, deBold CR, Shelton R, Orth DN. Psychiatric phenomenology in Cushing's disease. Pharmacopsychiat 1992; 25(4):192-198.
13. Dorn LD, Burgess ES, Friedman TC, Dubbert B, Gold PW, Chrousos GP. The longitudinal course of psychopathology in Cushing's syndrome after correction of hypercortisolism. J Clin Endocrinol Metab 1997; 82(3):912-919.
14. Sonino N, Fava GA, Raffi AR, Boscaro M, Fallo F. Clinical correlates of major depression in Cushing's disease. Psychopathology 1998; 31(6):302-306.
15. Kelly WF. Psychiatric aspects of Cushing's syndrome. QJM 1996; 89(7):543-551.
16. American Psychiatric Association . Diagnostic and Statistical Manual for Mental Disorders (DSM-IV). Fourth ed. Washington : American Psychiatric Association, 1994.

17. Boyd JH, Weissman MM. Epidemiology. In: Paykel ES, editor. Handbook of Affective Disorders. New York: Guilford Press, 1982: 109-125.
18. Pearson Murphy BE. Steroids and depression. J Steroid Biochem Molec Biol 1991; 38:537-559.
19. Arana GW, Baldessarini RJ, Ornsteen M. The dexamethasone suppression test for diagnosis and prognosis in psychiatry. Commentary and review. Arch Gen Psychiatry 1985; 42:1193-1204.
20. Miller KB, Nelson JC. Does the dexamethasone suppression test relate to subtypes, factors, symptoms, or severity? Arch Gen Psychiatry 1987; 44:769-774.
21. Thakore JH, Dinan TG. Cortisol synthesis inhibition: A new treatment strategy for the clinical and endocrine manifestations of depression. Biol Psychiatry 1995; 37:364-368.
22. Wolkowitz OM, Reus VI, Manfredi F, Ingbar J, Brizendine L, Weingartner H. Ketokonazole administration in hypercortisolemic depression. Am J Psychiatry 1993; 150:810-812.
23. Pearson Murphy BE, Ghadirian AM, Filipini D, Dhar V. Endocrine responses to antiglucocorticoid strategies in refractory depression. Biol Psychiatry 1994; 35:618.
24. Ghadirian AM, Engelsman F, Dhar V et al. Psychotropic effects of steroid synthesis inhibitors in refractory depression. Biol Psychiatry 1994; 35:618.
25. Wolkowitz OM, Reus VI, Manfredi F et al. Antiglucocorticoid medication effects on specific depressive symptoms. Biol Psychiatry 1994; 35:678-679.
26. Nemeroff CB. The role of corticotropin-releasing factor in the pathogenesis of major depression. Pharmacopsychiat 1994; 21:76-82.
27. Nemeroff CB, Widerlov E, Bissette G et al. Elevated concentrations of CSF corticotropin-releasing factor- like immunoreactivity in depressed patients. Science 1984; 226:1342-1344.
28. Geracioti TD, Orth DN, Ekhator NN, Blumenkopf B, Loosen PT. Serial cerebrospinal fluid corticotropin-releasing hormone concentrations in healthy and depressed subjects. J Clin Endocrinol Metab 1992; 74:1325-1330.
29. Kling MA, Roy A, Doran AR et al. Cerebrospinal fluid immunoreactive corticotropin-releasing hormone and adrenocorticotropin secretion in Cushing's disease and major depression: potential clinical implications [see comments]. J Clin Endocrinol & Metab 1991; 72(2):260-271.
30. Stewart MA, Drake F, Winokur G. Depression among medically ill patients. Dis Nerv Syst 1965; 26:479-485.
31. Schwab JJ, Bialow M, Brown JM, Holzer CE . Diagnosing depression in medical inpatients. Ann Intern Med 1967; 67:695-707.
32. Moffic HS, Paykel ES. Depression in medical in-patients. Brit J Psychiatry 1975; 126:346-353.
33. Abdel-Nasser AM, Abd El-Azim S, Taal E, El Badawy SA, Rasker JJ, Valkenburg HA. Depression and depressive symptoms in rheumatoid arthritis patients: an analysis of their occurrence and determinants. Br J Rheumatol 1998; 37(4):391-397.
34. Starkman MN, Schteingart DE. Neuropsychiatric manisfestations of patients with Cushing's syndrome: Relationship to cortisol and adrenocorticotropin hormone levels. Arch Intern Med 1981; 141:215-219.
35. Grigoriadis DE, Pearsall D, deSouza EB. Effects of chronic antidepressant and benzodiazepine treatment on corticotropin-releasing factor receptors in rat brain and pituitary. Neuropsychopharm 1988; 2:53-60.
36. Starkman MN, Schteingart DE, Schork MA. Depressed mood and other psychiatric manifestations of Cushing's syndrome: Relationship to hormone levels. Psychosom Med 1981; 43:3-18.
37. Cohen SI. Cushing's syndrome: A psychiatric study of 29 patients. Brit J Psychiatry 1980; 136:120-124.

38. Kelly WF, Checkley SA, Bender DA, Mashiter K. Cushing's syndrome and depression--a prospective study of 26 patients. Brit J Psychiatry 1983; 142:16-19.
39. Jeffcoate WJ, Silverstone J, Edwards C, Besser G. Psychiatric manifestation of Cushing's syndrome: response to lowering of plasma cortisol. Qtrly J Med 1979; 48:465-472.
40. Kelly WF, Checkley SA, Bender DA. Cushing's syndrome, tryptophan and depression. Brit J Psychiatry 1980; 136:125-132.
41. Sonino N, Fava GA. Psychosomatic aspects of Cushing's disease. Psychother Psychosom 1998; 67(3):140-146.
42. Voigt KH, Bossert S, Bretschneider S, Bliestle A, Fehm HL. Disturbed cortisol secretion in man: contrasting Cushing's disease and endogenous depression . Psychiatry Res 1985; 15(4):341-350.
43. Musselman DL, Evans DL, Nemeroff CB. The relationship of depression to cardiovascular disease: epidemiology, biology, and treatment. Arch Gen Psychiatry 1998; 55(7):580-592.
44. Cushing H. Psychic disturbances associated with disorders of the ductless glands. Am J of Insanity 1913; 69:965-990.
45. Gifford S, Gunderson JG. Cushing's disease as a psychosomatic disorder: A selective review of the clinical and experimental literature and a report of ten cases. Persp Biol Med 1970; 13:169-221.
46. Sonino N, Fava GA, Grandi S, Mantero F, Boscaro M. Stressful life events in the pathogenesis of Cushing's disease. Clin Endocrinol 1988; 29:617-623.
47. Isojarvi JI, Myllyla VV, Pakarinen AJ. Effects of carbamazepine on pituitary responsiveness to luteinizing hormone-releasing hormone, thyrotropin-releasing hormone, and metoclopramide in epileptic patients. Epilepsia 1989; 30(1):50-56.
48. Reus VI. Neuropsychiatric disturbance in endocrine disease. In: Nemeroff CB, Loosen PT, editors. Handbook of Clinical Psychoneuroendocrinology. New York : Guilford Press, 1988: 71-84.
49. Mazet P, Simon D, Luton JP, Bricaire H. [Psychic symptoms and personality of 50 patients with Cushing's syndrome (author's transl)]. Nouv Presse Med 1981; 10(31):2565-2570.
50. McEwen BS, Angulo J, Cameron H et al. Paradoxical effects of adrenal steroids on the brain: protection versus degeneration. Biol Psychiatry 1992; 31:177-199.
51. Kellner C, Rubinow D, Gold P, Post R. Relationship of cortisol hypersecretion and brain CT scan alterations in depressed patients. Psychiat Res 1983; 8:191-197.
52. Schlegel S, von Bardeleben U, Wiedemann K, Frommberger U, Holsboer F. Computerized brain tomography measures compared with spontaneous and suppressed cortisol levels in major depression. Psychoneuroendocr 1989; 14:209-216.
53. Bentson JR, Reza M, Winter J, Wilson G. Steroids and apparent cerebral atrophy on computed tomography scans. J Comput Assist Tomogr 1978; 2:16-23.
54. Okuno T, Ito M, Knoishi Y, Yoshioka MK, Nakano Y. Cerebral atrophy following ACTH therapy. J Comput Assist Tomogr 1980; 4:20-23.
55. Heinz ER, Martinez J, Haenggeli A. Reversibility of cerebral atrophy in anorexia nervosa. J Comput Assist Tomogr 1977; 1:415-418.
56. Zola-Morgan S, Squire LR, Amaral DG. Human amnesia and the medial temporal region:enduring memory impairment following a bilateral lesion limited to field CA1 of the hippocampus. J Neurosci 1986; 6:2950-2967.
57. Squire LR, Shimamura AP, Amaral DG. Memory and the hippocampus. In: Byrne J, Berry W, editors. Neural Modes of Plasticity. New York: Academic Press, 1989: 208-239.
58. Sapolsky RM, Uno H, Rebert CS, Finch CE. Hippocampal damage associated with prolonged glucocorticoid exposure in primates. J Neurosci 1990; 10:2897-2902.
59. Starkman MN, Gebarski SS, Berent S, Schteingart DE. Hippocampal formation volume, memory dysfunction, and cortisol levels in patients with Cushing's syndrome. Biol Psychiatry 1992; 32(9):756-765.

60. Axelson DA, Doraiswamy PM, McDonald WM et al. Hypercortisolemia and hippocampal changes in depression. Psychiatry Res 1993; 47:163-173.
61. Krishnan KR, Doraiswamy PM, Figiel GS et al. Hippocampal abnormalities in depression. J Neuropsych & Clin Neurosciences 1991; 3(4):387-391.
62. Bremner JD, Randall P, Scott TM et al. MRI-based measurement of hippocampal volume in patients with combat-related posttraumatic stress disorder. Am J Psychiatry 1995; 152(7):973-981.
63. Ling MHM, Perry PJ, Tsuang MT. Side effects of corticosteroid therapy. Arch Gen Psychiatry 1981; 38:471-477.
64. The Boston Collaborative Drug Surveillance Program. Acute adverse reactions to prednisone in relation to dosage. Clin Pharmacol Ther 1972; 13:695-698.
65. Lewis DA, Smith RE. Steroid-induced psychiatric syndromes. J Affect Dis 1994; 5:319-332.
66. Goolker p, Schein J. Psychic effects of ACTH and cortisone. Psychosom Med 1953; 15:589-613.
67. Wolkowitz OM, Rubinow D, Doran AR et al. Prednisone effects on neurochemistry and behavior. Preliminary findings. Arch Gen Psychiatry 1990; 47(10):963-968.
68. Price R. A Whole New Life. An Illness and a Healing. New York: Macmillan, 1994.
69. Varney NR, Alexander B, McIndoe JH. Reversible steroid dementia in patients without steroid psychoses. Am J Psychiatry 1984; 141:369-372.
70. Hall RCW, Popkin MK, Stickney SK, Gardner ER. Presentation of the steroid psychoses. J Nerv Ment Dis 1979; 167:229-236.
71. Reckart MD, Eisendrath SJ. Exogenous corticosteroid effects on mood and cognition: case presentations. Interntl J of Psychosomatics 1990; 37(1-4):57-61.

4

A RATIONAL APPROACH TO THE PATIENT SUSPECTED OF HAVING CUSHING'S SYNDROME

Michael Fowler M.D. and Lewis S. Blevins, Jr., M.D.
Division of Endocrinology, Diabetes & Metabolism, Department of Medicine, Vanderbilt School of Medicine, Nashville, TN, USA

In 1889 Sir William Osler detailed the case of a 31-year old man suffering from "an acute myxedematous condition, with tachycardia, glycosuria, melena, mania and death." In particular, Osler was impressed by the rapid weight gain and the "bloated" appearance of his patient's face (1, 2). Unfortunately, however, the constellation of symptoms and signs that we now attribute to glucocorticoid excess had not been described and the patient succumbed while being treated for hypothyroidism. In 1932, Harvey Cushing described the clinical features associated with basophilic tumors of the pituitary. Since then, the association has become well characterized and patients with features due to glucocorticoid excess are said to have Cushing's syndrome (3).

We now know that patients with Cushing's syndrome can have any one of a number of disease processes, all of which ultimately lead to the hypersecretion of cortisol and other adrenal steroids from the adrenal glands. Perhaps the greatest challenge to physicians evaluating patients suspected of having Cushing's syndrome is in determining whether patients have it or not. Once a diagnosis is proved, one must consider a broad differential diagnosis and identify the exact cause of the disorder in order to design an appropriate disease-specific management plan that will maximize the chance for therapeutic success. Based on our experience, conducting an ordered diagnostic evaluation and avoiding the temptation to take shortcuts in the evaluation process can minimize errors and increase the likelihood of a successful outcome. In this chapter we will provide an overview of our approach to the evaluation of patients suspected of having endogenous hypercortisolism.

The first step in the evaluation of patients with suspected endogenous hypercortisolism is to determine whether to proceed with screening tests or diagnostic tests. Screening tests, including the overnight 1 mg dexamethasone suppression test, determination of the 24-h urine free cortisol, and afternoon or evening serum cortisol levels, are designed to be sensitive but not specific. These tests are relatively inexpensive, can be performed on an outpatient basis, and are best employed to test large populations and individual patients when the pre-test probability of the disease is low. For example, in one study, three per-cent patients with poorly controlled diabetes mellitus had hypercortisolism but they lacked other overt manifestations of the disorder (4). It would be appropriate to evaluate this population with screening tests rather than with diagnostic tests in order to identify those patients who have the disorder.

Diagnostic tests, including the 24-h urine free cortisol, the dexamethasone-suppressed CRH stimulation test, and the formal low-dose dexamethasone suppression test, are designed to have the best combination of sensitivity and specificity. These tests are used to distinguish those with pathologic hypercortisolism from those who are normal and also those with pseudo-Cushing's syndrome. They are appropriate when the pre-test probability of the disorder, as estimated by the clinical evaluation and results of screening tests, is high (5). In many cases, the initial approach and selection of tests seems to be driven by clinical intuition rather than by evidence-based medicine. We would argue, however, that intuition is the foundation of the art of medicine and believe that it is usually based on valid unpublished observational experiences. For example, while there are no symptoms or signs alone that are highly specific for hypercortisolism, a constellation of features, including proximal myopathy, central redistribution of fat with weight gain, widened purplish striae, hirsutism, atrophy of the dermis, hypertension, and diabetes mellitus, usually heightens ones suspicion of the disorder. Diagnostic tests would naturally be selected in this context despite their higher overall complexity and costs. Normal diagnostic test results usually exclude a diagnosis of pathologic hypercortisolism. One must, however be mindful of the fact that some patients with Cushing's syndrome have normal 24-h urine free cortisol excretion rates and some even suppress glucocorticoid production normally during the formal low-dose dexamethasone suppression test (6). The dexamethasone-suppressed CRH stimulation test can be useful when test results are equivocal as it seems to better discriminate between affected and unaffected patients (7). When a diagnosis of Cushing's syndrome is strongly suspected in the setting of normal test results, one may choose to test periodically to identify those with cyclical or periodic hormonogenesis and those with progressive disease (8). In our experience, patients with true underlying Cushing's syndrome usually suffer from progression of their disorder during long-term follow-up. When diagnostic tests confirm a clinical suspicion of Cushing's syndrome, one should proceed to work through the differential diagnosis of the causes of hypercortisolism (Tables 1 and 2).

Table 1. ACTH-dependent causes of Cushing's Syndrome.

Pituitary	**Non-Pituitary**
Microadenoma	Medullary thyroid carcinoma
Macroadenoma	Thymic carcinoid
Corticotroph hyperplasia	Bronchial carcinoid
Primary	Small-cell bronchogenic carcinoma
Secondary	Pheochromocytoma
CRH-secreting	Islet-cell carcinoma
hammartoma	Others (ovary, colon)
Ectopic CRH-secreting	
tumor	

Table 2. ACTH-independent causes of Cushing's Syndrome.

Exogenous Glucocorticoids
- Therapeutic use
- Surreptitious use

Adrenocortical adenoma
Adrenocortical carcinoma
Bilateral adrenal hyperplasia
- McCune-Allbright syndrome
- Primary pigmented nodular adrenocortical disease
- Ectopic receptor expression
  - LH receptor
  - β-adrenergic receptor
  - GIP receptor
  - AVP receptor
- Idiopathic

Determination of whether the patient's hypercortisolism is ACTH-dependent is the most important step in working through the differential diagnosis. ACTH levels are either inappropriately normal or high when the hypercortisolism is due to pituitary tumors and neuroendocrine tumors that secrete ACTH (Table 1). ACTH levels are usually low, and often undetectable, in the setting of autonomous production of cortisol from one or both of the adrenal glands and in the setting of surreptitious ingestion of synthetic glucocorticoids (Table 2).

When ACTH levels are normal or high, one must differentiate between pituitary tumors and ectopic ACTH-secreting neuroendocrine tumors. This differentiation is one of the most difficult and crucial in clinical endocrinology and, despite having an arsenal of well-studied tests at our disposal, patients are occasionally misdiagnosed and treated inappropriately. Several tests have proved to be somewhat useful in distinguishing between patients with pituitary tumors and ectopic ACTH hypersecretion. These include the formal high-dose dexamethasone suppresion test, the overnight 8 mg high-dose dexamethasone suppression test, the metyrapone test, and the CRH stimulation test (9, 10, 11, 12). MRI of the sella with gadolinium contrast enhancement readily identifies patients with ACTH-secreting macroadenomas but fails to identify nearly one-half of ACTH-secreting microadenomas (13). Inferior petrosal sinus sampling with CRH administration (IPSS) has become the "gold standard" test to differentiate pituitary from ectopic sources of ACTH hypersecretion (14). The invasiveness, potential complications, and expenses of this procedure will undoubtedly be limiting factors in many cases. The choice of tests to localize the source of ACTH hypersecretion should be based on the evaluating physician's skill, familiarity, and level of confidence with these different modalities. We prefer to begin with an overnight 8mg high-dose dexamethasone suppresion test and an MRI of the sella. These tests are evaluated in the clinical context and if they do not support a diagnosis of pituitary-dependent ACTH hypersecretion, we recommend IPSS to distinguish between pituitary and ectopic ACTH secretion. Results consistent with pituitary-dependent ACTH secretion prompt us to refer patients for transsphenoidal surgery for a presumed pituitary tumor. Results indicative of a non-pituitary source of ACTH secretion prompt a thorough investigation for neuroendocrine tumors known to be associated with the syndrome.

When a diagnosis of ectopic ACTH hypersecretion is proved, one must conduct a thorough search in attempt to localize the responsible tumor (15). Indium-labeled pentreotide scintigraphy is often employed to localize the tumors responsible for the syndrome (16). Since most of these tumors are located in the chest, we also begin our evaluation with a spiral CT with contrast or an MRI of the chest. These studies identify most small cell lung carcinomas, bronchial carcinoids, and thymic carcinoids. Extending these imaging studies through the upper abdomen permits one to adequately screen for pheochromocytomas and pancreatic islet-cell carcinomas. Urinary studies for 5-HIAA and fractionated catecholamines and their metabolites are sometimes useful when bronchial carcinoid tumor and pheochromocytoma, respectively, are suspected. Palpation of the thyroid gland, fine needle aspiration of any nodules, and measurement of the serum calcitonin are usually sufficient to evaluate for the possibility of medullary thyroid carcinoma. Long-term follow-up with periodic imaging and medical therapy or adrenalectomy to control the hypercortisolism is sometimes necessary as some tumors escape detection for many years.

ACTH-independent hypercortisolism usually indicates autonomous hypersecretion of cortisol from one or both adrenal glands due to a tumor or adrenal hyperplasia. CT of the abdomen is the preferred imaging modality to evaluate the adrenal glands (17). Surgical excision, and rarely necessary, fine needle aspiration, are the procedures that best differentiate benign from malignant adrenocortical tumors. A panel of dynamic tests in conjunction with a thorough history and physical examination may determine the underlying cause of adrenal hyperplasia (18). If the adrenal glands are small on CT, surreptitious misuse and use or abuse of exogenous glucocorticoids must be considered (19). HPLC analysis of a 24-h urine sample may identify synthetic glucocorticoids and their metabolites.

Hypercortisolism can result in a number of serious and life-threatening disorders. Further, some responsible tumors are aggressive and can cause significant morbidity and mortality. Knowledge, understanding, and application of a logical and proper sequence of diagnostic studies will enable physicians to evaluate patients expediently while minimizing costs and diagnostic errors. This approach will ultimately lead to exact diagnoses and the development of treatment plans that will allow for the greatest chances of therapeutic success.

REFERENCES

1. Osler, W. "An Acute Myxoedematous Condition, With Tachycardia, Glycosuria, Melena, Mania and Death. J Nervous Mental Dis 26:65-71, 1899.
2. Altschule, MD. "A Near Miss – Osler's Early Description of Cushing's Syndrome With, Regrettably, No Post-Mortem Examination." N Engl J Med. 302: 1153-1155, 1980.
3. Cushing HW. The Pituitary Body and Its Disorders: Clinical States Produced by Disorders of the Hypophysis Cerebri. Philadelphia: JB Lippincott, 1912.
4. Leibowitz G, Tsur A, Chayen S, Salameh M, Itamar R, Cerasi E, Gross D. "Pre-clinical Cushing's Syndrome: An Unexpected Frequent Cause of Poor Glycaemic Control in Obese Diabetic Patients." Clin Endocrinol 44: 717-722, 1996.
5. Ross EJ and Linch DC. "Cushing's Syndrome- Killing Disease: Discriminatory Value of Signs and Symptoms Aiding Early Diagnosis." Lancet 646-649, 1982.

6. Newelle-Pross J, Trainer P, Besser M, Grossman A. The diagnosis and differential diagnosis of Cushing's Syndrome and Psuedo-Cushing's States. Endocr Rev 19: 647-672, 1998.
7. Yanovski JA, Cutler GB Jr. Chrousos GP, Nieman LK. The dexamethasone suppressed cortocotropin-releasing hormone stimulation differentiates mild Cushing's disease from normal physiology. J Clin Endocrinol Metab. 83: 348-352, 1998.
8. Hellman L, Weitzman ED, Roffwarg H, Fukushima D, Yoshida K, Gallagher T. Cortisol is Secreted Episodically in Cushing's Syndrome. J Clin Endocrinol Metab 30: 686-689.
9. Flack MR, Oldfield EH, Cutler GB Jr., Flack MR, Oldfield EH, Cutler GB, Zweig MH, Malley JD, Chrousos GP, Loriaux DL, Nieman LK. Urine free cortisol in the high-dose dexamethasone suppression test for the differential diagnosis of the Cushing's Syndrome. Ann Intern. Med 116:211-217, 1992.
10. Avergnious PC, Yanovski JA, Oldfield EH, Neiman LK, Cutler GB Jr. Metyrapone and dexamethasone suppression tests for the differential diagnosis of the adrenalcortocotropin-dependent Cushing's Syndrome: A Comparison. Ann Intern Med 121: 318-327, 1994.
11. Dichek HL, Nieman LK, Oldfield EH, Pass HI, Malley JD, Cutler GB Jr. A Comparison of the standard high-dose dexamethasone suppression test and the overnight 8-mg dexamethasone suppression test for the differential diagnosis of adrenocortocotropin-dependent Cushing's Syndrome. J Clin Endocrinol Metab. 78: 418-422, 1994.
12. Nieman LK, Chrousos GP, Oldfield EH, Avgerinos PC, Cutler GB, Loriaux DL. The ovine corticotropin-releasing hormone stimulation test and the dexamethasone suppression test in the differential diagnosis of Cushing's Syndrome. Ann Intern Med 105: 862-867, 1986.
13. Orth DN, Cushing's Syndrome. N Engl J Med 332: 791-802, 1995.
14. Oldfield EH, Doppman JL, Nieman LK, Chrousos GP, Miller DL, Katz DA, Cutler GB, Loriaux DL. Petrosal sinus sampling with and without corticotropin-releasing hormone for the differential diagnosis of Cushing's Syndrome. N Eng J Med 325: 897-905, 1991.
15. Doppman JL. The search for occult ectopic ACTH-producing tumors. The Endocrinologist 2: 41-46, 1992.
16. de Herder WW, Krenning EB, Malchoff CD, Hofland LJ, Riubi JC, Kwekkeboom DJ, Oei HY, Pols HA, Burnning HA, Nobels FR, Lamberts SW. Somatostatin receptor scintigraphy: It's value and tumor localization in patients with Cushing's Syndrome caused by ectopic cortocotropin or cortocotropin-releasing hormone secretion. Am J. Med 96:305-312, 1994.
17. Perry RR, Neiman LK, Cutler GB, Chrousos GP, Loriaux DL, Doppman JL, Travis WD, Norton JA. Primary adrenal causes of Cushing's Syndrome; Diagnosis and surgical management. Ann Surg 210:59-68, 1989.
18. Lacroix A, Mircescu H, Memat P. Clinical evaluation of the presence of abnormal hormone receptors in adrenal Cushing's Syndrome. The Endocrinologist 9: 9-15, 1999.
19. Quddusi S, Browne P, Hirsch IB. Cushing's Syndrome due to surreptitious glucocorticiod administration. JAMA 158: 294-296 ,1998.

5
EVALUATION OF CUSHING'S SYNDROME

John Newell-Price, M.D. and Mike Besser, M.D.
Northern General Hospital, Sheffield, UK and St Bartholomew's Hospital, London, UK.

OVERVIEW

The protean clinical manifestations of Cushing's syndrome (CS) are really only the starting point for biochemical evaluation. When florid clinical features are present diagnosis may be straightforward. In contrast, when physical symptoms and signs are subtle diagnosis may be far more taxing. The principal biochemical features of Cushing's syndrome are an excess endogenous integrated secretion of cortisol, loss of the normal feed-back of the hypothalamo-pituitary-adrenal axis (HPA) and disturbance of the normal circadian rhythm of cortisol secretion. The biochemical tests that are used in the diagnosis of Cushing's syndrome rely upon these disturbances of normal physiology. In the investigation of Cushing's syndrome initial biochemical tests of high *sensitivity* should be employed to avoid missing mild disease, although the diagnosis may latter be refuted by tests of higher *specificity*. Differential diagnosis can only be embarked upon once the diagnosis of Cushing's syndrome is firmly established. In this chapter we will review the biochemical tests used for diagnosis and differential diagnosis. Imaging tests are discussed in the relevant chapters.

ASSESSMENT OF CIRCADIAN RHYTHM

The loss of the normal circadian rhythm of cortisol secretion seen in patients with Cushing's syndrome was first reported by Doe et al in 1960 (1). The overlap in serum cortisol values in patients with Cushing's syndrome and normal subjects diminishes with clock time and this forms the basis of the midnight cortisol test. In our own series a single sleeping serum cortisol was greater than 50 nmol/L (1.8μg/dL) in 150 patients with subsequently confirmed Cushing's syndrome (three of whom had suppressed circulating cortisol levels on a 2mg/day 48-hour low-dose dexamethasone-suppression test), whilst in control subjects the sleeping midnight serum cortisol was less than 50 nmol/L (1.8 μg/dL) in all (2) (Figure 1). The test does, however, require in-patient admission for a period of 48 hours or more to avoid false positive responses due to the stress of hospitalization, and the blood sample needs to be drawn within 5-10 minutes of waking the patient. To avoid false positive results due to anticipation the patients should not be warned that the test is to be performed; if the patient is awake the test is not readily interpretable. In our experience the test affords no discriminatory power between any of the causes of Cushing's syndrome.

Similar blunting of the circadian rhythm may be seen in many patients with depressive illness (3) acute illness, heart failure and infection and may result in a false positive results (4). Discrimination between Cushing's syndrome and pseudo-Cushing's states has been reported with a serum cortisol value at midnight was greater than 207 nmol/L (7.5 μg/dL) in 94% of 97 patients with Cushing's syndrome, whilst it was less than this in 31 individuals with pseudo-Cushing's states (5). This difference in cut-off point may reflect that those studied in this latter report were awake at the time of sampling. It is our experience that many patients with time Cushing's syndrome will have *sleeping* midnight serum cortisol values below this level (Figure 1), and thus if this manner of sampling is employed a cut-off level above 50 nmol/L (1.8 μg/dL) is recommended. Late-night salivary cortisol estimation may be an alternative approach and has the advantage that admission to hospital may not be required (6) and may be particularly useful when investigating children (7,8).

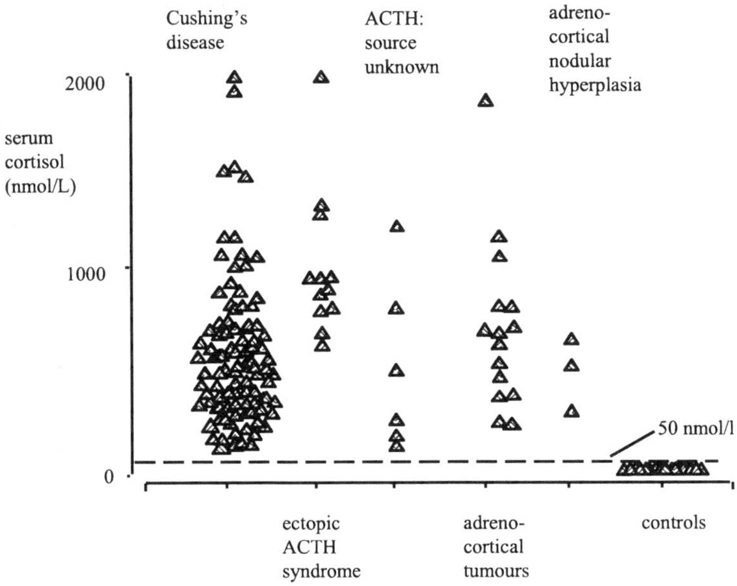

Figure 1. Sleeping serum midnight cortisol in 150 patients with Cushing's syndrome and controls. Serum cortisol <4550nmol/l=1.8 μg/dL (Reproduced with permission from Newell-Price et al., Clinical Endocrinology 1995, 43:545-550).

URINARY FREE CORTISOL

Urinary free cortisol (UFC) collection is a non-invasive procedure and is widely used as a screening test for the diagnosis of Cushing's syndrome. Twenty-four UFC measurements by radioimmunoassay (RIA) should reflect the integrated cortisol secretion, with a raised level being consistent with Cushing's syndrome. In most assays the upper normal range is between 220-330 nmol/24 hours (80 to 120 μg/24 hours). The majority of problems associated with this test relate to the adequacy of collection, although in some assays there may be cross-reactivity with exogenous glucocorticoids. Expressing the UFC over creatinine allows the adequacy of collection to be established and improves the specificity (9), but it should be noted that creatinine may vary with changes in lean body mass. Furthermore, creatinine is not subject to circadian variation like cortisol, so that this corrector has intrinsic error if used to correct for missed samples at differing times of the 24-hour period. 11% of 146 patients with Cushing's syndrome had at least one out four 24-hour collections with values within the normal range in one study (10). More recently, the lowest UFC level among multiple collections were normal in up to 15% of 333 patients with CS (11). Further, 'raised' 24-hour UFC levels have been documented in 40% of depressed in-patients (12), 50% of women with the polycystic ovarian syndrome (13), and by definition almost complete overlap in levels is seen in the differentiation from various causes of pseudo-Cushing's states (14), emphasizing the potential for diagnostic confusion. If factitious Cushing's syndrome is suspected high performance liquid chromatography may detect synthetic glucocorticoid in the urine (15). Overall, multiple UFC estimations have a high sensitivity, but relatively low specificity and thus if several UFC collections are normal Cushing's syndrome is highly unlikely.

LOW-DOSE DEXAMETHASONE TESTING

Since the original description by Liddle in 1960 (16) of the 48 hour 2 mg/day low-dose dexamethasone test (LDDST), this diagnostic tool has remained an important part of the evaluation of suspected Cushing's syndrome. A variety of regimens exist for the dexamethasone administration, and a range of diagnostic 'cut-offs' that classify responses have been reported.

The overnight test usually involves the oral administration of dexamethasone 1mg at 2300h or 2400h and then obtaining a serum cortisol sample at 0800h or 0900h the next morning (17). Because of its ease of administration as an outpatient, it has been widely advocated as a screening test. Since some patients with Cushing's disease demonstrate unusual suppression of serum cortisol to dexamethasone (18), a cut-off of a post-dexamethasone serum cortisol of 50 nmol/L (1.8 μg/dL) or less to exclude Cushing's syndrome is recommended (19). At this level false positive rates will be higher, but the main value of the overnight test is that of ease of outpatient-screening to exclude Cushing's syndrome, and thus this may be an acceptable price to pay for enhanced sensitivity.

In our own experience the original 48-hour 2mg/day dexamethasone suppression test, using serum cortisol measurements, in 150 individuals with proven Cushing's syndrome resulted in a sensitivity of 98% (three patients with histologically-proven Cushing's disease suppressed to less than 50 nmol/L) (2). The 1 mg overnight test has a specificity of 87.5% (20), whilst the reported specificity for the 2mg/day 48 hour test is 97-100% (21). It is, therefore, our routine practice to use the 'Liddle' 48 hour, 2mg/day dexamethasone suppression test (dexamethasone 0.5mg given strictly 6-

hourly), with serum cortisol measurement, in both in-patient and outpatient settings, since, with adequate written instructions, compliance is extremely high and the results reproducible.

Variable absorption or compliance with administration, of dexamethasone will influence the result of dexamethasone tests. For this reason simultaneous measurement of serum cortisol and dexamethasone has been advocated for the overnight test to ensure adequate plasma dexamethasone concentrations of 5.6 nmol/L (0.22 µg/dL) or greater and to confirm compliance (22). This does, however, require access to, and the cost of, a dexamethasone assay, and in general it is unnecessary although it may be particularly useful in cases of suspected malabsorption. An alternative is to use an intravenous infusion of dexamethasone (23, 24). Such tests are more complex to perform, but in certain circumstances their application may be of use.

Enhanced metabolism of dexamethasone may be caused by increased hepatic enzyme activity induced by drugs such as phenytoin, phenobarbitone, carbamezepine, and rifampicin, which may reduce dexamethasone concentration (25) and result in false positive responses to dexamethasone testing. Oral estrogens increase the cortisol binding globulin concentration in the circulation. Since RIA's measure total cortisol, false positive rates are seen in 50% of women on the oral contraceptive pill (26). It is our routine practice, where possible, and particularly in mild cases, to stop such estrogen-containing drugs and delay investigation for six weeks to allow the cortisol binding globulin to return to baseline. However, this may not be necessary in the case of trans-dermal estrogens.

CYCLICAL CUSHING'S SYNDROME

For unknown reasons some ACTH-secreting tumors of any type causing Cushing's syndrome may exhibit cyclical or intermittent secretion (27, 28), and this may be reflected by a history of variable and intermittent symptoms. In-patient admission, sometimes on repeated occasions, with sampling for sleeping midnight serum cortisol, is one means of getting around this diagnostic conundrum, and proceeding to further investigation if documented hypercortisolemia is present. If in-patient admission cannot be justified salivary cortisol estimations may be used to establish the diagnosis as an outpatient (29). Multiple and repeated 24 hour UFC collections may also prove useful. Clearly the diagnostic dimension of time is often needed to establish the diagnosis.

DIFFERENTIATION FROM PSEUDO-CUSHING'S STATES

The differentiation between mild Cushing's syndrome and pseudo-Cushing's syndrome can prove extremely difficult, and be a considerable challenge to the physician. A pseudo-Cushing's state may be defined as some or all of the clinical features that resemble true Cushing's syndrome together with some evidence of hypercortisolism, but resolution of the underlying primary condition results in the disappearance of the Cushing's-like state. Such findings may feature particularly in patients with depression and alcohol-induced pseudo-Cushing's syndrome.

In both Cushing's syndrome and pseudo-Cushing's states there is prevailing hypercortisolemia, and hence almost complete overlap between groups on basal 24-hour UFC collections (14). The dexamethasone-suppressed CRH test has been proposed as a test of high diagnostic accuracy to aid discrimination. The test involves the administration of CRH 100μg intravenously after a 48-hour, 2mg/day LDDST (14). In the original retrospective report a serum cortisol value at 15 minutes after injection of CRH of greater than 38 nmol/L (1.4 μg/dL) was seen in all patients with Cushing's syndrome, but in none with a pseudo-Cushingoid state, giving it a sensitivity and specificity of 100%. However, many serum cortisol RIA's will have poor precision at this level of cortisol, and care is needed in its application. Moreover, a prospective follow-up report on a further 98 patients resulted in a specificity of 96% and sensitivity of 98% since two patients were misclassified (30). It thus seems likely that this may prove to be a useful test in differentiating Cushing's syndrome from mild secondary hypercortisolism, but is still far from completely reliable.

In depressed patients, although there is often loss of suppression on a LDDST and a loss of the normal circadian rhythm of cortisol, there is usually a cortisol response to adequate insulin-induced hypoglycemia (plasma glucose <2.2 mmol/l (40mg/dl)), whilst such a response is seen in only 18% of patients with Cushing's syndrome (3, 31-33). In certain cases with pseudo-Cushing's syndrome associated with depression, the insulin tolerance test may be useful, although clearly overlap in responses exists (34).

Desmopressin, a long-acting analog of vasopressin (35), with relative specificity for the renal V_2 and V_{1b} (V3) receptors with little V_1-mediated pressor activity (36) has been reported for the purposes of discriminating between mild Cushing's syndrome and pseudo-Cushing's states. It has been shown previously to have no intrinsic *in vivo* ACTH-releasing characteristics when given as an infusion in normal volunteers (37). Moreover it appears to be free of the V_1 receptor-mediated pressor side-effects that preclude routine use of vasopressin. Following administration of desmopressin 10μg intravenously, each of 14 patients with Cushing's disease exhibited a rise in serum cortisol of 36% whilst such rises were seen in two out of 20 normal controls and four out of 11 patients with depression (38). No systematic ACTH responses to this peptide were observed in normal controls or depressed patients, whilst a rise was seen in all the patients with Cushing's disease, and as such the responses resemble those seen in this context following the CRH administration. Similarly, desmopressin has been shown to invoke ACTH responses in patients with mild Cushing's disease but not in those with obesity or pseudo-CS (39). It should be noted, however, that up to 15% of patients with Cushing's disease fail to respond to desmopressin (34) and thus in clinical practice the sensitivity of this test may be compromised. Moreover there are insufficient data on the responses seen in ectopic ACTH secretion and is unlikely that the test will be of use in ACTH-independent cases of CS.

Patients with alcohol-induced pseudo-Cushing's syndrome may cause diagnostic difficulty, but the sleeping midnight serum cortisol value has been shown to become undetectable within five days of supervised in-patient admission with abstinence, effectively excluding Cushing's syndrome (40). More recently both desmopressin and hexarelin stimulation tests (see below) have been shown not to elicit ACTH or cortisol responses in these patients, in contrast to patients with Cushing's disease (41).

ACTH-DEPENDENT VS. ACTH-INDEPENDENT

Only once the diagnosis of Cushing's syndrome is established may the differential diagnosis be entertained. The initial step is to establish if ACTH is detectable in the plasma, since if it is consistently undetectable the diagnosis of ACTH-independent Cushing's syndrome may be made, and focus should turn to adrenal imaging. It is, however, our experience that a few patients with pituitary-dependent Cushing's disease have the occasional undetectable plasma ACTH, as assessed by a conventional RIA with a lower limit of detection of 10 pg/ml (34). Thus, if measured by this means we would recommend the collection of several plasma ACTH collections to avoid this potential problem, or the measurement of ACTH after CRH stimulation. Nevertheless, we have not seen any patient with ACTH-dependent Cushing's syndrome that has had more than one or two ACTH levels of less than 10 pg/ml. Immunoradiometric assays (IRMA) provide increased speed of throughput, high reproducibility, and sensitivity for ACTH measurement.

Using IRMA, ACTH levels consistently below 5 pg/ml are found in patients with cortisol-producing adrenal adenomas, autonomous bilateral adrenal hyperplasia and Cushing's syndrome to the administration of exogenous glucocorticoids (42). A gray area does, however, exist and low-detectable ACTH levels need cautious interpretation and repeated measurement. Some clinicians advocate the use of the CRH test in these circumstances to confirm adrenal autonomy but we have not found this to be necessary. A detailed discussion on screening for the presence of aberrantly expressed G-protein coupled receptors in the adrenal cortex in patients with macronodular adrenal hyperplasia is beyond the scope of this chapter (43, 44).

DIFFERENTIAL DIAGNOSIS OF ACTH-DEPENDENT CUSHING'S SYNDROME

Overview

Corticotroph adenomas and tumors causing the ectopic ACTH syndrome are frequently small and thus difficult to visualize, and strenuous efforts at localization are required to allow correct management: these in turn rely heavily on biochemical testing to direct the imaging to the appropriate site. In most centers, during the investigation of ACTH-dependent Cushing's syndrome the likelihood that a patient has pituitary disease is usually between 85% and 90%. Biochemical testing is used in an attempt to improve upon the pre-test likelihood, and to direct the physician to the appropriate imaging and sampling modalities prior to formal management. Assuming that the default mode of treatment is pituitary surgery, peripheral non-invasive tests are best reported such that specificity is optimized (this inevitably being at the cost of reduced sensitivity) so that patients with ectopic ACTH syndrome do not undergo inappropriate pituitary surgery.

Basal testing

Plasma ACTH. Plasma ACTH levels tend to be higher in the ectopic ACTH syndrome than in Cushing's disease but there is a large overlap between values. Although the presence of pro-opiomelanocortin (POMC) precursors has been attributed to ectopic ACTH secretion, particularly when caused by small cell lung carcinoma (45), these peptides may also be detected in pituitary-dependent disease and the assays are not widely available (34). We do not routinely use such assays but in certain circumstances they may be of benefit.

Serum potassium

Serum potassium is virtually always low in the ectopic ACTH syndrome, and this may therefore be an extremely helpful discriminator, although up to 10% of patients with Cushing's disease exhibit hypokalemia (4). The reason for the hypokalemia appears to be the saturation of 11-β hydroxysteroid dehydrogenase by excessive cortisol, which under normal physiological circumstances protects the mineralocorticoid receptor from the effects of cortisol (46). However, this generally reflects the prevailing levels of cortisol rather than the specific etiology. In our experience we have seen only one patient with the ectopic ACTH syndrome (with a bronchial carcinoid) who did not have hypokalemia and associated alkalosis.

Ectopic co-secretion

In up to 70% of cases of occult ectopic tumors there may be co-secretion of one or more additional peptides which can offer a clue as to the presence of ectopic ACTH secretion. Measurement of such 'tumor markers', therefore, can be done and is helpful especially calcitonin, somatostatin, gastrin, pancreatic polypeptide, vasoactive intestinal peptide, glucagon, hCG-β, α-fetoprotein, α-subunit, neuron-specific enolase, growth hormone-releasing hormone, CRH and carcinoembryonic antigen (47, 48). The presence of an additional peptide provides stronger evidence for the ectopic ACTH syndrome, may also serve as a tumor marker during follow-up, and may occasionally be useful during venous sampling for localization (48).

DYNAMIC NON-INVASIVE TESTING

High dose dexamethasone testing (HDDST)

The HDDST has remained one of the main biochemical tools utilized in the differential diagnosis of ACTH-dependent Cushing's syndrome. The most commonly quoted cut-point consistent with Cushing's disease is a 50% suppression from basal pre-dexamethasone serum cortisol following administration of high-dose dexamethasone, although this cut-off has no intrinsic validity (49). Dexamethasone, 2mg 6-hourly for 48 hours is frequently used and is best performed using serum cortisol measurements. This test is somewhat cumbersome and some physicians use the 8 mg overnight dexamethasone suppression test with administration of a single 8 mg dose of dexamethasone orally at 2300h, with measurement of serum cortisol at 0800h before and after administration (50, 51). Overall, the 48-hour test is more accurate than the overnight test (34), and despite suggested improvements (52) the sensitivity and specificity of the test by itself remain below the pre-test likelihood since between 15-20% of patients with Cushing's disease fail to suppress using this criterion (34).

Part of the failure of suppression on high-dose dexamethasone testing in patients with Cushing's disease may relate to inadequate levels of circulating dexamethasone due to inadequate absorption, increased clearance or poor compliance (53). To circumvent these problems the use of a 5 hour intravenous infusion of dexamethasone at a rate of 1mg/h has been advocated, with suppression of serum cortisol by 50% or more being consistent with Cushing's disease (54). A slight modification of this test, with an infusion of dexamethasone for seven hours, demonstrated a fall in serum cortisol of 190 nmol/L or more in all of 90 patients with Cushing's disease, but only in two of 7 with the ectopic ACTH syndrome, giving a sensitivity of 100% and a specificity of 90% (55). Both false positive responders in this study were ectopic secretors of CRH, and since this is a rare cause of Cushing's syndrome more data on the responses seen in the ectopic ACTH syndrome are needed. A follow-up study revealed that the test performed less well with a sensitivity of 94% and specificity of only 63% for the diagnosis of Cushing's disease (56).

Because of this we would always consider the response in association with the responses seen following peripheral stimulation testing (see below), although some have called for the complete abandonment of the HDDST (57), and it appears that there may be some justification in such assertions since even supervised intravenous infusion tests can perform poorly.

Testing with Corticotropin-Releasing Hormone (CRH)

CRH has been extensively used in the differential diagnosis of ACTH-dependent Cushing's syndrome (34). It is important to note that the majority of reports documenting the use of this peptide in this context have utilized the ovine (oCRH) rather than human-sequence peptide (hCRH).

The test is performed with the patient in a rested, fasted, and recumbent state. Most test protocols take samples for plasma ACTH and serum cortisol at basal samples between -15 and 0 minutes, and stimulated samples at 15, 30, 45, 60 , 90, and 120 minutes following the intravenous administration of CRH 1μg/kg body weight, or a total dose of 100μg. The test is well tolerated, with side effects consisting of mild short-lived mild facial flushing and a metallic taste in the mouth.

Kaye and Crapo suggested that responses consistent with Cushing's were a rise from basal in peak serum cortisol of ≥20%, or a rise in peak plasma ACTH of ≥50% following the administration of CRH (58). Using these criteria for the plasma ACTH responses the test has a sensitivity of 86% and a specificity of 95%, whilst serum cortisol responses give an improved sensitivity of 91% and a similar specificity of 95%. In contrast, in the largest reported series from the NIH the responses to ovine oCRH, administered at 0800h, that best discriminated between pituitary and non-pituitary origins of ACTH secretion was a rise of 35% or more in the mean plasma ACTH concentrations at 15 and 30 minutes above the mean basal value at -5 and -1 minutes; this was seen in 93 of 100 patients with Cushing's disease, while a response less than this was observed in all 16 patients studied with the ectopic ACTH syndrome (13 of which had carcinoid tumors) giving a sensitivity of 93% and a specificity of 100% (59). The numbers of patients studied in this latter single study (59) is comparable in size to the total number analyzed by Kaye and Crapo (58), and yet differing response criteria are recommended. This may in part be explained by differences in study protocols, since the meta-analysis included studies utilizing evening testing, and in part by different assays employed.

The human sequence peptide has similar effects to the oCRH in normal individuals and patients with Cushing's disease. Some reports indicate that testing with hCRH is less accurate than when using oCRH (60). The responses are qualitatively similar, albeit with a quantitatively lower response to hCRH, in patients with Cushing's disease, Cushing's syndrome due to adrenal adenoma, and in obese and lean volunteers (61). For this reason pooled data in which both peptide sequences were utilized independently in different patients cannot be relied upon to set response criteria for each peptide (11). Published large series are needed that report the responses to hCRH that best discriminate between patients with the ectopic ACTH syndrome and Cushing's disease.

Combined analysis of all published series reveals that between 7% and 14% of all patients with Cushing's disease fail to respond to CRH if the best discriminating criteria are applied (34). Whilst most of the Cushing's disease non-responders exhibit suppression on a HDDST (49, 62), there are rare cases in which an ACTH-secreting bronchial carcinoid tumor may suppress on a HDDST and exhibit responsiveness to CRH (63). Such cases are very uncommon and analysis of the responses to high dose dexamethasone testing and CRH stimulation will often allow correct classification.

Testing with desmopressin

Desmopressin 10 µg iv does, however, cause the release of ACTH and cortisol in 84% of patients with Cushing's disease, but has a specificity that is worse than CRH testing (34). Several reports utilizing this peptide do not include responses from patients with the ectopic ACTH syndrome. Therefore, testing with desmopressin is inferior to testing with CRH in terms of sensitivity and specificity, although this peptide is cheaper and more easily available worldwide. A possible explanation for the relatively poorer specificity of the desmopressin test is the more common expression of the V_{1b} (or V_3) receptor in ACTH-secreting non-pituitary tumors (64, 65). Nevertheless some patients with Cushing's disease respond only to one peptide or the other (34) and thus in certain circumstances testing with desmopressin may be useful.

Testing with peptide combinations

Since 7-14% of patients with Cushing's disease do not respond to CRH, the use of this peptide in combination with other peptides has been assessed. Administration of 10U AVP in combination with 1µg/kg oCRH in patients with Cushing's disease resulted in a rise in serum cortisol of than 20% or more in 40 out of 41 patients, and a rise of 35% or more in plasma ACTH in all patients tested, a better response than that seen following administration of CRH alone (66). Thus, it has been suggested that this will be an improvement over the standard CRH test, although patients with the ectopic ACTH syndrome were not studied. Furthermore, vasoconstriction and an increase in blood pressure are important side-effects of testing with AVP, and thus caution is needed in patients with vascular disease. We have used desmopressin 10µg iv and hCRH 100µg iv in combination in 17 patients with Cushing's disease and 5 patients with the occult ectopic ACTH syndrome (age range 11-73y), all histologically-confirmed, and compared the response of serum cortisol and ACTH to those seen when testing with each peptide individually. An increase in serum cortisol of 39% or more was seen in every patient with Cushing's disease, but 29% or less in each of the five patients with the ectopic ACTH syndrome, thus giving 100% sensitivity and specificity in this small series (67) (Figure 2). Moreover, there are no associated adverse effects of the test, including no significant increases in blood pressure. The discrimination does, however, remain quantitative in nature, and it remains to be seen whether the co-administration of desmopressin and hCRH is an improvement over the standard CRH test.

Testing with hexarelin

Administration of the synthetic peptide hexarelin, a member of the growth hormone-releasing peptides (GHRPs) family, has recently been shown to have far greater ACTH and cortisol releasing effects than hCRH in ten patients with Cushing's disease (68). The responses seen were remarkable in terms of the absolute levels of stimulated serum cortisol, and even more impressively plasma ACTH, both of which were far higher than those seen even when testing with oCRH or hCRH in patients with Cushing's disease. In contrast, no ACTH rise was seen in two patients with the ectopic ACTH syndrome, whilst the cortisol response in one overlapped those of the Cushing's disease group. No ACTH or cortisol response was seen in five patients with cortisol-secreting adrenal adenomas. Use of a recombinant formulation of the natural ligand for this receptor 'ghrelin' (69) may possibly be useful in the future. However, GHRP receptors are expressed in ACTH-secreting non-pituitary tumors (70, 71). It seems likely, therefore, that with further experience more responders in the ectopic ACTH-syndrome group will be reported

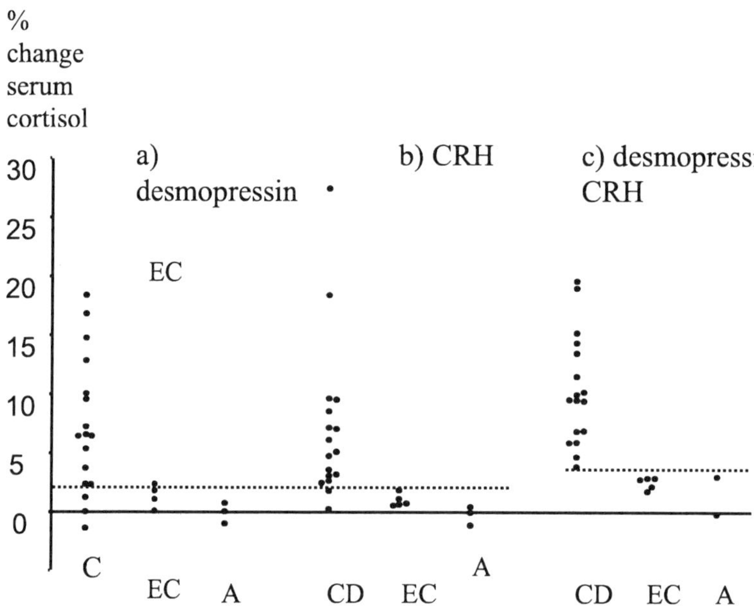

Figure 2 Effect of desmopressin 10μg i.v., hCRH 100 μg i.v., and desmopressin 10μg i.v. + hCRH 100 μg i.v. in combination, in patients with Cushing's disease (CD), Ectopic CTH secretion (EC). and adrenal Cushing's syndrome (A). (Reproduced with permission from Newell-Price et al., JCEM 1997, 82:176-181).

Invasive testing

1. Inferior petrosal sinus sampling for ACTH

Although in many circumstances peripheral biochemical tests will provide evidence of pituitary disease, the presence of a pituitary lesion on imaging, especially if less than 4mm in diameter, does not necessarily confirm functionality, since 10% of the general population harbor pituitary incidentalomas (72). Moreover, in at least 40-50% of cases of Cushing's disease no abnormality will be disclosed on pituitary imaging(34). Thus, venous sampling from the inferior petrosal sinuses and/or cavernous sinuses is now widely practiced, firstly to confirm or refute a central (pituitary) source of ACTH, especially when pituitary imaging is negative, and secondly to attempt to lateralize the site of a pituitary tumor to guide neurosurgical approaches.

A significant gradient between the pituitary (central) and peripheral values of plasma ACTH, obtained by simultaneous sampling, is indicative of Cushing's disease when there is prevailing hypercortisolaemia. Simultaneous bilateral inferior petrosal sampling is required as the drainage of the pituitary tends to have dominant and non-dominant drainage to the inferior petrosal sinuses and therefore unilateral sampling may miss a central source. A basal ratio of the values of plasma ACTH obtained from central and peripheral samples of greater than 2.0 is widely accepted as indicative of Cushing's disease (34). Although a baseline ratio of more than 2.0 is consistent with Cushing's disease, ACTH secretion is intermittent and a significant minority of patients with Cushing's disease have a ratio less than this on the basal samples. For this reason, after basal samples are obtained a stimulating agent to increase the sensitivity of the test. Both hCRH and oCRH have been used for this purpose with great success, and more recently desmopressin or desmopressin in combination with CRH (73, 74). Following administration of intravenous CRH 100µg, peripheral and simultaneous bilateral inferior petrosal sinus plasma ACTH samples are obtained at 3, 5, 10, and 15 minutes (in varying reports). In most studies a peak stimulated central to peripheral ratio of 3.0 or more, which usually occurs between 3 to 5 minutes post-CRH, is indicative of Cushing's disease. Metyrapone pretreatment has also been used with success to enhance the central to peripheral gradient (75), suggesting it may have a role when CRH is not available. Applying the criteria of a basal ratio of ≥2.0 or a CRH stimulated ratio of ≥ 3.0 to published series that have compared the results seen in Cushing's disease and the ectopic ACTH syndrome reveals an overall sensitivity of 96% and specificity of 100% (34) (although false positive responses have been documented-*see below*). BIPSS remains, however, an invasive technique requiring a high degree of skill and familiarity, and is thus best performed in centers with suitable experience. False negative results occur in 4% (34) and may relate to inaccurate catheter placement or rarely to hypoplastic inferior petrosal sinuses, whilst false positive results are extremely uncommon, and may result from cyclical secretion of ACTH, or theoretically by treatment with cortisol lowering agents resulting in the de-suppression of the normal corticotrophs which might then respond to CRH (76). Since this technique does not reliably distinguish between normal individuals, or those with pseudo-Cushingoid states, from Cushing's disease (77), it is essential to confirm the presence of hypercortisolism before performing the test. This is of particular relevance when considering complications. Although the test is well tolerated, with most patients experiencing slight discomfort in the ear whilst the catheters are being placed and small injections of i.v. contrast given, adverse effects when they do occur have included brain stem vascular damage (78-80). Such rare complications may relate to catheter design and might be avoided by the immediate cessation of the procedure, and catheter withdrawal, at the onset of the slightest neurological symptom (80). Heparinization of patients is recommended (80), and in our experience of over 130 procedures we have had no serious complications (73) and in our hands has a sensitivity and specificity of 97% and 100% respectively.

Use of this technique for lateralisation of a corticotroph adenoma within the sella is less reliable. An inter-sinus ratio of 1.4 or greater has been suggested as being consistent with the ipsilateral localization of a microadenoma (81). Using this ratio, analysis of reports documenting its use in the lateralisation of corticotroph microadenomas reveals that the diagnostic accuracy of inferior petrosal sinus sampling is 78% using findings at pituitary surgery as the 'gold standard' (34). CRH stimulation does not improve the accuracy of the localization in patients with Cushing's disease. If, however, the ACTH level in the non-dominant inferior petrosal sinus is less than three times the peripheral ACTH level, the accuracy improves to 83% (82). Occasionally, a reversal of lateralizing gradient is seen from the pre to the post-CRH values, and in this case the test cannot be relied upon for lateralisation (83). Although the lateralizing result may direct the surgeon to begin an initial examination of the pituitary gland on the side ipsilateral to the catheter gradient, a full exploration is required if 22% of tumors are not to be missed. Recommendations for ipsilateral hemihypophysectomy, in the absence of a clear tumor being visualized at operation, on the basis of the lateralizing data from inferior petrosal sinus sampling are hard to substantiate, since in over 20% of cases the tumor may be contralateral, although anecdotally this approach has been successful (84).

Normal individuals demonstrate unilateral gradients, despite anatomically symmetrical inferior petrosal sinuses (77). This may be caused by the normal pituitary having a 'dominant' side (85). In patients with Cushing's disease, however, the normal corticotrophs should be fully suppressed, and thus the existence of a dominant side to the pituitary should not affect the results. Asymmetric drainage has been demonstrated by gentle cavernous sinus venography prior to bilateral venous sampling from the inferior petrosal and cavernous sinuses (86). In nine out of 23 (39%) patients with Cushing's disease the drainage was asymmetric, with six of the 9 demonstrating drainage of both cavernous sinuses to the right petrosal sinus and no drainage to the left. Inferior petrosal sinus sampling and cavernous sampling correctly lateralized the tumor in all 12 with symmetric drainage, but in those with asymmetric drainage it was correct in basal and CRH-stimulated samples in only three and four out 9 respectively. As a whole the lateralisation was correct in 70%, which is similar to other published series. These data may explain some of the mis-lateralisation that occurs when performing BIPSS, and would suggest that the results for lateralisation cannot be relied upon in patients with Cushing's disease who have asymmetric drainage.

2. Cavernous sinus venous sampling

Teramoto and co-workers have suggested that sampling directly from the posterior cavernous sinuses, rather than the inferior petrosal sinuses, may improve diagnostic accuracy and obviate the need for the administration of CRH (87). However, in a comparison between this technique in the same 15 patients with Cushing's disease to that of basal and stimulated BIPSS, Doppman and co-workers found a false negative rate of 20% during cavernous sampling (88). More recently cavernous sinus sampling with CRH stimulation has been reported (89), and this appears to improve the diagnostic accuracy. Therefore this technique appears to carry few advantages for the assessment of centralization of an ACTH gradient. It may, however, be more accurate in localization of a microadenoma within the fossa with correct lateralization in 83% of all cases, and in 89% of cases with good catheter position and symmetric venous flow (89). The catheters are more expensive and in most centers, including ours, BIPSS remains the investigation most commonly performed.

Imaging

Pituitary

In ACTH-dependent Cushing's syndrome pituitary imaging is used to help identify and localize the position of a pituitary microadenoma, and hence to guide initial surgical exploration. Gadolinium-enhanced MRI offers a sensitivity for corticotroph microadenomas ranging from 50-60% (34). The majority of these have a hypointense signal on MRI, which fails to enhance with gadolinium. However, since in approximately 5% of pituitary microadenomas the tumor will take up the gadolinium contrast medium giving an isointense signal on MRI, pre-contrast images are essential (90). Furthermore, since incidental pituitary adenomas have been reported in up to 27% of post-mortem examinations (91), and in 10% of the 30-40y age group on MRI[72], any imaging result must be interpreted in the context of the biochemical evaluations as false positive results are possible (34). In some series, if a clear tumor is identified on MRI the chances that the position disclosed on imaging will correlate with the surgical findings are in the region of 75-98% (92, 93), and it has therefore been suggested that it is similar or superior to BIPSS for localization (92). In contrast, one large series reported that in only 52% of 41 cases did the MRI correlate with the surgical findings (94). Thus, a significant number of pituitary microadenomas are not visualized on MRI and pre-operative localization is not always possible by this technique. Furthermore, BIPSS is far from completely reliable for this purpose (discussed above). Dynamic MRI with very rapid sequence acquisition does not appear to improve diagnostic accuracy (95). In an effort to improve localization the use of intraoperative ultrasonography has been proposed (96, 97). In the latter study four out of 18 corticotroph microadenomas were clearly identified by preoperative MRI in a position that correlated with the surgical findings, whilst two further adenomas identified preoperatively were established to be at a different site at operation. In contrast, 13 of 18 were disclosed on intraoperative ultrasound as hyperechoic masses, with no complications of the technique which is, however, very operator-dependent. In this study no details were given of the results of BIPSS, but it seems from this small series that intraoperative ultrasound may prove useful in the localization of pituitary microadenomas, and further highlights the limitations of MRI for this purpose.

Adrenal

Adrenal imaging plays an important role in the diagnostic workup. In many circumstances a cortisol-secreting adrenal tumor will be obvious. The distinction between adrenal adenoma and carcinoma is based on the evidence of vascular invasion or metastases, but tumors greater than 6 cm in diameter on scanning should be regarded as malignant (4). In a review of 13 studies Fig et al (98) reported that all of 14 patients with adrenal carcinoma and all of 70 patients with adrenal adenoma causing Cushing's syndrome were correctly identified by computed tomography (CT) of the adrenal glands (98). Unfortunately, imaging is not always so clear-cut and some degree of nodularity of the glands may be apparent (99). Since a unilateral cortisol-secreting adrenal tumor will result in suppression of ACTH-secretion, the remaining ipsilateral and contralateral adrenal gland should be atrophic in appearance, and if any degree of hypertrophy is present the possibility of asymmetric macronodular hyperplasia should be considered (100). Careful scrutiny of abdominal imaging is required to avoid the inappropriate unilateral excision of bilateral disease masquerading as an adrenal adenoma. Rarely, ACTH-independent massive macronodular adrenal hyperplasia (weighing 69-149g) may be present on imaging with complete replacement of both adrenal glands, lack of a central to peripheral ACTH gradient on BIPSS, and an absence of an adenoma on MR imaging of the pituitary gland; in such cases bilateral adrenalectomy is indicated (90). However, macronodular hyperplasia may exist in Cushing's disease, although usually with not so dramatic appearances, and in all cases of ACTH-dependent Cushing's syndrome the adrenal glands may be bilaterally or unilaterally hyperplastic (100), with or without nodularity, although this may not be present in one third of cases (101). In such cases further detailed and careful biochemical evaluation is crucial.

Ectopic secretion

Small cell lung cancer and bronchial carcinoid tumors are the commonest source of ectopic ACTH secretion. Although the former is usually obvious the latter may prove extremely hard to localize (4, 34, 90, 102). High-resolution CT scans may reveal small bronchial carcinoid lesions inapparent on plain radiography (Figure 3), and since bronchial carcinoid tumors are usually 1 cm or less overlapping cuts of 1 cm or less should be employed (34).

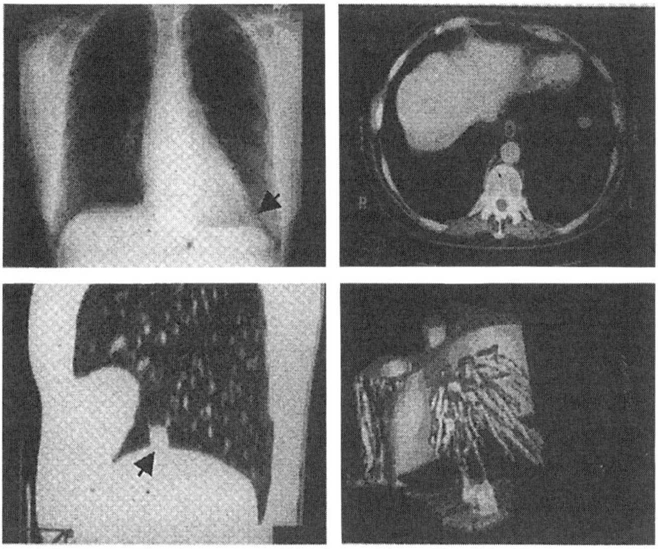

Figure 3. Plain radiograph (top left), axial (top right), saggital (bottom left) and 3-D reconstruction (bottom right) of a bronchial carcinoid tumor (arrows) abutting the diaphragm.

Small bronchial carcinoid tumors may, however, be confused with pulmonary vascular shadows. We have found that imaging the thorax in both supine and prone positions to be a simple and extremely effective means of resolving this diagnostic difficulty, since vascular shadows change and tumors do not. MRI seems to be an improvement over CT for this purpose, and results in improved discrimination: scanning in 10 patients with surgically-proven bronchial carcinoid tumors demonstrated high signal intensity on T2-weighted and short-inversion-time inversion recovery images in all, whilst CT scanning was equivocal in two (103). Thymic carcinoid tumors causing

Cushing's syndrome are generally larger than 2 cm and readily visualized by CT (104), although it should be noted that thymic remnant tissue is found 33 % of patients with active Cushing's syndrome, especially if young (105). Although the commonest site for an ectopic source of ACTH secretion is in the chest, many varied sites may ultimately be localized. Therefore, it may be necessary to perform extensive CT scanning of the abdomen to disclose, in particular, pancreatic islet cell tumors, intestinal carcinoid tumors, and pheochromocytomas (106). Islet cell tumors causing Cushing's syndrome are frequently large and have usually metastasized by the time of diagnosis and imaging (107), this presumably relating to the relatively late secretion of ACTH in the natural history of these tumors. A lesion disclosed on abdominal and chest imaging in the search for the source of ectopic ACTH or CRH secretion does not necessarily prove functionality. Furthermore, despite extensive imaging many tumors remain occult (4, 102, 106). Many carcinoid tumors, small cell lung cancers and medullary cell carcinomas of the thyroid express high numbers of high-affinity somatostatin receptors (108). It has therefore been proposed that the use of radiolabeled octreotide may be of use to confirm functionality (109) or disclose occult lesions not visualized by other means (110). In this latter series, although 8 of 10 patients with ectopic ACTH syndrome were disclosed on somatostatin receptor scintagraphy, all were apparent on conventional imaging and its use has thus been challenged since it did not reveal any lesions that were not disclosed by conventional imaging modalities (84). It remains to be seen whether improvements in this technique will allow disclosure of truly occult lesions.

CONCLUSIONS

Mild Cushing's syndrome may be difficult to distinguish from pseudo-Cushing's states, but use of some of the more recently described tests may help with this diagnostic conundrum. In the differential diagnosis of ACTH-dependent Cushing's syndrome invasive testing is the most accurate and useful, and some clinicians will opt to proceed directly to such testing once ACTH has been easily detected in the plasma. We rely heavily on the results from BIPSS, but since this is itself not perfect it is our routine practice to also perform non-invasive testing to build up an overall picture of the likely site of ACTH secretion. Figure 4 outlines a proposed scheme for investigation:

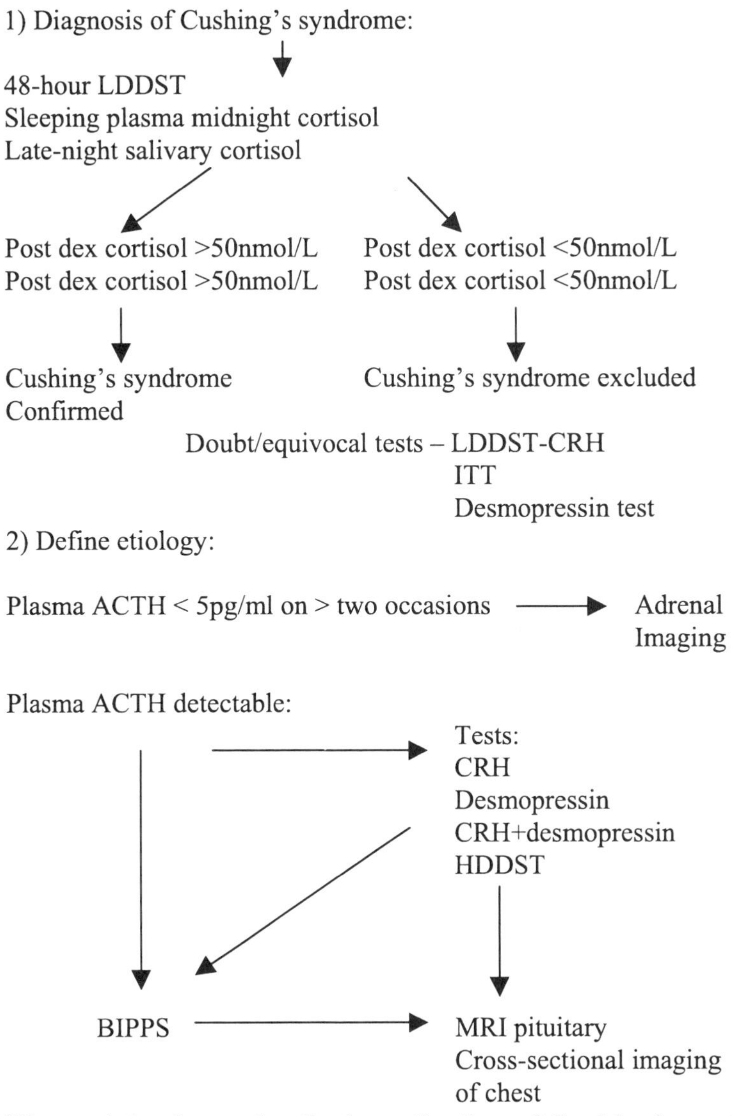

Figure 4 A scheme for the investigation of Cushing's syndrome

REFERENCES

1. Doe RP, Vennes JA, Flink EB. Diurnal variation of 17-hydroxycorticosteroids, sodium, potassium, magnesium, and creatinine in normal subjects and in cases of treated adrenal insufficiency and Cushing's syndrome. Journal of Clinical Endocrinology and Metabolism 1960; (20):253-65.
2. Newell-Price J, Trainer P, Perry L, Wass J, Grossman A, Besser M. A single sleeping midnight cortisol has 100% sensitivity for the diagnosis of Cushing's syndrome.

Clin.Endocrinol.(Oxf). 1995; 43(5):545-50.

3. Besser GM, Edwards CRW. Cushing's syndrome. Clinics in Endocrinology and Metabolism 1972; (1):451-90.
4. Trainer PJ, Grossman A. The diagnosis and differential diagnosis of Cushing's syndrome. Clin.Endocrinol.(Oxf). 1991; 34(4):317-30.
5. Papanicolaou DA, Yanovski JA, Cutler GB, Chrousos GP, Nieman LK. A single midnight cortisol measurement discriminates Cushing's syndrome from pseudo-Cushing's states. 1994.
6. Raff H, Raff JL, Findling JW. Late-night salivary cortisol as a screening test for Cushing's syndrome. Journal of Clinical Endocrinology & Metabolism 1998; 83(8):2681-6.
7. Gafni RI, Papanicolaou DA, Nieman LK. Nighttime salivary cortisol measurement as a simple, noninvasive, outpatient screening test for Cushing's syndrome in children and adolescents. Journal of Pediatrics 2000; 137(1):30-5.
8. Castro M, Elias PC, Quidute AR, Halah FP, Moreira AC. Out-patient screening for Cushing's syndrome: the sensitivity of the combination of circadian rhythm and overnight dexamethasone suppression salivary cortisol tests. Journal of Clinical Endocrinology & Metabolism 1999; 84(3):878-82.
9. Contreras LN, Hane S, Tyrrell JB. Urinary cortisol in the assessment of pituitary-adrenal function: utility of 24-hour and spot determinations. J.Clin.Endocrinol.Metab. 1986; 62(5):965-9.
10. Nieman LK, Cutler GBJr. The sensitivity of the urine free cortisol measurement as a screening test for Cushing's syndrome. 72nd Annual Meeeting. 1990:P822.
11. Invitti C, Giraldi FP, de Martin M, Cavagnini F. Diagnosis and management of Cushing's syndrome: results of an Italian multicentre study. Study Group of the Italian Society of Endocrinology on the Pathophysiology of the Hypothalamic-Pituitary-Adrenal Axis. Journal of Clinical Endocrinology & Metabolism 1999; 84(2):440-8.
12. Carroll BJ, Curtis GC, Davies BM, Mendels J, Sugerman AA. Urinary free cortisol excretion in depression. Psychol.Med. 1976; 6(1):43-50.
13. Cizza G, Nieman LK, Doppman JL *et al.* Factitious Cushing syndrome. J.Clin.Endocrinol.Metab. 1996; 81(10):3573-7.
14. Yanovski JA, Cutler GBJr, Chrousos GP, Nieman LK. Corticotropin-releasing hormone stimulation following low-dose dexamethasone administration. A new test to distinguish Cushing's syndrome from pseudo-Cushing's states. JAMA 1993; 269(17):2232-8.
15. Lin CL, Wu TJ, Machacek DA, Jiang NS, Kao PC. Urinary free cortisol and cortisone determined by high performance liquid chromatography in the diagnosis of Cushing's syndrome. J.Clin.Endocrinol.Metab. 1997; 82(1):151-5.
16. Liddle GW. Tests of pituitary-adrenal suppressibility in the diagnosis of Cushing's syndrome. J.Clin.Endocrinol.Metab. 1960; (20):1539-60.
17. Nugent CA, Nichols T, Tyler FH. Diagnosis of Cushing' syndrome-single dose dexamethasone. Archives of Internal Medicine 1965; 116:172-6.
18. Blethen SL, Chasalow FI. Overnight dexamethasone suppression test: normal responses and the diagnosis of Cushing's syndrome. Steroids 1989; 54(2):185-93.
19. Wood PJ, Barth JH, Freedman DB, Perry L, Sheridan B. Evidence for the low dose dexamethasone suppression test to screen for Cushing's syndrome--recommendations for a protocol for biochemistry laboratories. Ann.Clin.Biochem. 1997; 34(Pt 3):222-9.
20. Cronin C, Igoe D, Duffy MJ, Cunningham SK, McKenna TJ. The overnight dexamethasone test is a worthwhile screening procedure Clin.Endocrinol.(Oxf). 1990; 33(1):27-33.
21. Kennedy L, Atkinson AB, Johnston H, Sheridan B, Hadden DR. Serum cortisol concentrations during low dose dexamethasone suppression test to screen for Cushing's syndrome. Br.Med.J.(Clin.Res.Ed). 1984; 289(6453):1188-91.
22. Meikle AW. Dexamethasone suppression tests: usefulness of simultaneous measurement of serum cortisol and dexamethasone. Clin.Endocrinol.(Oxf). 1982; 16(4):401-8.

23. Abou Samra AB, Dechaud H, Estour B *et al.* Beta-lipotropin and cortisol responses to an intravenous infusion dexamethasone suppression test in Cushing's syndrome and obesity. J.Clin.Endocrinol.Metab. 1985; 61(1):116-9.
24. Atkinson AB, McAteer EJ, Hadden DR, Kennedy L, Sheridan B, Traub AI. A weight-related intravenous dexamethasone suppression test distinguishes obese controls from patients with Cushing's syndrome. Acta Endocrinol.(Copenh) 1989; 120(6):753-9.
25. Jubiz W, Meikle AW, Levinson RA, Mizutani S, West CD, Tyler FH. Effect of diphenylhydantoin on the metabolism of dexamethasone. N.Engl.J.Med. 1970; 283(1):11-4.
26. Tiller JW, Maguire KP, Schweitzer I *et al.* The dexamethasone suppression test: a study in a normal population. Psychoneuroendocrinology. 1988; 13(5):377-84.
27. Atkinson AB, Kennedy AL, Carson DJ, Hadden DR, Weaver JA, Sheridan B. Five cases of cyclical Cushing's syndrome. Br.Med.J.(Clin.Res.Ed). 1985; 291(6507):1453-7.
28. Bailey RE. Periodic hormonogenesis--a new phenomenon. Periodicity in function of a hormone-producing tumor in man. J.Clin.Endocrinol.Metab. 1971; 32(3):317-27.
29. Mosnier-Pudar H, Thomopoulos P, Bertagna X, Fournier C, Guiban D, Luton JP. Long-distance and long-term follow-up of a patient with intermittent Cushing's disease by salivary cortisol measurements. Eur.J.Endocrinol. 1995; 133(3):313-6.
30. Yanovski JA, Cutler GBJr, Chrousos GP, Nieman LK. Prospective evaluation of the dexamethasone-suppressed corticotropin-releasing hormone test in the differential diagnosis of Cushing's syndrome and pseudo-Cushing's states. 77th Annual Meeting. 1995:R 39-2.
31. James VH, Landon J, Wynn V, Greenwood FC. A fundamental defect of adrenocortical control in Cushing's disease. J.Endocrinol. 1968; 40(1):15-28.
32. Crapo L. Cushing's syndrome: a review of diagnostic tests. Metabolism 1979; 28(9):955-77.
33. Butler PW, Besser GM. Pituitary-adrenal function in severe depressive illness. Lancet 1968; 1(554):1234-6.
34. Newell-Price JDC, Trainer PJ, Besser GM, Grossman AB. The diagnosis and differential diagnosis of Cushing's syndrome and pseudo-Cushing's states. Endocrine Reviews 1998; 19(5):647-72.
35. Edwards CR, Kitau MJ, Chard T, Besser GM. Vasopressin analogue DDAVP in diabetes insipidus: clinical and laboratory studies. Br.Med.J. 1973; 3(876):375-8.
36. Sawyer WH, Acosta M, Manning M. Structural changes in the arginine vasopressin molecule that prolong its antidiuretic action. Endocrinology 1974; 95(1):140-9.
37. Gaillard RC, Riondel AM, Ling N, Muller AF. Corticotropin releasing factor activity of CRF 41 in normal man is potentiated by angiotensin II and vasopressin but not by desmopressin. Life Sci. 1988; 43(23):1935-44.
38. Malerbi DA, Fragoso MC, Vieira Filho AH, Brenlha EM, Mendonca BB. Cortisol and adrenocorticotropin response to desmopressin in women with Cushing's disease compared with depressive illness. J.Clin.Endocrinol.Metab. 1996; 81(6):2233-7.
39. Tsagarakis S, Vasiliou V, Kokkoris P, Stavropoulos G, Thalassinos N. Assessment of cortisol and ACTH responses to the desmopressin test in patients with Cushing's syndrome and simple obesity. Clinical Endocrinology 1999; 51(4):473-7.
40. Rees LH, Besser GM, Jeffcoate WJ, Goldie DJ, Marks V. Alcohol-induced pseudo-Cushing's syndrome. Lancet 1977; 1(8014):726-8.
41. Coiro V, Volpi R, Capretti L, Caffarri G, Chiodera P. Desmopressin and hexarelin tests in alcohol-induced pseudo-Cushing's syndrome. Journal of Internal Medicine 2000; 247(6):667-73.
42. Raff H, Findling JW. A new immunoradiometric assay for corticotropin evaluated in normal subjects and patients with Cushing's syndrome. Clin.Chem. 1989; 35(4):596-600.
43. Lacroix A, Bolte E, Tremblay J *et al.* Gastric inhibitory polypeptide-dependent cortisol hypersecretion- -a new cause of Cushing's syndrome [see comments]. N.Engl.J.Med.

1992; 327(14):974-80.
44. Lacroix A, Tremblay J, Rousseau G, Bouvier M, Hamet P. Propanolol therapy for ectopic beta-adrenergic receptors in adrenal Cushing's syndrome. New England Journal of Medicine 1997; 337:1429-34.
45. Stewart PM, Gibson S, Crosby SR *et al.* ACTH precursors characterize the ectopic ACTH syndrome. Clin.Endocrinol.(Oxf). 1994; 40(2):199-204.
46. Stewart PM, Walker BR, Holder G, O'Halloran D, Shackleton CH. 11 beta-Hydroxysteroid dehydrogenase activity in Cushing's syndrome: explaining the mineralocorticoid excess state of the ectopic adrenocorticotropin syndrome. J.Clin.Endocrinol.Metab. 1995; 80(12):3617-20.
47. Howlett TA, Rees LH. Is it possible to diagnose pituitary-dependent Cushing's disease?. Ann.Clin.Biochem. 1985; 22(Pt 6):550-8.
48. Howlett TA, Price J, Hale AC *et al.* Pituitary ACTH dependent Cushing's syndrome due to ectopic production of a bombesin-like peptide by a medullary carcinoma of the thyroid. Clin.Endocrinol.(Oxf). 1985; 22(1):91-101.
49. Grossman AB, Howlett TA, Perry L *et al.* CRF in the differential diagnosis of Cushing's syndrome: a comparison with the dexamethasone suppression test. Clin.Endocrinol.(Oxf). 1988; 29(2):167-78.
50. Tyrrell JB, Findling JW, Aron DC, Fitzgerald PA, Forsham PH. An overnight high-dose dexamethasone suppression test for rapid differential diagnosis of Cushing's syndrome. Ann.Intern.Med. 1986; 104(2):180-6.
51. Bruno OD, Rossi MA, Contreras LN *et al.* Nocturnal high-dose dexamethasone suppression test in the aetiological diagnosis of Cushing's syndrome. Acta Endocrinol.(Copenh) 1985; 109(2):158-62.
52. Flack MR, Oldfield EH, Cutler GBJr *et al.* Urine free cortisol in the high-dose dexamethasone suppression test for the differential diagnosis of the Cushing syndrome. Ann.Intern.Med. 1992; 116(3):211-7.
53. Meikle AW, Lagerquist LG, Tyler FH. Apparently normal pituitary-adrenal suppressibility in Cushing's syndrome: dexamethasone metabolism and plasma levels. J.Lab.Clin.Med. 1975; 86(3):472-8.
54. Croughs RJ, Docter R, Jong FHde. Comparison of oral and intravenous dexamethasone suppression tests in the differential diagnosis of Cushing's syndrome. Acta Endocrinol.(Copenh) 1973; 72(1):54-62.
55. Biemond P, de Jong FH, Lamberts SW. Continuous dexamethasone infusion for seven hours in patients with the Cushing syndrome. A superior differential diagnostic test. Ann.Intern.Med. 1990; 112(10):738-42.
56. van den Bogaert DP, de Herder WW, de Jong FH, Biemond P, van der LelyA. J. , Lamberts SW. The continuous 7-hour intravenous dexamethasone suppression test in the differential diagnosis of ACTH-dependent Cushing's syndrome. Clinical Endocrinology 1999; 51(2):193-8.
57. Aron DC, Raff H, Findling JW. Effectiveness versus efficacy: the limited value in clinical practice of high dose dexamethasone suppression testing in the differential diagnosis of adrenocorticotropin-dependent Cushing's syndrome. J.Clin.Endocrinol.Metab. 1997; 82(6):1780-5.
58. Kaye TB, Crapo L. The Cushing syndrome: an update on diagnostic tests [see comments]. [Review] [78 refs]. Ann.Intern.Med. 1990; 112(6):434-44.
59. Nieman LK, Oldfield EH, Wesley R *et al.* A simplified morning ovine corticotropin-releasing hormone stimulation test for the differential diagnosis of adrenocorticotropin-dependent Cushing's syndrome. J.Clin.Endocrinol.Metab. 1993; 77(5):1308-12.
60. Nieman LK, Cutler GBJr, Oldfield EH, Loriaux DL, Chrousos GP. The ovine corticotropin-releasing hormone (CRH) stimulation test is superior to the human CRH stimulation test for the diagnosis of Cushing's disease. J.Clin.Endocrinol.Metab. 1989; 69(1):165-9.

61. Trainer PJ, Faria M, Newell-Price J *et al.* A comparison of the effects of human and ovine corticotropin- releasing hormone on the pituitary-adrenal axis. J.Clin.Endocrinol.Metab. 1995; 80(2):412-7.
62. Hermus AR, Pieters GF, Pesman GJ *et al.* The corticotropin-releasing-hormone test versus the high-dose dexamethasone test in the differential diagnosis of Cushing's syndrome. Lancet 1986; 2(8506):540-4.
63. Malchoff CD, Orth DN, Abboud C, Carney JA, Pairolero PC, Carey RM. Ectopic ACTH syndrome caused by a bronchial carcinoid tumor responsive to dexamethasone, metyrapone, and corticotropin- releasing factor. Am.J.Med. 1988; 84(4):760-4.
64. De Keyzer Y, Lenne F, Auzan C *et al.* The pituitary V3 vasopressin receptor and the corticotroph phenotype in ectopic ACTH syndrome. J.Clin.Invest. 1996; 97(5):1311-8.
65. Dahia PL, Ahmed-Shuaib A, Jacobs RA *et al.* Vasopressin receptor expression and mutation analysis in corticotropin-secreting tumors. J.Clin.Endocrinol.Metab. 1996; 81(5):1768-71.
66. Dickstein G, DeBold CR, Gaitan D *et al.* Plasma corticotropin and cortisol responses to ovine corticotropin-releasing hormone (CRH), arginine vasopressin (AVP), CRH plus AVP, and CRH plus metyrapone in patients with Cushing's disease. J.Clin.Endocrinol.Metab. 1996; 81(8):2934-41.
67. Newell-Price J, Perry L, Medbak S *et al.* A combined test using desmopressin and corticotropin-releasing hormone in the differential diagnosis of Cushing's syndrome. J.Clin.Endocrinol.Metab. 1997; 82(1):176-81.
68. Ghigo E, Arvat E, Ramunni J *et al.* Adrenocorticotropin- and cortisol-releasing effect of hexarelin, a synthetic growth hormone-releasing peptide, in normal subjects and patients with Cushing's syndrome. J.Clin.Endocrinol.Metab. 1997; 82(8):2439-44.
69. Kojima M, Hosoda H, Date Y, Nakazato M, Matsuo H, Kangawa K. Ghrelin is a growth-hormone-releasing acylated peptide from stomach. Nature 1999; 402(6762):656-60.
70. De Keyzer Y, Lenne F, Bertagna X. Widespread transcription of the growth hormone-releasing peptide receptor gene in neuroendocrine human tumors. Eur.J.Endocrinol. 1997; 137(6):715-8.
71. Korbonits M, Jacobs RA, Aylwin SJ *et al.* Expression of the hormone secretagogue receptor in pituitary and other neuroendocrine tumors. J.Clin.Endocrinolol.Metab. 1998, 83(10):3624-30.
72. Hall WA, Luciano MG, Doppman JL, Patronas NJ, Oldfield EH. Pituitary magnetic resonance imaging in normal human volunteers: occult adenomas in the general population. Ann.Intern.Med. 1994; 120(10):817-20.
73. Kaltsas GA, Giannulis MG, Newell-Price JD *et al.* A critical analysis of the value of simultaneous inferior petrosal sinus sampling in Cushing's disease and the occult ectopic adrenocorticotropin syndrome. Journal of Clinical Endocrinology & Metabolism 1999; 84(2):487-92.
74. Tsagarakis S, Kaskarelis IS, Kokkoris P, Malagari C, Thalassinos N. The application of a combined stimulation with CRH and desmopressin during bilateral inferior petrosal sinus sampling in patients with Cushing's syndrome. Clinical Endocrinology 2000; 52(3):355-61.
75. Cuneo RC, Lee W, Harper J *et al.* Metyrapone pre-treated inferior petrosal sinus sampling in the differential diagnosis of ACTH-dependent Cushing's syndrome. Clin.Endocrinol.(Oxf). 1997; 46(5):607-18.
76. Yamamoto Y, Davis DH, Nippoldt TB *et al.* False-positive inferior petrosal sinus sampling in the diagnosis of Cushing's disease. Report of two cases. J.Neurosurg. 1995; 83(6):1087-91.
77. Yanovski JA, Cutler GBJr, Doppman JL *et al.* The limited ability of inferior petrosal sinus sampling with corticotropin-releasing hormone to distinguish Cushing's disease from pseudo-Cushing states or normal physiology. J.Clin.Endocrinol.Metab. 1993; 77(2):503-9.
78. Seyer H, Honegger J, Schott W *et al.* Raymond's syndrome following petrosal sinus

sampling. Acta Neurochir.(Wien). 1994; 131(1-2):157-9.
79. Sturrock ND, Jeffcoate WJ. A neurological complication of inferior petrosal sinus sampling during investigation for Cushing's disease: a case report. J.Neurol.Neurosurg.Psychiatry 1997; 62(5):527-8.
80. Miller DL, Doppman JL, Peterman SB, Nieman LK, Oldfield EH, Chang R. Neurologic complications of petrosal sinus sampling. Radiology 1992; 185(1):143-7.
81. Oldfield EH, Chrousos GP, Schulte HM *et al.* Preoperative lateralization of ACTH-secreting pituitary microadenomas by bilateral and simultaneous inferior petrosal venous sinus sampling. N.Engl.J.Med. 1985; 312(2):100-3.
82. Oldfield EH, Doppman JL, Nieman LK *et al.* Petrosal sinus sampling with and without corticotropin-releasing hormone for the differential diagnosis of Cushing's syndrome [published erratum appears in N Engl J Med 1992 Apr 23;326(17): 1172] [see comments]. N.Engl.J.Med. 1991; 325(13):897-905.
83. Miller DL, Doppman JL, Nieman LK *et al.* Petrosal sinus sampling: discordant lateralization of ACTH- secreting pituitary microadenomas before and after stimulation with corticotropin-releasing hormone. Radiology 1990; 176(2):429-31.
84. Doppman JL. Somatostatin receptor scintigraphy and the ectopic ACTH syndrome- -the solution or just another test? Am.J.Med. 1994; 96(4):303-4.
85. Kalogeras KT, Nieman LK, Friedman TC *et al.* Inferior petrosal sinus sampling in healthy subjects reveals a unilateral corticotropin-releasing hormone-induced arginine vasopressin release associated with ipsilateral adrenocorticotropin secretion. J.Clin.Invest. 1996; 97(9):2045-50.
86. Mamelak AN, Dowd CF, Tyrrell JB, McDonald JF, Wilson CB. Venous angiography is needed to interpret inferior petrosal sinus and cavernous sinus sampling data for lateralizing adrenocorticotropin-secreting adenomas. J.Clin.Endocrinol.Metab. 1996; 81(2):475-81.
87. Teramoto A, Yoshida Y, Sanno N, Nemoto S. Cavernous sinus sampling in patients with adrenocorticotrophic hormone-dependent Cushing's syndrome with emphasis on inter- and intracavernous adrenocorticotrophic hormone gradients. Journal of Neurosurgery 1998; 89(5):762-8.
88. Doppman JL, Nieman LK, Chang R *et al.* Selective venous sampling from the cavernous sinuses is not a more reliable technique than sampling from the inferior petrosal sinuses in Cushing's syndrome. J.Clin.Endocrinol.Metab. 1995; 80(8):2485-9.
89. Graham KE, Samuels MH, Nesbit GM *et al.* Cavernous sinus sampling is highly accurate in distinguishing Cushing's disease from the ectopic adrenocorticotropin syndrome and in predicting intrapituitary tumor location. Journal of Clinical Endocrinology & Metabolism 1999; 84(5):1602-10.
90. Findling JW, Doppman JL. Biochemical and radiologic diagnosis of Cushing's syndrome. [Review]. Endocrinol.Metab.Clin.North Am. 1994; 23(3):511-37.
91. Burrow GN, Wortzman G, Rewcastle NB, Holgate RC, Kovacs K. Microadenomas of the pituitary and abnormal sellar tomograms in an unselected autopsy series. N.Engl.J.Med. 1981; 304(3):156-8.
92. de Herder WW, Uitterlinden P, Pieterman H *et al.* Pituitary tumour localization in patients with Cushing's disease by magnetic resonance imaging. Is there a place for petrosal sinus sampling? Clin.Endocrinol.(Oxf). 1994; 40(1):87-92.
93. Barrou Z, Abecassis JP, Guilhaume B *et al.* Magnetic resonance imaging in Cushing disease. Prediction of surgical results. Presse Med. 1997; 26(1):7-11.
94. Buchfelder M, Nistor R, Fahlbusch R, Huk WJ. The accuracy of CT and MR evaluation of the sella turcica for detection of adrenocorticotropic hormone-secreting adenomas in Cushing disease. AJNR.Am.J.Neuroradiol. 1993; 14(5):1183-90.
95. Tabarin A, Laurent F, Catargi B *et al.* Comparative evaluation of conventional and dynamic magnetic resonance imaging of the pituitary gland for the diagnosis of Cushing's disease. Clinical Endocrinology 1998; 49(3):293-300.

96. Doppman JL, Ram Z, Shawker TH, Oldfield EH. Intraoperative US of the pituitary gland. Work in progress. Radiology 1994; 192(1):111-5.
97. Ram Z, Shawker TH, Bradford MH, Doppman JL, Oldfield EH. Intraoperative ultrasound-directed resection of pituitary tumors. J.Neurosurg. 1995; 83(2):225-30.
98. Fig LM, Gross MD, Shapiro B *et al.* Adrenal localization in the adrenocorticotropic hormone- independent Cushing syndrome. Ann.Intern.Med. 1988; 109(7):547-53.
99. Samuels MH, Loriaux DL. Cushing's syndrome and the nodular adrenal gland. [Review]. Endocrinol.Metab.Clin.North Am. 1994; 23(3):555-69.
100. Doppman JL, Miller DL, Dwyer AJ *et al.* Macronodular adrenal hyperplasia in Cushing disease. Radiology 1988; 166(2):347-52.
101. Sohaib SA, Hanson JA, Newell-Price JD *et al.* CT appearance of the adrenal glands in adrenocorticotrophic hormone-dependent Cushing's syndrome. AJR. American Journal of Roentgenology 1999; 172(4):997-1002.
102. Wajchenberg BL, Mendonca BB, Liberman B *et al.* Ectopic adrenocorticotropic hormone syndrome [published erratum appears in Endocr Rev 1995 Aug;16(4):507]. [Review] [238 refs]. Endocr.Rev. 1994; 15(6):752-87.
103. Doppman JL, Pass HI, Nieman LK *et al.* Detection of ACTH-producing bronchial carcinoid tumors: MR imaging vs CT. AJR.Am.J.Roentgenol. 1991; 156(1):39-43.
104. Doppman JL, Pass HI, Nieman LK *et al.* Corticotropin-secreting carcinoid tumors of the thymus: diagnostic unreliability of thymic venous sampling. Radiology 1992; 184(1):71-4.
105. Hanson JA, Sohaib SA, Newell-Price J *et al.* Computed tomography appearance of the thymus and anterior mediastinum in active Cushing's syndrome. Journal of Clinical Endocrinology & Metabolism 1999; 84(2):602-5.
106. Doppman JL, Nieman L, Miller DL *et al.* Ectopic adrenocorticotropic hormone syndrome: localization studies in 28 patients. Radiology 1989; 172(1):115-24.
107. Doppman JL, Nieman LK, Cutler GBJr *et al.* Adrenocorticotropic hormone--secreting islet cell tumors: are they always malignant? Radiology 1994; 190(1):59-64.
108. Reubi JC, Kvols LK, Waser B *et al.* Detection of somatostatin receptors in surgical and percutaneous needle biopsy samples of carcinoids and islet cell carcinomas. Cancer Res. 1990; 50(18):5969-77.
109. Phlipponneau M, Nocaudie M, Epelbaum J *et al.* Somatostatin analogs for the localization and preoperative treatment of an adrenocorticotropin-secreting bronchial carcinoid tumor. J.Clin.Endocrinol.Metab. 1994; 78(1):20-4.
110. de Herder WW, Krenning EP, Malchoff CD *et al.* Somatostatin receptor scintigraphy: its value in tumor localization in patients with Cushing's syndrome caused by ectopic corticotropin or corticotropin-releasing hormone secretion [see comments]. Am.J.Med. 1994; 96(4):305-12.

6
ASSAY METHODS IN CUSHING'S SYNDROME

Wendell Nicholson, BS
Robert J. Workman, MD
Division of Endocrinology, Diabetes and Metabolism, Department of Medicine, Vanderbilt School of Medicine, Nashville, TN, USA.

The successful and accurate laboratory evaluation of Cushing's syndrome is heavily dependent upon the availability of assays which are specific and sufficiently sensitive to detect the analytes of interest not only under normal physiologic conditions but also especially in situations where these analytes are suppressed either due to natural causes (e.g. suppression of pituitary ACTH by a functional adrenal adenoma) or to pharmacologic testing (e.g. dexamethasone suppression). Immunoassays (radioimmunoassay [RIA], enzyme-linked immunosorbent assay [ELISA], immunoradiometric assay [IRMA] etc) have several advantages over other methods, including relatively low cost, speed, large capacity, and usual ability to measure substances directly (with the notable exception of urinary steroids), thus avoiding extraction methods and the possible need to correct for variable recoveries, . The quality or characteristics of the primary antisera and the purity and specific activity of the tracer or other detection method contribute significantly to the sensitivity and specificity of such assays. HPLC is well adapted for small analytes such as steroids and has the potential of detecting synthetic as well as endogenously produced steroids (1-3). The finding of unusually low cortisol levels in a situation where elevated levels would be expected suggests exposure to synthetic steroids. Sample extraction is mandatory for all samples, including serum, plasma, saliva and urine. Substances which interfere with HPLC methods are due primarily to coelution with the substance of interest. For example, carbamazepine and digoxin elute with cortisol in many solvent systems (4,5).

Recent changes in the medical insurance industry have had a major impact on the way medicine is practiced, including the choice of laboratory. In endocrinology it is crucial to have a reliable laboratory and to be familiar with the characteristics and limitations of the assays available. This information may not be readily accessible. One approach to overcome these limitations, especially if an unanticipated or uninterpretable result is obtained, is to perform tests by totally different methodologies to confirm the validity of any result. Examples of this practice include obtaining both a urinary free cortisol (UFF; by initial extraction and either RIA or HPLC) and 17-hydroxycorticosteroids (17OHCS; based upon the Porter and Silber method dependent upon the reaction of steroids possessing a dihydroxy acetone side chain at the 17 position with acidic phenylhydrazine; the resultant yellow chromogen is measured at 410 nm). There ought to be a correlation between the two assays. This is not a perfect solution. If a subject were taking steroids surreptitiously such as prednisone, the RIA and 17-hydroxysteroids might both be elevated. If cortisol by HPLC is available, however, a peak eluted at the expected position for prednisolone should be detected; prednisone is rapidly converted to prednisolone *in vivo* by hepatic11 β-hydroxysteroid dehydrogenase, type I. A low or normal UFF with elevated 17OHCS may be observed due to interference with the Porter and Silber reaction due to color development detected at 410 nm; spurious large increases are observed with Librium (chlordiazepoxide) and paraldehyde; moderate increases occur with colchicine and ketonuria. There are also factors, such as phenothiazine, pentazocine, propoxyphene and bile salts, which interfere with various steps of the Porter and Silber reaction to produce a factitiously low result. However, if the antibody employed in the RIA is non-specific, it is possible that the UFF will be high but the 17OHCS will be normal. Measurement of 17-ketosteroids is particularly important if there is any evidence of virilization in females in association with Cushingoid features; this observation often is an indication of adrenal carcinoma rather than a benign pituitary or adrenal lesion. 17-ketosteroids traditionally are measured by the Zimmerman reaction involving reaction of steroid extracts with m-dinitrobenzene to produce a chromogen detected at 520 nm. Cephalosporins, erythromycin, meprobamate, penicillin, spironolactone and testosterone produce falsely high levels.

Steroids are usually measured in serum, urine or saliva. The peptide hormones are best assayed on plasma samples which have been kept cold and which have been centrifuged at high speed to remove all cellular particles, including platelets. The conditions and time of day under which samples are collected are important in interpreting results. Pulsatile release of hormones, a normal or disturbed diurnal variation in secretion, and the effects of stress, concurrent illness, posture, salt and fluid balance are some of the factors which impact results. It is often helpful to obtain timed or multiple samples to account for these factors. Twenty-four hour urine collections integrate steroid excretion over an entire day. A simple and convenient method to obtain multiple samples involves collection by the subject of saliva at various time intervals. This is easily accomplished employing a disposable plastic suction bulb which is placed between the teeth and buccal mucosa. The saliva can be stored without a preservative in the refrigerator. Since the levels of steroids are lower in saliva than in serum or plasma, employment of saliva assumes the availability of assays with sufficient sensitivity to detect the analytes of interest.

Interference with steroid assays by exogenous steroids

Corticosteroids, especially synthetic steroids, are widely available; and when they are used improperly, can induce exogenous Cushing's syndrome (6,7). Although not a corticosteroid, Megestrol has been reported to cause Cushing's syndrome (8,9). Steroids may be taken unwittingly. Chinese herbal medications and herbal teas from India and Pakistan may be enriched with corticosteroids. Steroid preparations are produced in oral, nasal, aerosol, ophthalmic, topical and injectable formulations. Absorption can occur across mucosal membranes and the skin. If steroids are taken which cross-react in cortisol assays (such as prednisone in RIAs), a laboratory profile emerges with elevated cortisol and suppressed ACTH, a constellation suggestive of hypercortisolism of adrenal origin. In this case, if the individual is unaware of or denies use of a corticosteroid, a simple way to rule out endogenous overproduction of cortisol by the adrenals is to assay urinary corticosterone (7). Under most conditions the secretory rate of corticosterone is about 10% that of cortisol. Therefore, if urinary cortisol is elevated, the urinary corticosterone should be increased proportionately. However, if an exogenous steroid is taken, such as cortisol itself or prednisone, the cortisol may be elevated but the corticosterone will be suppressed. If the steroids taken do not cross-react with cortisol, then both ACTH and cortisol will be suppressed. In this situation, all urinary steroids, including normal metabolites, should be suppressed. It usually is impractical to try to identify which steroid has been taken. An abnormal or unknown peak on HPLC might provide a clue (1). A cortrosyn stimulation test might help support the diagnosis. Clinically the individual appears Cushingoid, but urinary steroids are low and serum cortisol cannot be stimulated appropriately by ACTH.

Urinary free cortisol

The precise quantification of UFF presents a challenge, and there is no perfect method (10-13). The most accurate techniques frequently are laborious, lengthy and require expensive equipment and reagents. However, these methods may be well suited for research and for large, centralized reference laboratories. From a theoretical standpoint the main problem with these methods, which all require extraction and sample purification before the final analytical step, is accurately calculating recoveries. There are various approaches to solving this problem, including using tritiated cortisol (14) or other synthetic steroids such as methylprednisolone (4). In the latter case the recovery of the synthetic steroid and cortisol may differ, necessitating a correction factor. On the other hand, simpler methods depending primarily on immunoassay tend to overestimate UFF, leading to false positives in the evaluation of subjects suspected of having Cushing's syndrome and to normal values in individuals who actually might have reduced adrenocortical function. Poor antibody specificity is probably the main reason for the spuriously elevated cortisol levels. However, there are immunoassays available with good specificity which, when combined with some relatively simple extraction methods, will yield results comparable to those obtained with other techniques (7,10,13,14).

The gold standard for determining UFF is isotope dilution gas chromatography-mass spectroscopy. According to Ulick et al (15) the normal range for UFF is 53-93 nmol (19.1-33.5 μg) per day. This range can be compared with the results of other methods. In the unusual case where the range is lower, it can be assumed that recoveries from extraction procedures are unacceptably low or variable. If the range is higher, then interfering substances are being detected. In many instances the interference is from a cross-reacting steroid (10,13) (e.g. in an immunoassay) or from a coeluting drug (e.g. in HPLC). However, often the source of interference is not known and is presumably not due to steroids. Certain immunoassays are prone to matrix effects. Proteins, lipids, solvents, salts and probably other substances can interfere with antibody-antigen interactions.

Commercially available methods will invariably provide some cross-reactivity data. It may be useful to perform cross-reactivities if information on substances of interest is lacking or if methods are modified. For example, Morineau et al (16) describe a prepurification of urine on commercially available solid phase cartridges. They evaluated several chemically distinct cartridges and concluded that diol columns were the best based upon the criterion of efficiency in separating tritiated analogs of cortisol and 11-desoxycortisol. This criterion does not address interference by non-steroidal compounds. Also, among the manufacturers of diol columns, those from Waters and Amersham seemed to perform better than those from Baker. Following the solid phase extraction, which involves several washing and elution steps, the extracts were assayed by RIA.

Dexamethasone: Assay and Application

Dexamethasone (Decadron) is a synthetic corticosteroid which is approximately 30 times more potent than cortisol by weight, which has a duration of action 3-5 times longer than cortisol and which possesses essentially no mineralocorticoid activity. Since the early 1970's various types of assays have been available to measure dexamethasone in blood. A great deal of pharmacokinetic data was generated. However, much of this data suffered from the shortcoming of insensitivity of the dexamethasone assays. In order to obtain accurate and meaningful results, either large, supraphysiologic doses of dexamethasone were administered or large quantities of blood were sampled to extract and concentrate the dexamethasone. Radioimmunoassay is ideally suited to dexamethasone measurement since very specific and sensitive antibodies are now available (17). Dexamethasone is sufficiently different in structure from cortisol or the two most commonly used synthetic steroids, Prednisone (Prednisolone) and Medrol, that cross-reactivity is usually not an issue. Tritiated dexamethasone is commercially available. It is relatively simple to prepare radioiodinated dexamethasone, which significantly increases the sensitivity of the assay.

The most common diagnostic application of dexamethasone is in the dexamethasone suppression test (DST), which is widely employed 1) in endocrinology to detect and confirm the presence of Cushing's syndrome and to distinguish pseudo-Cushing's (e.g. due to chronic alcoholism) (18) and AIDS (19) from true Cushing's syndrome and 2) in psychiatry to confirm the diagnosis of endogenous depression. Complicating factors have been the lack of a standardized protocol for performing the DST and the lack of correlating dexamethasone serum levels with the cortisol levels in either blood or urine. To achieve the maximum amount of information from a DST it is necessary to obtain simultaneous dexamethasone and cortisol blood levels at specific time points after the dose of dexamethasone or to confirm blood dexamethasone levels in the middle or at the end of timed urine collections for cortisol (20). It is necessary to obtain this data for the following reasons: 1) different individuals metabolize dexamethasone at different rates, which may be unpredictable (21,22) or which may be correlated with body weight (23), i.e. the heavier an individual, the larger the volume of distribution of the drug within the body and the longer the half-life; 2) a number of medications or drugs markedly accelerate the clearance of dexamethasone, resulting in a much lower dexamethasone level than predicted; examples of such substances include alcohol, phenobarbital and other barbiturates, the benzodiazepines, dilantin, carbamazepine (and probably other anti-epileptics), rifampin and mitotane. The phenomenon of drug interference is illustrated in Figure 1. Sixteen of seventeen subjects who were taking one or more of the substances mentioned above and were undergoing overnight testing with dexamethasone were correctly identified as having an inadequate dexamethasone suppression due to concomitant drug exposure.

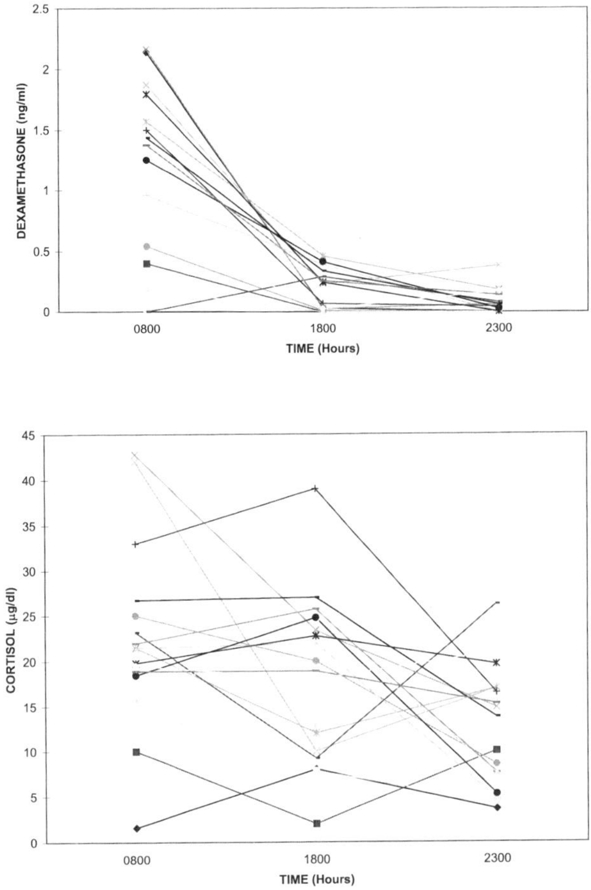

Figure 1. Plasma cortisol and dexamethasone levels in patients taking one or more substances which accelerate the clearance of dexamethasone. All individuals were given 1.0 mg dexamethasone at midnight and plasma levels of dexamethasone and cortisol were measured at 0800, 1800, and 2300 hours the following day.

In spite of these limitations, the DST can be performed in a consistent and reliable fashion and yield valuable results. Assuming an individual is not taking any of the medications listed above, the DST can be performed as an overnight screening test or in the conventional two or four day test, which is usually employed mainly to confirm Cushing's syndrome and to help determine the etiology [examples of difficulty in interpreting the conventional DST without dexamethasone levels are illustrated by Ashcraft et al (24)]. In the overnight DST 1.0 mg of dexamethasone is given orally between 2200 and 2400 hours and blood is drawn 8 hours later for both cortisol and dexamethasone. If the dexamethasone is administered parenterally (im or iv), the dose must be adjusted to deliver 1.0 mg of dexamethasone itself because the parenteral preparations are either salts or esters, which contribute to the weight of material in the vial. An individual with a normal pituitary-adrenal axis will suppress the A.M. cortisol to <3.5 μg/dl with a concomitant dexamethasone >3.5 ng/ml. Usually the dexamethasone level will not exceed 12 ng/ml and if a very high dexamethasone level is noted, it is possible that the person is a slow metabolizer of dexamethasone and will need to be retested with lower doses. Conversely, if low dexamethasone levels are observed, the test may be uninterpretable due to lack of suppression of cortisol. If substances which accelerate dexamethasone metabolism (such as anti-epileptics) cannot be safely stopped, then the DST should be repeated with higher doses of dexamethasone until the serum sample obtained eight hours following administration contains >3.5 ng/ml of dexamethasone. There is nothing sacred about the dose of dexamethasone employed, and doses as high as 32-64 mg have been necessary in some instances. These high doses are very useful if Cushing's syndrome due to either primary adrenocortical disease or ectopic ACTH is suspected. If very high serum dexamethasone levels are not associated with suppression of the serum cortisol, Cushing's disease of pituitary origin can be ruled out with a high degree of confidence. Fortunately, patients who are taking corticosteroids surreptitiously are not as smart as the assays. The exact intent of administering different doses of dexamethasone should not be explained to the patient, who will not be able to adjust the dose of exogenous steroid appropriately to attain the anticipated response.

In the conventional DST 0.5 mg of dexamethasone is given orally every 6 hours for 8 doses, and serial 24 hour urines for either 17OHCS and/or UFF are collected. A normal response correlates with the following results on the second day: dexamethasone 4-10 ng/ml 6 hours after last dose, 17OHCS < 3.0 mg/gram of creatinine and UFF < 5 μg/gram of creatinine. Individuals with Cushing's syndrome of any etiology should not suppress. During the next two days dexamethasone is administered at a dose of 2.0 mg every 6 hours, and the same tests are repeated as described above. Patients with Cushing's syndrome due to a pituitary adenoma (Cushing's disease) should suppress their urinary steroids, and the dexamethasone level should range from 25-45 ng/ml 6 hours after the last dose. In subjects with Cushing's syndrome due to either an autonomous adrenal tumor (adenoma or carcinoma) or to ectopic ACTH from an occult neoplasm (e.g., small cell lung carcinoma, bronchial carcinoid or an unsuspected pheochromocytoma), the urinary steroids will not suppress or change in spite of attaining adequate levels of dexamethasone in the circulation. Without the dexamethasone values, however, you never can be certain if some other factor such as poor absorption, non-compliance, abnormal metabolism or drug interference were responsible for the lack of suppressibility. An example of a low and high dose dexamethasone test with concomitant serum cortisol and dexamethasone levels in an individual with pituitary Cushing's disease is shown in Figure 2. This individual was cured by transsphendoidal hypophysectomy.

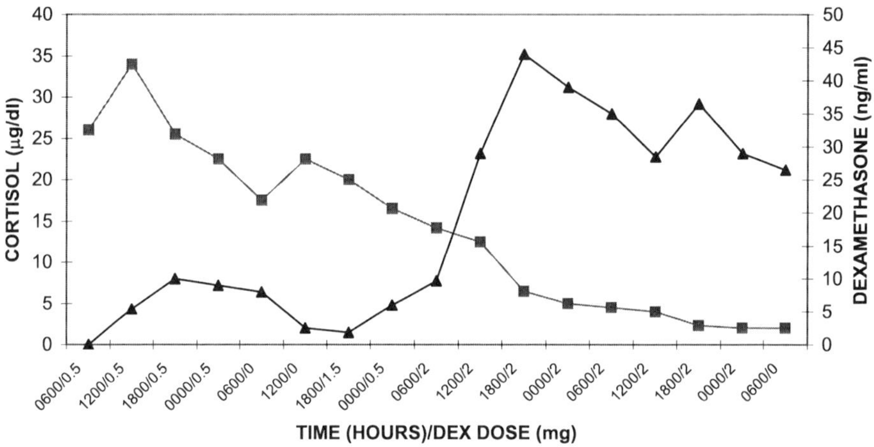

Figure 2. Plasma cortisol and dexamethasone in a patient with Cushing's disease undergoing successive low and high dose dexamethasone testing. On the second day of the low dose test, the 0.5mg doses of dexamethasone were missed at 0600 and 1200 hours. At 1800 hours, 1.5 mg of dexamethasone was administered in an attempt to compensate for the missed doses. (■-■cortisol; ▲-▲ dexamethasone)

In psychiatry similar considerations exist, especially since the psychiatric patient population is often exposed to medications for treatment or to drugs of abuse which may alter (usually accelerate) dexamethasone metabolism. Several psychiatric disorders (e.g., chronic alcohol abuse, endogenous depression, anorexia nervosa) are commonly associated with abnormalities of the hypothalamic-pituitary-adrenal axis such as elevated plasma cortisol levels and loss of the normal diurnal variability of cortisol so that the possibility of Cushing's syndrome is frequently entertained in the differential diagnosis. Whenever a DST is performed in a patient with mental illness, it is essential to obtain a dexamethasone level with the cortisol (25,26).

There are also other medical conditions in which Cushing's syndrome needs to be excluded. For example, chronic renal failure is associated with many endocrine disturbances, including the frequent observation of elevated serum cortisols due to the decreased renal clearance of cortisol and many of its metabolites, which potentially may cross-react in cortisol assays. In this situation dexamethasone suppression testing properly done and associated with determination of the dexamethasone levels achieved will unequivocally clarify whether Cushing's syndrome is present (17). The pharmacokinetics of dexamethasone generally are not affected by chronic renal failure. Due to the delayed clearance of endogenous corticosteroids in this disease, an overnight DST is usually unsatisfactory. The conventional 48 hour low and high dose DST should allow sufficient time for cortisol and its metabolites to clear, but each phase could be prolonged to three or four days if necessary. One group has circumvented this problem by assaying the serum cortisol after paper chromatography (14), but this is a cumbersome method not generally available and which necessitates some calculation of recoveries. Urinary steroid determinations are misleading and useless in assessing adrenal function in chronic renal failure (27).

Finally, there are therapeutic situations where dexamethasone is preferentially employed in association with other medications which accelerate the clearance of dexamethasone. Therefore, it is impossible to know if the dosage of dexamethasone is correct. In trauma to or in surgery on the central nervous system traditionally large doses of dexamethasone are administered to prevent edema of neural tissues without causing generalized fluid retention because dexamethasone is devoid of mineralocorticoid activity. It is a very common practice simultaneously to use anti-epileptics to prevent seizure activity. In the treatment of adrenal carcinoma o,p'DDD (mitotane) is one of the most effective drugs. Frequently dexamethasone replacement is given concomitantly because it will not interfere with cortisol assays in blood or urine, which may be utilized to follow the response to therapy. Since mitotane is another agent known to accelerate dexamethasone metabolism, however, it is highly likely the patient will be underreplaced and be in a state of chronic adrenal insufficiency. The anorexia and nausea which often have been attributed to a side effect of mitotane may, in fact, be due to Addisonism. Therefore, in both neurosurgical cases and in therapy of adrenal carcinoma where dexamethasone is employed in combination with mitotane or possibly chemotherapy it may be helpful for better patient management to determine the dexamethasone levels.

ACTH immunoassay

Once the diagnosis of Cushing's syndrome has been confirmed on the basis of cortisol measurements under basal conditions and following dexamethasone suppression testing, the next step in the differential diagnosis is to determine whether the hypercortisolism is ACTH-dependent or ACTH-independent. This can be accomplished by assay of several plasma samples for ACTH. Consistently below normal or undetectable plasma ACTH would indicate that the hypercortisolism is ACTH-independent. On the other hand, the continued presence of ACTH in the plasma of patients with Cushing's syndrome necessitates identifying the source of the ACTH secretion.

Human pituitary ACTH is synthesized as part of a 241 amino acid precursor, pro-opiomelanocortin (POMC) (28) (Figure 3). Pituitary POMC processing takes place in mature dense core granules and is accomplished by the protein convertases, PC1 and PC2 (29,30). Pulse-chase experiments in the mouse have shown that the first cleavage of POMC takes place at the carboxyl terminus of ACTH (31), which in man would yield the corresponding ACTH precursor, $POMC_{1-150}$, or pro-ACTH and the 89 amino acid β-endorphin (βEND) precursor, $POMC_{153-241}$, or β-lipotropin (βLPH) (Figure 3). The initial cleavage of POMC as well as the subsequent cleavage of pro-ACTH to yield murine ACTH, $POMC_{112-150}$ in man (Figure 3), is preferentially accomplished by PC1 (29). Further processing by PC2 produces joining peptide (JP), α-melanocyte stimulating hormone (αMSH), and βEND (29,30), corresponding to $POMC_{79-109}$, $POMC_{112-124}$, and $POMC_{211-241}$, respectively in man (Figure 3).

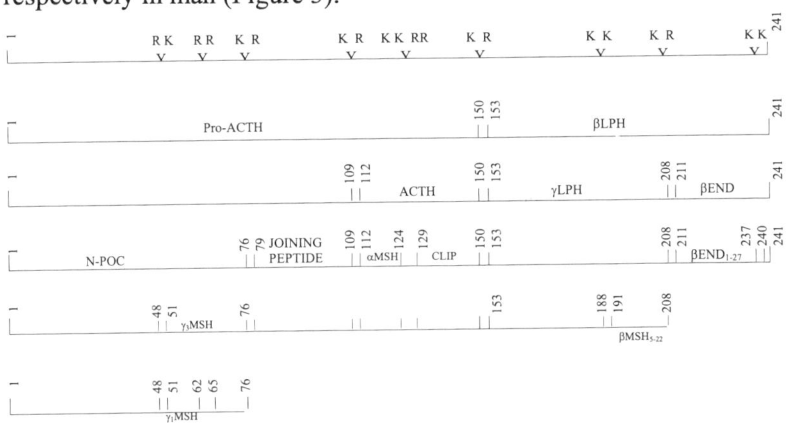

Figure 3. Schematic representation of human POMC, based on the sequence proposed by Takanishi et al (28), with 9 potential cleavage sites, basic amino acid pairs, identified (R = arginine; K = lysine).

Since 1966 several immunoassays for measuring ACTH in human plasma have been reported. Initially, these immunoassays were competition assays consisting of radioactive ACTH for detection and a single antiserum which usually recognized a single ACTH epitope. Each assay was characterized, to some degree, with respect to specificity as it relates to 1) the cross-reactivity of the ACTH antibody with selected structurally related peptides. i.e. those POMC maturation products containing the heptapeptide core sequence conserved in $POMC_{53\text{-}59}$ ($\gamma_3MSH_{3\text{-}9}$), $POMC_{115\text{-}121}$ ($ACTH_{4\text{-}10}$), and $POMC_{197\text{-}203}$ ($\beta LPH_{45\text{-}51}$ or $\beta MSH_{7\text{-}13}$) (Figure 3), which might also be co-secreted with ACTH, and 2) the effects of human plasma itself on the ACTH RIA.

Initially, the development of antibodies with the required specificity for the human ACTH sequence, i.e. one whose epitope spanned $POMC_{124\text{-}129}$ or $ACTH_{13\text{-}18}$, a likely site for proteolytic cleavage to produce two steroidogenically inert fragments of ACTH (32), was left to chance as only native ACTH was available for immunization (33-36). Later, immunizing with $ACTH_{1\text{-}24}$, a synthetic fragment which contains the entire steroidogenic portion of the molecule (32), appeared to favor the production of ACTH antibodies of the desired specificity (37). In either case the antibodies selected for these ACTH RIAs seldom cross-reacted with the other POMC maturation products which share the heptapeptide core sequence. Since no reference standards for human POMC and pro-ACTH have been established , the cross-reactivity of the ACTH precursors in these ACTH RIAs has never been systematically evaluated.

The lack of ACTH RIA specificity caused by human plasma and known as the matrix effect is easily identified. Its presence is apparent when 1) dilutions of a human plasma sample do not generate an RIA competition curve which is parallel to that generated by the ACTH reference standard, making an accurate interpolation of sample ACTH impossible, and/or 2) plasma from individuals expected to have low concentrations of ACTH appears to contain significant amounts of ACTH (37). The cause of the matrix effect in the ACTH RIA is unknown, but is thought to involve inhibition of labeled ACTH-antibody binding by plasma proteins (37). This non-specific reduction in specific antibody binding to the radioactive ACTH is incorrectly interpreted as ACTH activity. Thus far, attempts to alleviate the matrix effect in the plasma ACTH RIA have included partial purification and concentration of sample ACTH prior to assay (34-36) or diluting both the reference standard and plasma sample with either hormone-free human plasma (33) or solutions of hormone-free animal plasma protein fractions (37).

RIAs with the requisite sensitivity and specificity to study plasma ACTH over the wide range of levels encountered in patients with Cushing's syndrome are listed in Table 1. The well-known overlap in plasma ACTH between patients with pituitary dependent Cushing's syndrome and either normal volunteers or patients with the ectopic ACTH syndrome is readily apparent. In one study (37), the expected trend toward suppressed levels of plasma ACTH in patients with Cushing's syndrome of adrenal origin was demonstrated at the same time, underscoring the need for repetitive measurements of plasma ACTH in this situation since, on one occasion, one of these patients apparently had a normal plasma ACTH level.

Table 1. Plasma ACTH[a] in normal subjects and in patients with Cushing's Syndrome as measured by radioimmunoassay

NORMAL SUBJECTS	PATIENTS WITH CUSHING'S SYNDROME			REF.
	PITUITARY	ADRENAL	ECTOPIC ACTH	
<1 - 17 (70)			14 ->220 (8)	33
6 - 15 (10)	12, 86 (2)		46, 56 (2)	34
7 - 34 (24)	16 - 100 (3)			35
2 - 12 (36)	11 - 28 (5)		28, 5500 (2)	36
2 - 6 (17)	5 - 21 (8)	<1 - 2[b] (7)	28 - 180 (4)	37
3 - 4[c,d]	2 - 120[d]	2 - 6 (12)	4 - 160 (7)	38

a. Immunoreactive plasma ACTH concentration (pM) in samples collected between 0800 – 1000 hours. The number of normal subjects or patients studied prior to treatment is indicated in parenthesis.
b. Plasma ACTH was < 1.1 pM in 6 of 7 patients.
c. "Lower normal limit."
d. Number of subjects not reported.

Since 1984 attempts to circumvent some of the difficulties inherent in the RIA of human plasma ACTH have centered on development of two antibody-site immunoradiometric assays (IRMA) (39) by using two antibodies to separate strategic ACTH epitopes (40). One antibody is made radioactive for detection, and the other, which binds to a different ACTH epitope, is immobilized to facilitate separation of the radioactive antibody-ACTH complex from the non-complexed antibody radioactivity. The amount of radioactive antibody-ACTH complex bound to the immobilized antibody is directly proportional to the amount of ACTH present. In general, these ACTH IRMAs, which are technically easier to perform than RIAs, equal or exceed ACTH RIA sensitivity without resorting to prolonged incubations. Matrix composition has virtually no effect on the ACTH IRMA, eliminating the need for sample preparation prior to assay (41). They are thought to be specific for $ACTH_{1-39}$; however, cross-reactivity with the ACTH precursors, POMC and pro-ACTH, was apparent but indeterminate in the two assays tested thus far (41,42). Cross-reactivity of POMC and pro-ACTH in the former assay was reassessed and found to be <1% and 10%, respectively (43). Although no cross reactivity has been reported with POMC maturation products, a marked decrease (>80%) in plasma ACTH as measured by ACTH-IRMA coincided with the dramatic increase (>400 fold) in plasma ACTH as measured by RIA following the intravenous injection of 250 μg $ACTH_{1-24}$ into normal subjects, whose endogeneous ACTH levels had been increased following the overnight administration of 2.5 g metyrapone (44). The authors suggested that this artifactual decrease in plasma ACTH as measured by IRMA was caused by the large amount of injected $ACTH_{1-24}$ binding only one of the two IRMA antibodies, the radiolabeled $ACTH_{1-17}$ antibody in this case, and thus making it unavailable to bind the endogeneous $ACTH_{1-39}$ and create a signal. They concluded that a thorough examination of cross-reactivity of selected POMC maturation products in the ACTH IRMA should not only include testing these peptides in the absence of $ACTH_{1-39}$ but also in the presence of known amounts of the native peptide.

Studies of plasma ACTH employing immunoradiometric assays in patients with Cushing's syndrome are listed in Table 2. As with RIA, plasma ACTH levels obtained by IRMA in patients with pituitary-dependent Cushing's syndrome overlap the levels found in both normal subjects and patients with the ectopic ACTH syndrome However, the distinction between ACTH-independent and ACTH-dependent Cushing's syndrome tends to be more apparent when plasma ACTH is determined by IRMA.

Recently, the two antibody-site immunoassay format for ACTH has been modified to include methods of detection other than radioactivity. These modifications include covalently coupling the detection antibody to 1) either a specific enzyme which then reacts with an appropriate substrate to yield a chromogen which can be measured spectrophotometrically (the enzyme linked immunosorbent assay or ELISA) or 2) to an appropriate chemiluminescent substrate which can later be induced to emit light (the immunochemiluminescence assay or ICLA). The salient features of these immunoassays include 1) an extended shelf-like for the non-radioactive detection antibody compared to the 60-day T/2 of the [^{125}I]-labeled detection antibody of the IRMA, 2) improved sensitivity, at least with the ICLA and 3) avoidance of hazardous radioisotopes in the work place. The final assessment of the utility of these non-isotopic two antibody-site immunoassays for the measurement of plasma ACTH in patients with Cushing's syndrome will have to await report of studies similar to those listed for the ACTH IRMA listed in Table 2.

Immunoassay of CRH

Since the isolation (51) and characterization (52) of corticotrophin releasing hormone (CRH), the major physiological regulator of the hypothalamic-pituitary-adrenal axis (52), clinical scientists have investigated the role of this hypothalamic peptide in the etiology of Cushing's syndrome. Human CRH occurs mostly in neurons in the parvocellular region of the paraventricular nucleus where it is contained mainly in 100-150nm secretory granules (53). While extra-hypothalamic concentrations of CRH-like material have been identified in numerous normal human tissues and fluids, the relationship between extra-hypothalamic CRH and pituitary ACTH secretion remains unclear (53). Human CRH is synthesized as part of a 196 amino acid precursor, prepro-CRH (54) and is contained in prepro-$CRH_{154\text{-}194}$ (Figure 4). Full ACTH releasing activity resides in prepro-$CRH_{157\text{-}194}$ ($CRH_{4\text{-}41}$)(52).

Table 2. Plasma ACTH[a] in normal subjects and in patients with Cushing's Syndrome as measured by immunoradiometric assay.

NORMAL SUBJECTS	PATIENTS WITH CUSHING'S SYNDROME			REF.
	PITUITARY	ADRENAL	ECTOPIC ACTH	
2 - 11 (28)	4 - 15 (7)			45
5 - 11 (6)	12 ± 4 (6)		41 ± 14 (5)	46
2 - 12 (45)	4 - 24 (10)	2 - 3[b] (5)	17 - 160 (3)	47[C]
2 - 14 (45)	2 - 21 (9)	<1 (5)	18, 24 (2)	47[C]
4 - 14 (30)	10, 16 (2)		100 (1)	48[C]
4 - 12 (30)	10, 18 (2)		160 (1)	48[C]
6 - 15 (30)	17, 22 (2)		140 (1)	48[C]
<1 - 11 (20)	8 - 23 (7)	<1 (1)	24 - 140 (9)	41
2 - 22 (94)	5 - 18 (9)	<1 (2)	14 - 200 (14)	49
2 - 12[d]	6 - 24 (10)	<1- 2[e] (5)	22 - 36 (2)	50
1 - 4[d,f]	3 - 60[g]	<1- 7[h] (28)	11 - 110 (16)	38

a. Immunoreactive plasma ACTH concentration (pM) in samples collected between 0800-1000 hours. The number of normal subjects or patients studied prior to treatment is indicated in parenthesis.
b. Plasma ACTH was < 1.9 pM in 4 of 5 patients.
c. Plasma ACTH in same samples as measured by different immunoradiometric assays.
d. Number of subjects not reported
e. Plasma ACTH was ≤ 1pM in 4 of 5 patients.
f. "Lower normal limit."
g. Number of patients not reported.
h. Plasma ACTH was < 2 pM in 19 of 28 patients.

Figure 4. Schematic representation of human prepro-CRH, based on the sequence proposed by Shibahara et al (52), with 3 potential cleavage sites, basic amino acid pairs, indicated. The presence of the dipeptide, Gly-Lys, as the carboxyl terminus of prepro-CRH (prepro-CRH$_{195\text{-}196}$) implies that the carboxyl terminal isoleucine (prepro-CRH$_{194}$) of human CRH is amidated [Shibahara et al (52)].

Attempts to identify the role of CRH in the etiology of Cushing's syndrome have been concentrated on RIA of plasma CRH, before and during dynamic testing, in all forms of the disease. Of special interest have been cases of ACTH-dependent Cushing's syndrome associated with unregulated neoplastic production and secretion of CRH-like material with and without concomitant production of ACTH-like material by the tumor. The former clinical situation has been referred to as the ectopic ACTH-CRH syndrome and the latter simply as the ectopic CRH syndrome (55). Both types mimic the ectopic ACTH syndrome with elevated levels of plasma ACTH and lack of cortisol suppression after high doses of dexamethasone. Furthermore, when CRH produced by the tumor results in corticotrope hyperplasia and Cushing's syndrome, localization of ACTH production by inferior petrosal sinus sampling might be taken as evidence that the patient has Cushing's disease (56).

There are two factors which complicate the measurement of CRH in human peripheral blood: 1) The presumed relatively high concentration of CRH in hypothalamic blood is very dilute by the time it reaches the periphery (53); and 2) human plasma contains a specific, high affinity, high capacity binding protein for CRH (CRH-BP) (57). The net effect of these two complications is to render the results of RIAs of CRH in unaltered human plasma uninterpretable (57,58). RIAs for CRH are based on competition between reference standard or sample CRH and radioactive CRH for a limited number of specific binding rites contained in a fixed amount of CRH antibody. Antibody-bound radioactivity is inversely proportional to the amount of reference or sample CRH added to the reaction. While the unconcentrated level of CRH in human plasma is frequently below the detection limits of CRH RIAs, an equally important consideration in establishing assay specificity is the requirement that, prior to RIA, plasma CRH must be separated from CRH-BP and prevented from re-associating with this binding protein. CRH-BP has the capacity to bind not only the CRH present in the plasma sample but to also bind a portion of the radio-labeled CRH added to the CRH RIA system, leaving less radioactivity available for binding to the CRH antibody. In this assay system, any reduction in CRH-antibody-bound radioactivity resulting from plasma sample CRH-BP binding of the tracer, is incorrectly interpreted as immunoreactive CRH (57,58).

The problems of lack of sensitivity and specificity in the plasma CRH RIA have been addressed simultaneously by employing methods of sample preparation which irreversibly separate CRH from CRH-BP and permit several-fold concentration of the peptide from a given plasma sample. These methods include partial purification of CRH by methanol or acetone extraction (58-60), absorption to and elution from octadecyl silylsilica (61), silicic acid (61), or Vycor glass (62), and immunoaffinity chromatography (63, 64). Results of plasma CRH RIAs in patients with Cushing's syndrome, based on these methods of sample preparation, are listed in Table 3. Apparent discrepancies in the role of plasma CRH in patients with disorders of hypothalamic–pituitary–adrenal function have been noted and doubt has been expressed concerning the relevance of measuring plasma CRH under these conditions (53). Since the review by Orth in 1992 (53), at least six additional studies involving plasma CRH measurements in patients with Cushing's syndrome have been reported (55,56,66,69,70,72) (Table 3). To date, 5 of 10 patients with either the ectopic CRH or the ectopic ACTH-CRH syndrome have been shown to have higher than normal concentrations of CRH in their plasma (55,59,65,67,69) (Table 3). With the two exceptions of patients in reference 62 (Table 3), plasma CRH levels in all other forms of Cushing's syndrome overlapped those in normal subjects.

Table 3. Plasma CRH[a] in normal subjects and patients with Cushing's Syndrome as measured by radioimmunioassay.

NORMAL SUBJECTS	PATIENTS WITH CUSHING'S SYNDROME					REF.
	PITUITARY	ADRENAL	ECTOPIC			
			ACTH	ACTH-CRH	CRH	
<2[b]					26[c] (1)	55
<4 - 23 (46)	<4 - 12 (24)	<4 (2)	<4 - 57[d] (3)			62
<1[b]					6[c] (1)	59
<8[b]				59[e] (1)		65
<1 - 2 (26)	<1, 1 (2)	<1 (2)				66
<4 - 23	(46)			57[f]	(1)	67
<1 - 2[b]				2[f] (1)		68
<1 (4)	<1	(10)		2[g]	(1)	69
					<2[h] (1)	70
<1 - 2 (33)	<1 (5)	<1 (4)				64
				16[g]	(1)	71
1 ± 0.1 (12)	<1 (5)	<1 (6)	3[d], 4[d]	(2)		63
					6[k] (1)	60
2[m] ± 0.8 (11)	2[m] ± 0.2 (25)	2[m] ± 0.1 (7)	2[m,d] ± 0.1	(17)		72
				2[g]	(1)	56

a. Immunoreactive plasma CRH (pM) in samples collected between 0800-1000 hours. The number of normal subjects or patients studied prior to treatment is indicated in parenthesis.
b. Number of subjects not reported.
c. Tumor immunohistochemically positive for CRH and negative for ACTH.
d. No tissue CRH or ACTH data reported.
e. Tumor contained CRH and βEND.
f. Tumor contained CRH and ACTH.
g. Tumor immunohistochemically positive for CRH and ACTH.
h. Tumor immunohistochemically positive for CRH, ACTH not reported.
k. Tumor contained CRH, ACTH was not detectable.
m. Simultaneous dominant inferior petrosal sinus levels for CRH were 2 ± 0.8, 2 ± 0.1 and 2 ± 0.1 pM respectively.

Utilizing the two antibody-binding site immunoassay system, either in the form of an IRMA or an ELISA to measure human plasma CRH, appears to circumvent the problems of specificity encountered with the RIA, while maintaining sensitivity without partially purifying and concentrating the sample CRH prior to assay. The plasma CRH IRMA system (73) consists of radiolabeled immunoaffinity purified rabbit anti-prepro $CRH_{189-194}$ (CRH_{36-41}) immunoglobulin for detection and guinea pig anti-prepro-$CRH_{154-174}$ (CRH_{1-20}) serum which is allowed to form a complex with CRH and the detection antibody. The two-antibody-CRH complex is removed from the reaction by allowing the guinea pig antibody portion of the complex to react with sheep anti-guinea pig immunoglobulin coupled to solid phase Dynospheres. The amount of radioactivity precipitated is directly proportional to the amount of CRH present. This CRH IRMA has a sensitivity of <1 pM. Basal levels of plasma CRH in normal subjects ranged from <1-6 pM. Plasma CRH remained in this range, <1-5 pM, in 17 patients with Cushing's syndrome. Three of the 17 had the ectopic ACTH syndrome and the etiology of the disease in the other 14 was not stated (73).

The plasma CRH ELISA system (74) contains protein G-Sepharose affinity-purified sheep anti-CRH immunoglobulin immobilized in the wells of a 96-well microtiter plate where it reacts with added reference standard CRH or plasma CRH. Rabbit anti-CRH serum is added to form a sheep anti-CRH immunoglobulin-CRH-rabbit anti-CRH immunoglobulin complex. Detection is accomplished by reacting goat anti-rabbit immunoglobulin-horseradish peroxidase with the rabbit anti-CRH immunoglobulin portion of the complex. This complex is then reacted with an appropriate substrate for the horseradish perioxidase, tetramethylbenzidine, to form a stable dark blue precipitate. Addition of sulfuric acid stops color development, solubilizes the precipitate, and produces a bright yellow chromogen which can be measured spectrophotometrically. Absorbance at 450 nm is directly proportional to the amount of CRH present in the well. This CRH ELISA has a sensitivity of 0.1 fmole in the reaction well, but the volume of reference standard CRH or plasma sample added to the reaction was not stated. The authors did indicate that plasma CRH was undetectable in early pregnancy (74).

Immunoassay of ACTH precursors

The hypercortisolism in a significant number of patients with Cushing's syndrome is due to the unregulated synthesis and secretion of ACTH-like materials by tumors arising from non-pituitary tissue – the ectopic ACTH syndrome (75). Clinically, this condition can be difficult to distinguish from pituitary-dependent Cushing's syndrome due to the overlapping circulating ACTH levels in both conditions (43) and the apparent sensitivity of some ectopic ACTH producing tumors to regulation by glucocorticoids (76). Before the concept was articulated that pituitary ACTH is synthesized as part of a large precursor molecule and that post translational processing of this precursor produces ACTH as well as a number of other biologically active peptides (Figure 3), investigators began a search for any alterations in ACTH synthesis and secretion that might characterize the ectopic ACTH syndrome.

Initial attempts to characterize the ectopic ACTH syndrome by identifying alterations in POMC synthesis and secretion were based on chromatographic isolation of the products of interest from extracts of the tumor and subsequent identification of each by RIA (77-83). Chromatographic separation of POMC from intermediate precursors and final cleavage products was necessary because these peptides are indistinguishable by RIA. The antibodies employed in these studies had been developed to POMC maturation products and crossreactivity with the respective precursor(s) was suggested by the finding of immunoreactive material in the chromotographic effluent with an apparent molecular weight(s) larger than the reference material. Plasma was studied less frequently than tissue because the ratio of plasma protein to plasma POMC-related peptide content was frequently too high to permit analysis with the chromatographic systems employed. In the eight cases of the ectopic ACTH syndrome studied (77,80,81) alterations in tumor POMC synthesis were assumed to be responsible for the ACTH-like material(s) of considerable but indeterminate molecular size found in the plasma.

Recent attempts to characterize the ectopic ACTH syndrome by identifying alterations in POMC synthesis and secretion have been directed toward more specific immunological measurement of ACTH precursors and selected cleavage products in plasma from all patients with ACTH-dependent Cushing's syndrome. To date, these immunoassays have been the two antibody-site immunoradiometric assays (IRMA) where one antibody to a specific POMC peptide epitope is made radioactive for detection and the other, which binds a different POMC peptide epitope, is immobilized to facilitate separation of the radioactive antibody–POMC peptide complex from non-complexed antibody radioactvity. The amount of radioactive antibody-POMC peptide complexed with the immobilized antibody is directly proportional to the amount of POMC-peptide present in the sample. With one exception the two antibody-site POMC precursor immunoassays offer a high degree of specificity. However, they employ antibodies which recognize strategic precursor epitopes which are also contained in the maturation products of the precursor. These maturation products, which frequently circulate in high concentrations in some cases of ACTH-dependent Cushing's syndrome, are able to bind a portion of the limited number of sites contained in the fixed amount of antibody used in the IRMA. This maturation product–antibody binding results in fewer antibody sites available for precursor binding. Any reduction in precursor-antibody binding is incorrectly interpreted as reduced POMC precursor immunoreactivity.

Data from two laboratories, utilizing plasma IRMAs of POMC (84) or POMC plus pro-ACTH (85) in ACTH dependent Cushing's syndrome, have been reported. Raffin-Sanson, et al (84) assembled an IRMA for human POMC with antisera obtained from Nichols Institute Diagnostics, USA. The radioactive or detection antibody recognized $POMC_{112\text{-}128}$ ($ACTH_{1\text{-}17}$) (Figure 3), and the immobilized antibody reacted with an unspecified sequence of amino acids located in the carboxyl terminus of βEND ($POMC_{211\text{-}241}$)(Figure 3). The reference material, prepared by the authors, was human POMC semi-purified from culture medium harvested from the cell line, DMS-79. This cell line originated from pleural fluid collected from a patient with small cell anaplastic carcinoma of the lung and was subsequently shown to produce POMC (86). Human POMC, the only known POMC-related peptide that would be measured by this IRMA (Figure 3), was undetectable in the plasma of 17 normal subjects, 4 patients with Addison's disease, whose plasma ACTH levels ranged from 80-230 pM, and 34 patients with Cushing's disease due to pituitary microadenoma. Seven of 8 Cushing's disease patients with highly invasive pituitary macroadenomas had elevated levels of POMC in their plasma. Plasma POMC levels were elevated in 16 of 20 patients with the ectopic ACTH syndrome. Each of the four patients with the ectopic ACTH syndrome and undetectable POMC in their plasma had typical bronchial carcinoids.

White and Gibson (85) used an IRMA, developed in house, which measured both POMC and pro-ACTH equally (87). The detection antibody reacted with $POMC_{121\text{-}129}$ ($ACTH_{10\text{-}18}$) and the immobilized antibody recognized $POMC_{51\text{-}62}$ (γ_1MSH)(87). The reference material was human POMC, semi-purified from medium collected from an invasive human corticotrope adenoma known to produce POMC in culture (87). Human POMC plus pro-ACTH (Figure 3), the only two known POMC-related peptides that would be measured by this IRMA, ranged from 10-100 pM in the plasma of 27 patients with Cushing's disease and from 140-18,000 pM in 48 patients with the ectopic ACTH syndrome. Twelve of the patients with the ectopic ACTH syndrome had carcinoid tumors and their plasma POMC plus pro-ACTH levels ranged from 150-1,300 pM in contrast to plasma POMC alone which was undetectable in the 4 bronchial carcinoids from the previous study (84). These authors suggested that the failure to detect carcinoid ACTH precursors when only POMC was measured could be related to the differences in the incidence and possible level of PC1 and PC2 expression between carcinoids and other tumors. They pointed out that the incidence of PC1 and

PC2 expression is high in bronchial carcinoids (88) in which POMC processing has been reported to be more complete (83). Expression of these processing enzymes is restricted in small cell lung carcinoma cell lines (89) in which POMC cleavage is virtually nonexistent (90).

Immunoassay of Chromogranin A

Chromogranin A (CgA), an acidic glycoprotein found in secretory granules of a number of neuroendocrine cells, has been examined as a marker for neuroendocrine neoplasia (91). These tumors, including carcinoids, medullary thyroid carcinomas, and small lung cancers, have been associated with high plasma levels of CgA and can also be responsible for the ectopic ACTH syndrome. In a survey of 30 cases of Cushing's syndrome, Nobels et al (92) found normal plasma CgA levels (≤ 220μg/L) in all 5 patients with Cushing's syndrome of adrenal origin, 12 of 15 patients with pituitary-dependent disease, and in 3 patients with Cushing's syndrome associated with bronchial carcinoids. Elevated levels of plasma CgA (380-14,000 μg/L) were found in the remaining 7 cases of the ectopic ACTH syndrome which included one additional bronchial carcinoid. These authors pointed out that the markedly elevated levels of plasma CgA could be explained by tumor bulk, as all seven cases showed widespread metastases. They concluded that plasma CgA measurement was of little value in making an early diagnosis of Cushing's syndrome due to ectopic production of ACTH and/or CRH by small neuroendocrine tumors.

Immunoassay of atrial natriuretic factor

The role of atrial natiuretic factor (ANF), a peptide synthesized in and secreted from the heart and involved in the regulation of water and salt homeostasis and cardiovascular function (93), in ACTH secretion is controversial (94). While all other forms of Cushing's syndrome were not studied, plasma levels of ANF were higher in patients with Cushing's disease than in normals (25-68 vs 9-23 pg/mL [95] and 15± 0.8 vs 10.8 ± 0.6 pM [94]). In 1994 Okada et al (96) described what was reported to be the first case of Cushing's syndrome associated with ectopic production of ANF. In a patient with ACTH-dependent Cushing's syndrome, normal plasma CRH and a concomitant thymic carcinoid, samples of venous blood approaching the tumor showed progressively increasing levels of plasma ANF from 120-620 pg/mL, while plasma ACTH in the same samples remained constant but elevated. Plasma ACTH and ANF returned to normal after the tumor was removed, and the tumor was immunohistochemically positive for ANF and negative for ACTH.

REFERENCES

1. Lin C-L, Wu T-J, Machacek DA, Jiang N-S, Kao, PC. 1997 Urinary free cortisol and cortisone determined by high performance liquid chromatography in the diagnosis of Cushing's syndrome.. J Clin Endocrinol Metab 82:151-155.
2. Quddusi S, Browne P, Toivola B, Hirsch IB. 1998 Cushing syndrome due to surreptitious glucocorticoid administration. Arch Intern Med. 158:294-296
3. Tsuruoka S, Sugimoto K, Fujimura A. 1998 Drug induced Cushing's syndrome due to surreptitious glucocorticoid administration. Therapeutic Drug Monitoring. 20:387-389.
4. Turpeinen U, Markkanen H, Valimaki M, Stenman U-H. 1997 Determination of urinary free cortisol by HPLC. Clin Chem. 43:1386-1391 .
5. Findling JW, Pinkstaff SM, Shaker JL, Raff H, Nelson JC. 1988 Pseudohypercortisoluria: Spurious elevation of urinary cortisol due to carbamazepine. The Endocrinologist 8:51-54 .
6. Findlay CA, Macdonald JF, Wallace AM, Geddes N, Donaldson MDC. 1998 Childhood Cushing's syndrome induced by betamethasone nose drops, and repeat prescriptions. Brit Med J. 317: 739-740 .
7. Workman RJ, Nicholson WE, McCammon DK. 1995 Factitious hypercortisoluria. J Clin Endocrinol Metabol 80:3050-3051 .
8. Padmanabhan S, Rosenberg AS. 1998 Cushing's syndrome induced by megestrol acetate in a patient with AIDS. Clin Infectious Dis. 27:217-218 .
9. Shotliff K, Nussey SS. 1997 Medroxyprogesterone acetate induced Cushing's syndrome. Brit J Clin Pharmacol. 44:304 .
10. Lee C, Goeger DE. 1998 Interference of 6β-hydroxycortisol in the quantitation of urinary free cortisol by immunoassay and its elimination by solid phase extraction. Clin Biochem 31:229-233 .
11. Murphy BEP, Okouneff LM, Klein GP, Ngo SC. 1981 Lack of specificity of cortisol determinations in human urine. J Clin Endocrinol Metab. 53:91-99
12. Nahoul K, Patricot M-C, Moatti J-P, Revol A. 1992 Determination of urinary cortisol with three commercial immunoassays. J Steroid Biochem Molec Biol. 43:573-580 .
13. Sapin R, Schlienger JL, Gasser F, Pradignac A, Grucker D. 1998 Improved specificity of a new direct assay for urinary cortisol: application in corticoid treated patients. Clin Chem Lab Med. 36:855-858 .
14. Van Herle AJ, Birnbaum JA, Slomowitz LA, Mayes D, Chandler DW, Rosenblit PD, Nissenson A. 1998 Paper chromatography prior to cortisol RIA allows for accurate use of the dexamethasone suppression test in chronic renal failure. Nephron 80:79-84
15. Ulick S, Chan CK, Wang JZ. 1991 Measurement of 4 urinary C-18 oxygenated corticosteroids by stable isotope dilution mass fragmentography. J Steroid Biochem Molec Biol. 38:59-66..
16. Morineau G, Gosling J, Patricot M-C, Soliman J, Boudou P, Halnak AA, LeBrun G, Brerault J-L, Julien R, Villette J-M, Fiet J. 1997 Convenient chromatographic prepurification step before measurement of urinary cortisol by radioimmunoassay. Clin Chem 43:786-793 .
17. Workman RJ, Vaughn WK, Stone WJ. 1986 Dexamethasone suppression testing in chronic renal failure: pharmacokinetics of dexamethasone and demonstration of a normal hypothalamic-pituitary-adrenal axis. J. Clin. Endocrinol. Metabol. 63:741-746 .
18. Veldman RG, Meinders AE. 1996 On the mechanism of alcohol-induced pseudo-Cushing's syndrome. Endo. Rev. 17:262-268
19. Hirsch MS, Klibanski A. 1998 Editorial response: What price progress? Pseudo-Cushing's syndrome associated with anti-retroviral therapy in patients with human immunodeficiency virus infection. Clin. Inf. Diseases 27:73-75 .
20. Meikle AW. 1982 Dexamethasone suppression tests: usefulness of simultaneous measurement of plasma cortisol and dexamethasone. Clin. Endocrinol. 16:401-408 .

21. Meikle AW, Lagerquist LG, Tyler FH. 1975 Apparently normal pituitary-adrenal suppressibility in Cushing's syndrome: dexamethasone metabolism and plasma levels. J. Lab. Clin. Med. 86: 472-478 .
22. Meikle AW, Clarke DH, Tyler FH. 1976 . Cushing's syndrome resulting from a low dose of dexamethasone. JAMA 235:1592-1593
23. Tseui SE, Moore RG, Ashley JJ, McBride WG. 1979 Disposition of synthetic glucocorticoids. I. Pharmacokinetics of dexamethasone in healthy adults. J. Pharmacokinetics and Biopharmaceutics 7:249-264 .
24. Ashcraft MW, Van Herle AG, Vener SL, Geffner DL. 1982 . Serum cortisol levels in Cushing's syndrome after low- and high-dose dexamethasone suppression. Ann. Int. Med. 97:21-26 .
25. Arana GW, Reichlin S, Workman RJ, Hasser R, Shader RI. 1988 DST diagnostic utility for depression is enhanced by expressing serum cortisol as a function of serum dexamethasone (Dexamethasone Suppression Index, DSI). Amer J Psychiatry 14:707-711.
26. Arana GW, Workman RG, Baldessarini RJ. 1984 . Association between low plasma levels of dexamethasone and elevated levels of cortisol in psychiatric patients given dexamethasone. Amer. J. Psychiatry. 141: 1619-1620 .
27. Issa BG, Page MD, Read G, John R, Douglas-Jones A, Scanlon MF. 1999 Undetectable urinary free cortisol concentrations in a case of Cushing's disease. Eur J Endocrinol 140:148-151 .
28. Takahashi H, Teranishi Y, Nakanishi S, Numa S. 1981 Isolation and structural organization of the human corticotropin-β-lipotropin precursor gene. FEBS Lett 135:97-102.
29. Benjannet S, Rondeau N, Day R, Chrétien M, Seidah NG. 1991 PC1 and PC2 are proprotein convertases capable of cleaving proopiomelanocortin at distinct pairs of basic residues. Proc Nat Acad Sci USA 88:3564-3568.
30. Thomas L, Leduc R, Thorne BA, Smeekens SP, Steiner DF, Thomas G. 1991 Kex2-like endoproteases PC2 and PC3 accurately cleave a model prohormone in mammalian cells: Evidence for a common core of neuroendocrine processing enzymes. Proc Natl Acad Sci USA 88:5297-5301.
31. Ney RL, Ogata E, Shimizu N, Nicholson WE, Liddle GW. 1964 Structure-function relationships of ACTH and MSH analogues. Excerpta Medica International Congress Series No. 83:1184-1191.
32. Eipper BA, Mains RE. 1980 Structure and biosynthesis of pro-adrenocorticotropin/endorphin and related peptides. Endocr Rev 1:1-27.
33. Berson, SA, Yalow RS. 1968 Radioimmunoassay of ACTH in plasma. J Clin Invest 47:2725-2751.
34. Demura H, West CD, Nugent CA, Nakagawa K, Tyler FH. 1966 A sensitive radioimmunoassay for plasma ACTH levels. J Clin Endocrinol 26:1297-1302.
35. Krieger DT, Liotta AS, Suda T, Goodgold A, Condon E. 1979 Human plasma immunoreactive lipotropin and adrenocorticotropin in normal subjects and in patients with pituitary-adrenal disease. J Clin Endocrinol Metab 48:566-571.
36. Landon J, Greenwood FC. 1968 Homologous radioimmunoassay for plasma-levels of corticotrophin in man. Lancet 1:273-276.
37. Nicholson, WE, Davis DR, Sherrell BJ, Orth DN. Rapid radioimmunoassay for corticotropin in unextracted human plasma. Clin Chem 30:256-265.
38. Invitti C, Giraldi FP, de Martin M, Cavagnini F. 1999 Diagnosis and management of Cushing's syndrome: results of an Italian multicentre study. J Clin Endocrinol Metabol. 84:440-448.
39. Hodgkinson SC, Allolio B, Landon J, Lowry PJ. 1983 Development of a non-extracted 'two-site' immunoradiometric assay for corticotropin utilizing extreme amino- and carboxy-terminally directed antibodies. Biochem J 218:703-711.
40. Miles LEM, Hales CN. 1968 Labelled antibodies and immunological assay systems. Nature 219:186-189.

41. White A, Smith H, Hoadley M, Dobson SH, Ratcliffe JG. 1987 Clinical evaluation of a two-site immunoradiometric assay for adrenocorticotrophin in unextracted human plasma using monoclonal antibodies. Clin Endocrinol 26:41-52.
42. Fukata J, Naitoh Y, Usui T, Nakaishi S, Nakai Y, Imura H. 1989 Two-site immunoradiometric assay for adrenocorticotrophin: A cautionary study about the reactivity to its precursor molecules. Endocrinol Jpn 36:155-161.
43. Stewart PM, Gibson S, Crosby SR et al. 1994 ACTH precursors characterize the ectopic ACTH syndrome. Clin Endocrinol 40:199-204.
44. Raff H, Findling JW, Wong J. 1989 Short loop adrenocorticotropin (ACTH) feedback after ACTH- (1-24) injection in man is an artifact of the immunoradiometric assay J Clin Endocrinol Metab 69:678-680.
45. Kertesz G, Bourcier B, Cailla H, Jean F. 1998 Immunoradiometric assay of succinylated corticotropin: An improved method for quantification of ACTH. Clin Chem 44:78-85.
46. Raff H, Findling JW. 1989 A new immunoradiometric assay for corticotropin evaluated in normal subjects and patients with Cushing's syndrome. Clin Chem 35:596-600.
47. Rosano TG, Demers LM, Hillam R, Dybas MT, Leinung M. 1995 Clinical and analytical evaluation of an immunoradiometric assay for corticotropin. Clin Chem 41:1022-1027.
48. Van Rijn JLML, Van Landeghem BAJ, Haima P, Goldschmidt HMJ. 1996 Evaluation of ACTH immunoradiometric assays. Clin Biochem 29:93-95.
49. Zahradnik R, Brennan G, Hutchison JS, Odell WD. 1989 Immunoradiometric assay of corticotropin with use of avidin-biotin separation. Clin Chem 35:804-807.
50. Ghigo E, Arvat E, Ramunni J, et al. 1997 Adrenocorticotropin- and cortisol-releasing effect of hexarelin, a synthetic growth hormone-releasing peptide, in normal subjects. J Clin Endocrinol Metabol. 82:2439-2444.
51. Vale W, Spiess J, Rivier C, Rivier J. 1981 Characterization of 41-residue ovine hypothalamic peptide that stimulates secretion of corticotropin and β-endorphin. Science 213:1394-1397.
52. Vale W, Rivier C, Brown MR, et al. 1983 Chemical and biological characterization of corticotropin releasing factor. Recent Prog Horm Res 39:245-270.
53. Orth, DN. 1992 Corticotropin-releasing hormone in humans. Endocr Rev 13:164-191.
54. Shibahara S, Morimoto Y, Furutani Y, et al. 1983 Isolation and sequence analysis of the human corticotropin-releasing factor precursor gene. EMBO J 2:775-779.
55. Auchus RJ, Mastorakos G, Friedman TC, Chrousos GP. 1994 Corticotropin-releasing hormone production by a small cell carcinoma in a patient with ACTH-dependent Cushing's syndrome. J Endocrinol Invest 17:447-452.
56. Young J, Deneux C, Grino M, Oliver C, Chanson P, Schaison G. 1997 Pitfall of petrosal sinus sampling in a Cushing's syndrome secondary to ectopic adrenocorticotropin-corticotropin releasing hormone (ACTH-CRH) secretion. J Clin Endocrinol Metab 83:305-308.
57. Orth DN, Mount CD. 1987 Specific high-affinity binding protein for human corticotropin-releasing hormone in normal human plasma. Biochem Biophys Res Commun 143:411-417.
58. Ellis MJ, Livesey JH, Donald RA. 1988 Circulating plasma corticotropin-releasing factor-like immunoreactivity. J Endocrinol 117:299-307.
59. Fjellestad-Paulsen A, Abrahamsson P-A, Bjartell A, et al. 1988 Carcinoma of the prostate with Cushing's syndrome. Acta Endocrinol 119:506-516.
60. Tourniaire J, Rebattu B, Conte-Devolx B, et al. 1988 Syndrome de Cushing secondaire à la production ectopique de CRF par un carcinome médullaire du corps thyroïde. Ann Endocrinol (Paris) 4:61-67.
61. Sasaki A, Liotta AS, Luckey MM, Margioris AN, Suda T, Krieger DT. 1984 Immunoreactive corticotropin-releasing factor is present in human maternal plasma during the third trimester of pregnancy. J Clin Endocrinol Metab 59:812-814.
62. Cunnah D, Jessop DS, Besser GM, Rees LH. 1987 Measurement of circulating corticotrophin-releasing factor in man. J Endocrinol 113:123-131.

63. Suda T, Tomori N, Yajima F, et al. 1985 Immunoreactive corticotropin-releasing factor in human plasma. J Clin Invest 76:2026-2029.
64. Sasaki A, Sato S, Murakami O, et al. 1987 Immunoreactive corticotropin-releasing hormone present in human plasma may be derived from both hypothalamic and extrahypothalamic sources. J Clin Endocrinol Metab 65:176-182.
65. Gerl H, Knappe G, Rohde W, Stahl F, Wolff H, Martin H. 1990 Cushing-Syndrom bei CRF-produzierendem mediastinalem Karzinoid. Dtsch Med Wochenschr 115:332-336.
66. Hashimoto K, Nishioka T, Numata Y, Ogasa T, Kageyama J, Suemaru S. 1993 Plasma levels of corticotropin-releasing hormone in hypothalamic-pituitary-adrenal disorders and chronic renal failure. Acta Endocrinol 128:503-507.
67. Jessop DS, Cunnah D, Millar JGB, et al. 1987 A phaeochromocytoma presenting with Cushing's syndrome associated with increased concentrations of circulating corticotropin-releasing factor. J Endocrinol 113:133-138.
68. Nawata H, Higuchi K, Ikuyama S, et al. 1990 Corticotropin-releasing hormone- and adrenocorticotropin-producing pituitary carcincma with metastases to the liver and lung in a patient with Cushing's disease. J Clin Endocrinol Metab 71:1068-1073.
69. O'Brien T, Young WF Jr, Davila DG, et al. 1992 Cushing's syndrome associated with ectopic production of corticotrophin-releasing hormone, corticotrophin and vasopressin by a phaeochromocytoma. Clin Endocrinol 37:460-467.
70. Puchner MJA, Lüdecke DK, Valdueza JM, et al. 1993 Cushing's disease in a child caused by a corticotropin-releasing hormone-secreting intrasellar gangliocytoma associated with an adrenocorticotropic hormone-secreting pituitary adenoma. Neurosurgery 33:920-925.
71. Schteingart DE, Lloyd RV, Akil H, et al. 1986 Cushing's syndrome secondary to ectopic corticotropin-releasing hormone-adrenocorticotropin secretion. J Clin Endocrinol Metab 63:770-775.
72. Yanovski JA, Nieman LK, Doppman JL, et al. 1998 Plasma levels of corticotropin-releasing hormone in the inferior petrosal sinuses of healthy volunteers, patients with Cushing's syndrome, and patients with pseudo-Cushing states. J Clin Endocrinol Metab 83:1485-1488.
73. Linton EA, McLean C, Nieuwenhuyzen Kruseman AC, Tilders FJ, Van der Veen EA, Lowry, PJ. 1987 Direct measurement of human plasma corticotropin-releasing hormone by "two-site" immunoradiometric assay. J Clin Endocrinol Metab 64:1047-1053.
74. Behan DP, Khongsaly O, Liu X-J, et al. 1996 Measurement of corticotropin-releasing factor (CRF), CRF-binding protein (CRF-BP), and CRF/CRF-BP complex in human plasma by two-site enzyme-linked immunoabsorbant assay. J Clin Endocrinol Metab 81:2579-2586.
75. Liddle GW, Nicholson WE, Island DP, Orth DN, Abe K, Lowder SC. 1969 Clinical and laboratory studies of ectopic humoral syndromes. Recent Prog Horm Res 25:283-314.
76. Aron DC, Raff H, Findling JW. 1997 Effectiveness *versus* efficacy: The limited value in clinical practice of high dose dexamethasone suppression testing in the differential diagnosis of adrenocorticotropin-dependent Cushing's syndrome. J Clin Endocrinol Metab 82:1780-1785.
77. Yalow RS, Berson SA. 1971 Size heterogeneity of immunoreactive human ACTH in plasma and in extracts of pituitary glands and ACTH-producing thymoma. Biochem Biophys Res Commun 44:439-445.
78. Orth DN, Nicholson WE, Mitchell WM, Island DP, Liddle GW. 1973 Biologic and immunologic characterization and physical separation of ACTH and ACTH fragments in the ectopic ACTH syndrome. J Clin Invest 52:1756-1769.
79. Ratcliffe JG, Scott AP, Bennett HPJ, et al. 1973 Production of a corticotrophin-like intermediate lobe peptide and of corticotrophin by a bronchial carcinoid tumour. Clin Endocrinol 2:51-55.
80. Ratter SJ, Gillies G, Hope J, et al. 1983 Pro-opiocortin related peptides in human pituitary and ectopic ACTH secreting tumours. Clin Endocrinol 18:211-218.

81. Hale AC, Besser GM, Rees LH. 1986 Characterization of pro-opiomelanocortin-derived peptides in pituitary and ectopic adrenocorticotrophin-secreting tumours. J Endocrinol 108:49-56.
82. Bertagna X, Lenne F, Comar D, et al. 1986 Human β-melanocyte-stimulating hormone revisited. Proc Natl Acad Sci USA 83:9719-9723.
83. Vieau D, Massias J-F, Girard F, Luton J-P, Bertagna X. 1989 Corticotrophin-like intermediary lobe peptide as a marker of alternate pro-opiomelanocortin processing in ACTH-producing non-pituitary tumours. Clin Endocrinol 31:691-700.
84. Raffin-Sanson M-L, Massias J-F, Dumont C, et al. 1996 High plasma proopiomelanocortin in aggressive adrenocorticotropin-secreting tumors. J Clin Endocrinol Metab 81:4272-4277.
85. White A, Gibson S. 1998 ACTH precursors: Biological significance and clinical relevance. Clin Endocrinol 48:251-255.
86. Bertagna XY, Nicholson WE, Sorenson GD, Pettengill OS, Mount CD, Orth DN. 1978 Corticotropin, lipotropin, and β-endorphin production by a human nonpituitary tumor in culture: Evidence for a common precursor. Proc Natl Acad Sci USA 75:5160-5164.
87. Crosby SR, Stewart MF, Ratcliffe JG, White A. 1988 Direct measurement of the precursors of adrenocorticotropin in human plasma by two-site immunoradiometric assay. J Clin Endocrinol Metab 67:1272-1277.
88. Scopsi L, Gullo M, Rilke F, Martin S, Steiner DF. 1995 Proprotein convertases (PC1/PC3 and PC2) in normal and neoplastic human tissues: Their use as markers of neuroendocrine differentiation. J Clin Endocrinol Metab 80:294-301.
89. Creemers JWM, Roebroek AJM, Van de Ven WJM. 1992 Expression in human lung tumor cells of the proprotein processing enzyme PC1/PC3. FEBS Lett 300:82-88.
90. Crosby SR, Stewart MF, Farrell WE, Gibson S, White A. 1990 Comparison of ACTH and ACTH precursor peptides secreted by human pituitary and lung tumour cells *in vitro*. J Endocrinol 125:147-152.
91. Deftos LJ. 1991 Chromogranin A: Its role in endocrine function and as an endocrine and neuroendocrine tumor marker. Endocr Rev 12:181-187.
92. Nobels FRE, de Herder WW, Kwekkeboom DJ, et al. 1994 Serum chromogranin A in the differential diagnosis of Cushing's syndrome. Eur J Endocrinol 131:589-593.
93. De Bold, AJ. 1986 Atrial natriuretic factor: An overview. Fed Proc 45:2081-2085.
94. Opocher G, Rocco S, Carpenè G, et al. 1990 Atrial natriuretic peptide in Cushing's disease. J Endocrinol Invest 13:133-137.
95. Colao A, Pivonello R, Ferone D, et al. 1998 Effect of corticotrophin-releasing hormone on arginine vasopressin and atrial natriuretic factor in patients with Cushing's disease. Clin Endocrinol 49:77-84.
96. Okada S, Ohshima K, Mori M. 1994 The Cushing syndrome induced by atrial natriuretic peptide-producing thymic carcinoid. Ann Intern Med 121:75-76.

7
ACTH SECRETING PITUITARY ADENOMAS

Marie-Laure Raffin-Sanson, M.D. [1,3]
Yves de Keyzer, M.D. [3]
Xavier Bertagna, M.D. [2,3]
(1) Service de Médecine Interne 2, Hôpital Ambroise Paré, Boulogne/seine, France
(2) Clinique des Maladies Endocriniennes, Hôpital Cochin,
(3) Institut Cochin de Génétique Moléculaire, Faculté Cochin, Université Paris V - René Descartes, Paris, France.

The proposition that the pituitary was responsible for the clinical features of Cushing's disease was convincingly expressed for the first time in Harvey Cushing's classic monograph of 1932: The basophil adenomas of the pituitary body and their clinical manifestations (pituitary basophilism). Cushing was recognizing that:

"...striking clinical effects might be produced by minute, symptomatically predictable (pituitary) adenomas. So, it is the degree of secretory activity of an adenoma, which may be out of all proportion to its dimension, that evokes the recognizable symptom complex in all hypersecretory states."

he was still wondering however,

"...if the polyglandular features of the disorder are partly due, as premised, to a secondary hyperplasia of adrenal cortex..."

Much uncertainty remained, at that time, on the mechanism of this disorder. Today, much progress has been done. Yet...

CUSHING'S SYNDROME: PATHOPHYSIOLOGY, EPIDEMIOLOGY

Cushing's syndrome, which refers to the clinical and biological manifestations of excessive glucocorticoïd secretion, may result from various causes. In Cushing's disease, adrenocorticotropic hormone (ACTH) oversecretion by a pituitary corticotroph adenoma induces bilateral adrenocortical hyperplasia and excessive production of cortisol, androgens, and mineralocorticoids, which altogether provoke the clinical and biological features of the disease.

Cushing's disease is the most frequent cause of Cushing's syndrome in adult. In most series, its prevalence is approximately 70%, with a clear female preponderance, the sex ratio ranging from 3/1 to 10/1 (1-4). In our series of 809 adult patients with spontaneous Cushing's syndromes Cushing's disease accounts for 68% of the cases, and the female to male ratio is 2.8. Distribution of ages at diagnosis shows a peak in adult females between 25 and 45 yo. In children, things are different with a sex ratio of 1 and a preponderance of primary adrenal causes, Cushing's disease accounting for about 30% of the cases. Children with this condition are usually older than nine.

Cushing's disease is rare. Its true incidence which varies with age and sex, is difficult to evaluate, ranging roughly from 0.1 to 1 new cases per 100,000 and per year.

NORMAL CORTICOTROPH SECRETION

Corticotroph cell differentiation

To discuss the pathogenesis of corticotroph tumors, some understanding of normal corticotroph differentiation is required.

The anterior pituitary gland arises embryologically from Rathke's pouch, which develops as an invagination of the oral ectoderm, while the posterior lobe originates from the ventral hypothalamus. Interactions between the two tissues are essential for their differentiation. The cells from Rathke's pouch will proliferate and differentiate into specific cell lineage.

The corticotrophs are the first hormone-producing cells of the pituitary to reach terminal differentiation. Expression of proopiomelanocortin (POMC), the polypeptide precursor to ACTH, starts in corticotrophs at about day 12.5 of embryonic development in the mouse (5). The three Prop 1- and Pit 1- dependent lineages give rise, a few days later, to TSH positive thyrotrophs, GH positive somatotrophs and prolactin expressing lactotrophs

(6-8). Finally, the GATA 2 and SF-1 dependant gonadotroph appear around day 16.5 (9, 10).

The pituitary differentiation is directed by a complex set of factors, expressed at distinct and highly specific phases of pituitary development. Some of these factors are tissue (i.e. pituitary) specific, others are lineage restricted (11, 12).

Ptx 1, 2 and Lhx 3, 4 belong to the first group. Inactivation of Ptx 2, Lhx 3 or Lhx 4 blocks pituitary development at the Rathke's pouch stage, before the onset of cell differentiation except for corticotrophs (13-16). Ptx1 inactivation does not seem to have important consequences for pituitary embryogenesis, perhaps because it is rescued by Ptx 2.

Ptx 1 and 2 are expressed early in the anterior ectoderm that will give rise to Rathke's pouch. Their expression is maintained throughout pituitary development and in adult cell lineages (15, 17-19). Because Lhx 3 and 4 are expressed later, after the Rathke's pouch formation, and Lhx 3 transcription is dependent on the presence of Ptx 1 and 2 (20), those two last factors appear like upstream regulators required for progression of pituitary development. In the adult pituitary, by contrast, Ptx1 appears to be recruited for cell specific transcription of the POMC gene (21).

Determinants of corticotroph differentiation are largely unknown but seem to act at an early stage of pituitary development and appear different from those involved in the differentiation of the other pituitary lineages:

- when the Rathke's pouch is isolated from the infundibulum, no pituitary lineages can develop but the corticotroph cells (22, 23).

- as already stressed, mice deficient in Ptx2, Lhx3 or Lhx4 lack all pituitary cells but corticotrophs (13, 15). This also has been shown in man for Lhx3 (16).

- transgenic mice expressing pituitary directed LIF expression driven by the alpha GSU promoter show a deficit of all pituitary lineages except corticotroph cells which are hyperplastic (24).

Neuro D1 (25) may contribute to differentiation of the corticotroph lineage, as its expression closely precedes that of POMC in those cells (26). LIF, also expressed during embryogenesis, may favor corticotroph proliferation (24).

Above all, the transcription factor Tpit, highly specific for corticotroph and melanotroph lineages, recently demonstrated its importance in corticotroph orientation. Tpit belongs to the T box family of transcription factors and induces differentiation of undifferentiated cells into POMC expressing cells. Moreover, Tpit gene mutations were found in patients with isolated congenital corticotroph insufficiency, thus confirming the role of this gene in corticotroph development (27).

Mechanism of ACTH biosynthesis

The overall mechanism of POMC gene expression is schematically shown on Figure 1.

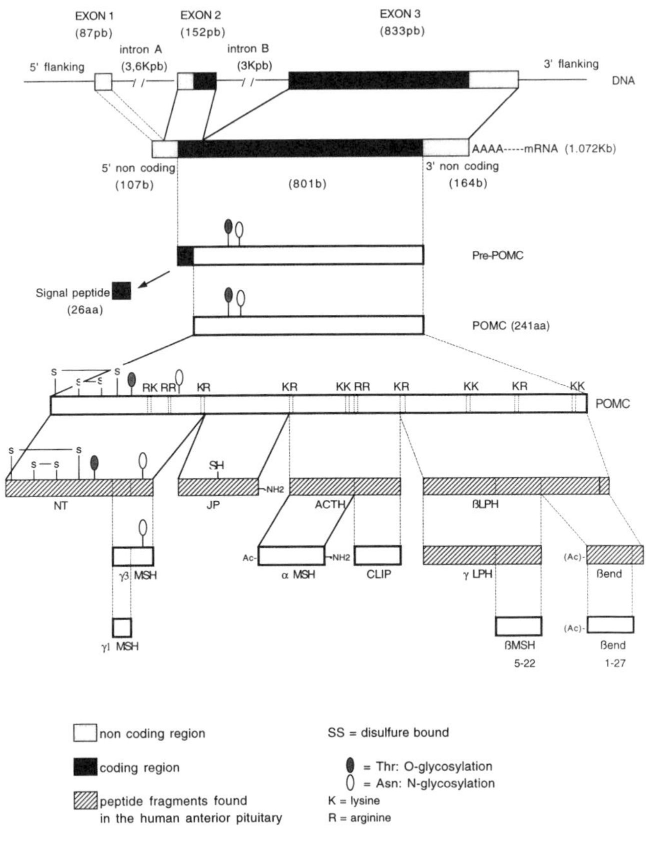

Figure 1. Schematic view of the human proopiomelanocortin (POMC) gene expression.

A single POMC gene by haploid genome is present on the distal region of chromosome 2 (2p23-25). It consists of three exons; coding regions are located within the second and third exons. After the splicing of the primary transcript, a mature messenger RNA (mRNA) of 1072 nucleotides (nt) is generated and a poly (A) + tail of about 200 nt is added. A prePOMC molecule is first translated starting with a 26 amino acid signal peptide necessary for the translocation of the nascent protein through the membrane of the rough endoplasmic reticulum.

Within the Golgi apparatus and the secretory granules the POMC molecule undergoes a series of proteolytic cleavages and chemical transformations which together result in the maturation or processing of the precursor (28): proteolysis occurs at pairs of basic amino acids. Among the 9 potential cleavage sites of the human POMC only 4 are used in the anterior pituitary, generating the N-Terminal fragment (29, 30) the Joining peptide (31-33), ACTH (34-36), ß-Lipotropin (ßLPH), and a small amount of gamma LPH and ß-Endorphin (ßend) (36, 37). Other chemical transformations include glycosylation of the N-Terminal fragment (38), C-terminal amidation of the Joining peptide (32, 33, 39, 40), and partial phosphorylation of ACTH on Ser^{31} (41, 42).

An alternate mode of POMC processing takes place in the intermediate lobe of the pituitary releasing smaller peptides such as ß-Melanocyte Stimulating Hormone (ß MSH), Corticotrophin-like intermediate lobe peptide (CLIP), and α-MSH (43, 44). It does not normally occur in the human pituitary where the intermediate lobe is only fully present in the fetus (45). However, this type of proteolysis also takes place in hypothalamic POMC neurons of arcuate nucleus where alpha MSH and CLIP are produced.

POMC gene expression is observed in many normal non-pituitary tissues (46, 47); it occurs at a very low level and predominantly through an alternate mode of gene transcription (48-50), generating very little POMC peptides if any (51-53). It is assumed that the highly predominant if not the sole source of circulating ACTH and POMC peptides in man, under normal circumstances, is the anterior pituitary corticotroph cell.

The coordinate proteolysis of POMC and the equimolar secretion of the various POMC peptides has two implications: any of the non-ACTH POMC peptide can be assayed in blood as a marker of the overall pituitary corticotroph activity; a specific pattern of POMC peptides is associated with the pituitary corticotroph, any qualitative abnormality suggesting either a non pituitary source (54-56) or an acquired processing abnormality in some aggressive - and poorly differentiated - pituitary macroadenomas (cf § 4.2).

Nuclear factors involved in the control of POMC transcription in pituitary

Regulatory elements controlling POMC gene expression in pituitary are contained within a 500bp fragment of the proximal POMC promoter (fig 2).

Pituitary-specific expression.

Binding sites for the factors required for tissue specific expression are present in the central and distal parts of this promoter fragment. Three elements binding transcription factors of a different family each, play a central role in pituitary specific POMC expression: an E box specific for bHLH heterodimers containing NeuroD1 (called CUTE complexes); a binding site for Pitx1 (58) a binding site for Tpit, flanking that of Pitx1 (59).

Pitx1 although not restricted to pituitary corticotrophs, plays a central role since it is able to interacts with both Tpit and the CUTE complex and activate POMC promoter in a synergistic fashion (58, 60). The cooperation between Pitx 1 ant Tpit, which is only expressed in pituitary corticotrophs and a small subset of dopaminergic neurons of the CNS, appear to be essential for POMC gene expression but also for the initiation of corticotroph differenciation during pituitary development (59).

The CUTE complex seems to play a less decisive role. In transfected cells it can interact with Pitx1 and synergistically induce POMC transcriptional activity, however, it is only transiently expressed during pituitary development, and *in vivo* its absence does not prevent Pitx1/Tpit-induced POMC expression (59).

Regulation of transcription.

CRH action is mediated by two separate elements that bind orphan nuclear receptors of the Nur family in the proximal and distal parts of the promoter (61, 62). Although the positive actions of CRH and its second messenger cAMP on POMC transcription are well established (63-65), their molecular basis are not totally understood.

The classical CRE (cAMP responsive element) motif has not been identified in the POMC promoter. Instead, CRH regulation appear to involve several regions of the promoter, including two separate elements binding orphan nuclear receptors of the Nur family in the proximal and distal parts of the promoter (61, 66, 67). In fact, in addition to stimulate cAMP production, CRH also induces the expression of Nur factors in corticotrophs.

The proximal site, a classic NBRE site, overlaps with a sequence known to mediate the negative action of glucocorticoïds, referred to as the nGRE, and the interplay between Nur factors and the glucoicorticoid receptor (GR) may underlie the antagonistic effects of CRH and glucocorticoïds (61).

The distal Nur binding response element (NurRE) is unique to the POMC gene (62). NurRE plays a major role in POMC induction by CRH; this action is mediated in part by an increased binding of unusual, Nur heterodimers to the NurRE site.

Glucocorticoid regulation: The negative regulation of POMC by glucocorticoïds requires the binding of the glucocorticoid receptor to the nGRE POMC promoter (68, 69). The interplay between GR and Nur binding on the nGRE site is not the only mechanism mediating glucocorticoïds inhibition of POMC transcription and glucocorticoid may act by multiple mechanism involving GR DNA binding as well as protein-protein interactions at different sites of the promoter. For instance, GR can interact with the unusual Nur complexes formed on the NurRE through protein-protein interactions and modulate transcriptional activity induced by CRH (67).

LIF regulation: LIF action on POMC transcription is mediated by activation of JAK-STAT pathway, leading to the activation of STAT1/3 factors and subsequent binding to POMC promoter after nuclear translocation (70).

Several potential (degenerated) STAT binding sites are present in the POMC promoter, including two overlapping NurRE that have been shown to bind STAT 1/3 proteins (71) however the precise molecular mechanisms of LIF action remain poorly undershood.

PITUITARY ACTH OVERSECRETION IN TRANSGENIC MICE : MODELS FOR CUSHING'S DISEASE ?

Transgenic animals often provide important tools in the analysis of the different stages of the development of tumors. In fact, study of this process from patient samples is often restricted by the lack of availability of tissue specimens, especially in pituitary. Thus many transgenic/knockout models have been generated, modifying the expression of candidate genes belonging to the regulatory pathway of corticotroph secretion or proliferation. Some of them developed clinical and biological features of Cushing's syndrome and/or POMC expressing pituitary adenomas.

ACTH oversecretion without pituitary adenoma

ACTH originates from the anterior lobe

LIF expression driven by alpha GSU promoter: These mice show clinical and biological features of hypercortisolism with thin skin, truncal obesity with increased intra abdominal fat, high corticosterone levels, incompletely suppressed by dexamethasone administration. They also exhibit dwarfism with very low IGF1 serum level, infertility because of central hypogonadotropism. Pituitary was hypotrophic with a dramatic decrease of somatotroph, lactotroph, gonadotroph, and thyrotroph cells while corticotroph lineages were hyperplastic. Thus, in the developing pituitary, LIF inhibits gonadotroph, somatotroph, thyrotroph and lactotroph lineages, but induces corticotroph development. Corticotroph hyperplasia causes a state of persistant glucocorticoid excess after birth and in adult animals (24).

CRH overexpression: The CRH gene was expressed under the control of the metallothionein promoter in transgenic mice (72, 73). Distribution revealed that transgene expression was primarily restricted to cells that expressed the endogenous CRH gene. The preservation of tissue specific expression of the transgene suggests the presence of elements located within the CRF intron, coding, or 3'-flanking regions, contributing to the cell type specificity of CRH gene expression. These mice show many features suggestive of Cushing's disease, such as excess fat accumulation, muscle atrophy, thin skin, and alopecia.

Endocrine abnormalities involving the hypothalamic-pituitary-adrenal included elevated ACTH and corticosterone plasma levels, reaching ten times the concentrations observed in wild type animals. Transgenic mice also showed an anxiogenic like behaviour, as in animals under chronic stress, and an impaired immune response linked to quantitative and qualitative alteration in B cell activity.

Both endocrine and behavioral features of these animals were counteracted by injection of a CRH antagonist.

These findings indicated that chronic production of excess CRH resulted in sustained stimulation of pituitary corticotroph cells, with elevated ACTH and glucocorticoid production, a condition that leads to the development of Cushing's syndrome. However, no tumors were seen in the mice pituitaries.

V3R overexpression: Transgenic mice were generated, that overexpressed the vasopressin V3 receptor under the control of rat POMC promoter sequences. The transgene was efficiently transcribed in both corticotrophs and melanotrophs. Circulating levels of ACTH were similar to those of controls but corticosterone levels were moderately increased. The levels of POMC mRNA in pituitary and of CRH and AVP mRNA in hypothalamus were comparable to those of control mice, thus indicating a partial resistance to glucocorticoid, considering elevated corticosterone levels. No clinical signs of Cushing's syndrome were observed, but a reduced anxiety like behaviour, suggesting an adaptative mechanism that modulated the activity of brain structures involved in stress response *(personal data).*

GR null mice: Three different models of glucocorticoid receptor KO have been described.

Generation of GR-deficient mice was obtained by gene targeting (74). Most of the pups died soon after birth, due to severe lung dysplasia. Also, neonatal induction of gluconeogenic enzymes in the liver was impaired. Feedback regulation via the hypothalamic-pituitary-adrenal axis was severely impaired resulting in elevated levels of plasma ACTH (15-fold) and corticosterone (2.5-fold). Extensive hyperplasia of the cortical zone of the adrenal was observed, with increased expression of key cortical steroid biosynthetic enzymes, such as side-chain cleavage enzyme, steroid 11 beta-hydroxylase, and aldosterone synthase. In contrast, adrenal medulla was atrophic.

Since the GR can influence transcription either through DNA-dependent and or independent mechanisms, the same group attempted to assess the relative importance of these two modes of (75). A point mutation was introduced into the GR by gene targeting using the Cre/loxP system. It specifically impaired dimerization and therefore interaction with DNA while functions that do not require DNA binding (i e cross talk with other transcription factors), were preserved. In contrast to GR-/- mice, these mutants were viable, revealing the in vivo relevance of DNA-independent activities of the GR. Corticosterone levels and expression of the side chain cleavage enzyme were up regulated in those mice, but to a lesser extent than in GR-/- mice. ACTH plasma levels were identical to those in wild type animals, despite the several fold increase of POMC mRNA expression and ACTH immunoreactivity in corticotroph cells. The level of CRH peptide in the median eminence was unaffected. Thus, glucocorticoid feedback in the HPA axis is a complex network, involving both DNA - dependent and independent mechanism.

The Cre Lox system allows specific disruption of the glucorticoid gene receptor in selected tissues. Loss of GR function in the nervous system (76*)* whereas GR expression is preserved in pituitary and peripheral tissues, impaired hypothalamus-pituitary-adrenal (HPA)-axis regulation. CRH was up regulated in the PVN and POMC gene expression was increased. ACTH plasma levels were identical to those observed in wild type animals, whereas corticosterone plasma levels were increased, leading to Cushing's syndrome. Circadian rhythm of secretion however was present. Conditional mutagenesis of GR in the nervous system also provides genetic evidence for the importance of GR signalling in emotional behaviour because mutant animals show impaired behavioural response to stress and reduced anxiety.

ACTH originates from the intermediate lobe

Dopaminergic receptor D2 deficient mice (77): Analysis of D2R KO mice showed unexpected ACTH overproduction whereas physiological products of POMC processing in the intermediate lobe are alpha MSH and beta endorphin.

Melanotroph secretion is under the negative control of dopamine which, in addition to controlling POMC gene expression, also regulates its processing. In the absence of dopaminergic receptors, the convertase PC1 is inappropriately upregulated in the intermediate lobe and ACTH is produced in addition to alpha MSH. As melanotroph cells do not express glucocorticoid receptors, ACTH production is not under the negative feedback of corticosteroids. Excessive corticosteroid secretion results, with subsequent hypertrophy of adrenal glands and clinical features of Cushing's syndrome *(*77). Why ACTH is not entirely converted into CLIP and alpha MSH by the convertase PC2 also present in melanotroph cells remains uncompletely explained.

These results were not observed in another model of D2 KO mice (78), perhaps because of a different genetic background.

KO 7B2: The protein 7B2 has been implicated in activation of prohormone convertase 2 (PC2) (79-82), one of the two neuroendocrine processing enzymes responsible for POMC processing in melanotrophs and hypothalamus.

7B2 null mice (83) have no demonstrable PC2 activity, are deficient in processing pancreatic islet hormones, and display hyperproinsulinemia, hypoglucagonemia and hypoglycemia. They show markedly elevated ACTH and corticosterone plasma levels, with adrenocortical hyperplasia. They die before 9 weeks of life from severe pituitary Cushing's disease, arising from intermediate lobe ACTH hypersecretion, whereas hypothalamic secretion of CRH and anterior lobe synthesis of POMC are suppressed.

Thus, a processing defect in the intermediate pituitary lobe can result in excessive ACTH secretion and clinical features of Cushing's. However, it is noteworthy that PC2 null mice (PC2-/-), in which such an abnormal POMC processing was expected, do not exhibit excessive glucocorticoïd secretion. *(84)*.

POMC expressing pituitary adenomas… of the intermediate lobe

Melanotroph tumors induced by a POMC simian virus 40 large T antigen transgene: Mice harboring a transgene composed of POMC gene promoter sequences (nucleotides-706 to +64) upstream of the simian virus (SV) 40 early gene encoding large T antigen developed large POMC-expressing pituitary tumors. Histologically the tumors arose from the intermediate lobe, contained POMC peptides. POMC processing in the pituitary tumors was indistinguishable from that in normal mouse intermediate lobe melanotrophs (85).

P27/Kip1 $^{-/-}$ mice: P27/Kip1gene is a member of the cyclin kinase inhibitor family of proteins. The P27 protein specifically inhibits the formation of cyclins / cyclin dependent kinases complexes, required for the progression from phase G1 to phase S of the cell cycle. These complexes mediate the phosphorylation of the Rb1 protein (86-92). P 27 KO mice develop multiorgan hyperplasia, excessive growth and pituitary tumors originating from the intermediate lobe melanotroph cells. The intermediate lobes are hyperplastic in those mice whereas anterior and posterior lobe are relatively normals. About half of the P27-/- animals develop pituitary tumors originating from the intermediate lobe, by 12 weeks of age. Melanotroph tumoral cells nucleus display abnormal features. Cellular sheets are irregular and abnormally thick, with papillary protrusion without invasion or metastasis; the tumors tend to form cystic structures. The abnormal tissue stained positively for alpha MSH, ACTH and beta endorphin. However, no clinical signs of glucocorticoid excess was observed in those mice (93-95).

Retinoblastoma (Rb1) protein deficient mice: The Rb1 gene product plays an essential role in regulating the progression of the cell cycle. In its dephosphorylated form, it blocks the progression from G1 to S phase. Rb -/- mice die during embryogenesis. All animals with one inactive Rb1 allele developed pituitary tumor of the intermediate lobe. Rb +/- mice develop more aggressive pituitary tumor than p27-/- mice, with hemorrhaging, necrosis and invasion into adjacent brain structures. The wild type Rb allele is lost in tumoral cells. However, development of these adenocarcinoma of the pituitary in Rb +/- mice is observed only after a long latency period after the loss of the wild type allele in melanotroph cells, suggesting the need of another molecular event (96-98).

Mice lacking one Rb allele and deficient for p27 (ie Rb+/- p27-/- mice) developed pituitary adenocarcinomas that were more aggressive and appeared earlier than in either Rb +/- or p27 -/- mice, therefore suggesting than Rb and p27 cooperate to suppress melanotroph tumor development (99).

P18 KO mice: Similar to the Rb and p27 KO mice, animals deficient in the cell cycle inhibitor p18 also develop pituitary tumors originating from the intermediate lobe. Intermediate lobe hyperplasia or adenoma was found in 9 to 12 months old mice. Anterior lobe was also enlarged but in a lesser extend. In older mice, melanotroph tumors progressed, compressing but not invading the anterior lobe. The cells stained strongly for POMC derived peptides. Plasma ACTH and corticosterone levels have not been determined, but no features of Cushing's syndrome were observed. Mice lacking both p27 and p18 developed rapidly invasive pituitary tumors from the intermediate lobe. This synergistic response in double null animals suggest that p18 and p27 work in a cooperative manner to suppress tumorigenesis in mice intermediate lobe (100).

POMC expressing pituitary adenomas... of the anterior lobe

Transgenic mice expressing the T antigen of polyoma virus under the control of the early promoter develop pituitary tumors with ACTH immunoreactivity. Corticotroph microadenomas are observed at 9 months of age with increased ACTH plasma levels, adrenal hyperplasia and abnormal weight gain. When these tumors are reimplanted in null mice, the transplantated animals show still higher ACTH plasma levels than knocked out mice, early corticosterone increase and clinical features of glucocorticoid excess with skin atrophy, alopecia, cervical fat deposit (101, 102).

From mice to human beings: the limits of animal models

These models have provided valuable tools for a better understanding of the pathways involved in physiological HPA regulation. However, regarding the pathogenesis of Cushing's disease, there is a lack of correspondance between the murine models an human disease. Most models of ACTH oversecretion in mice involve corticotroph hyperplasia and not adenoma as observed in human Cushing's disease (with the exception of the non specific introduction of T antigen of polyoma virus under the control of the early promoter in transgenic mice).

Moreover, most POMC expressing pituitary adenomas in mice originated from the intermediate lobe, whereas this structure is vestigial in humans. Even if these models represent invaluable tools for the understanding of specific gene functions, none of them has provided us with a clue for the understanding of pathophysiology of human corticotroph adenomas.

ACTH SECRETION IN CUSHING'S DISEASE

In most cases, if not all, Cushing's disease is caused by a corticotroph adenoma. This assertion is grounded on the frequence with which such adenomas are found at surgery, and the clinical, biochemical and histologic evidence of a suppressed CRH hypothalamic secretion.

These adenomas are usually small and indolent tumors, and may be difficult to visualise by MRI. In some cases, adenomas are large and show invasive features. Corticotroph carcinomas have even been described.

The typical corticotroph microadenoma

Prevalence

Since the late 1970s many groups have reported the high frequency of pituitary microadenomas found at surgery in patients who were systematically subjected to sellar exploration by the transsphenoidal route, whether or not a pituitary tumor had been suspected by prior X-ray, CT-Scan, or -more recently- MRI imaging. As a rule such tumors are found in more than 80% of the cases (103-111). Although small -and "silent"- corticotroph tumor are sometimes found at autopsy of non-Cushing's patients, the prevalence of such adenomas is definitively higher in patients with Cushing's disease (112).

Pathology

The basophilic adenomas of Cushing's disease have variable sizes, with a mean around 5 mm (106, 113, 114). Some are found outside the pituitary fossa and develop from the uppermost part of the pituitary gland (pars tuberalis), above the diaphragma being out of reach of the neurosurgeon via the transsphenoidal route, and several cases of Cushing's disease have been reported with an ectopic corticotroph adenoma (115-119).

The classic basophilic adenoma is not encapsulated and composed of compact cords of more or less homogeneous cells. The granule content is responsible for its basophilic and PAS staining properties: the latter is now-explained since non-ACTH POMC peptides (the N-Terminal fragment) are glycosylated (38). Occasionally the paucity of the granule content explains that some adenomas appear chromophobe at the light microscope. In some adenomas (120) tumor cells show characteristic features of the Crooke's cell (121) as depicted in the normal corticotroph of patients treated with corticosteroids: a ring-shaped homogeneous dense hyaline area constitutes an amorphous zone which repells the granules to the margin of the cell and close to the nucleus; ultrastructural studies show that it is made of filaments (122).

Immunocytochemistry has recently provided the ultimate means to recognize corticotroph cells by the specific immunodetection of their content (123, 124). The sensitivity of the method sometimes allows the detection of an immune signal in what appeared to be a chromophobe adenoma (124). Many adenomas will, unsurprisingly, react with different antisera directed against different POMC fragments, some will respond only to a given antiserum. Although this type of observation may point to some peculiar mode of POMC processing in a particular tumor which would not generate a generally accessible epitope to the antibody, as has been described for example in endorphin adenomas, it should be kept in mind that different anti-sera may show variable sensitivities. More recently the specific recognizition of POMC RNA by in situ hybridization has been achieved in human corticotroph adenomas.

The peri-adenomatous tissue shows a variable density of corticotroph cells, with frequent and typical Crooke's cells (122). The co-existence of corticotroph hyperplasia and adenoma has been reported (124-127).

ACTH production is excessive but qualitatively normal

Numerous studies performed in vitro show that in the vast majority of corticotroph adenomas the products of POMC gene transcription and of POMC processing are identical with those in the normal human anterior pituitary.

The gene transcription, shows no gross abnormality and the POMC transcripts in pituitary tumors are similar to those in the normal pituitary (129-132): A 1200 nt POMCmRNA is the highly predominant -if not sole-transcript. A small percentage (<5%) of transcripts result from an alternate mode of RNA splicing adding 30 nt at the 5' end of the second exon. It has no implication on the open reading frame which is not modified. Fewer than 1% of transcripts result from the use of an upstream promoter at -369 nt (133).

The N-Terminal fragment (134), the Joining Peptide (32), authentic $ACTH_{1-39}$ (35, 43, 55, 135), ß-LPH and variable amounts of gamma LPH and ß-Endorphin (135, 136, 137) are the normal end-products of POMC processing found both in tumor extracts and in culture media. A somewhat higher proportion of ß-Endorphin over ßLPH, -and gamma LPH over ßLPH- has been reported (129, 130). Yet the recruitement of proteolytic sites which are not normally activated in the normal pituitary is not observed and peptides like CLIP and alpha-MSH are neither formed nor released. This general finding support the use of highly specific IRMAs for plasma ACTH detection as a valid and significant means to evaluate patients with Cushing's disease (138). Thus tumor POMC peptides, including ACTH, usually show no peculiar or unexpected molecular forms in contrast with what is often found when POMC expression occurs in a non-pituitary tumor (Fig. 2). Yet, highly aggressive pituitary tumors and silent corticotroph adenomas, occasionally secrete abnormal POMC fragments (see further)...

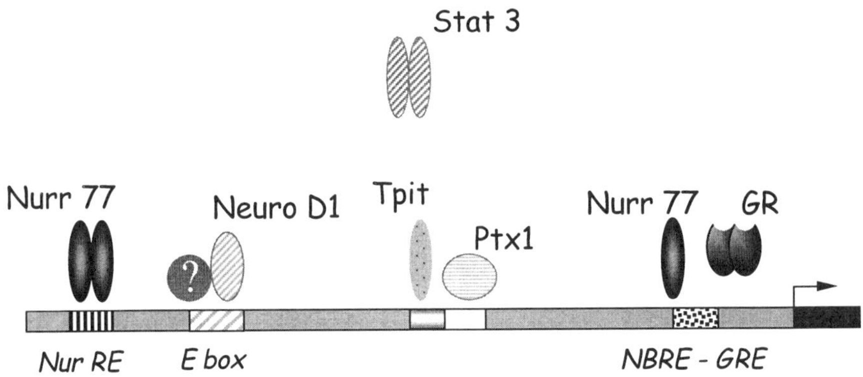

Figure 2. Schematic view of the proximal POMC promoter region and nuclear factors involved in the control of POMC transcription in pituitary.

ACTH secretion is dysregulated, but not autonomous

Increased sensitivity to secretagogs CRH, AVP, hexarelin

A vast majority of patients with Cushing's disease (ca 85%) have an exagerated ACTH response to the administration of CRH, when compared to normal subjects (139, 140). This is most likely explained by increased expression of CRH receptors in the tumoral cells as assessed by studies at the mRNA level (141).

This is the basis of the current use of the CRH test in the investigation of patients with Cushing's syndrome. It is a further evidence of the relative resistance to glucocorticoid action at the tumor level, since excess plasma cortisol normally suppresses the ACTH response to CRH.

Another physiological ACTH secretagog, vasopressin, also induces an exaggerated ACTH response in patients with Cushing's disease (142). Quite interestingly –and somewhat unexpectedly- the V2 agonist desmopressin induces an even higher response in most (~85%) patients (143). The recent characterization of the "pituitary-specific" vasopressin receptor (V3 or V1b) has provided some explanation (144): we -and others- have shown that desmopressin was also a V3 agonist (145, 146). Yet, the higher response to desmopressin than to vasopressin remains unexplained: tissue studies by RT PCR suggest that V2 receptor is also expressed in corticotroph tumors but its precise cellular localization has not been established (147).

Hexarelin, and possibly other realted peptides of the GHRP family, are weak stimulators of the ACTH response in normal subjects (148). Recent studies have shown that hexarelin induced an exaggerated ACTH release in patients with Cushing's disease (149). This phenomenon is most likely explained by the demonstration of high expression of GHRP receptors in corticotroph tumors (150).

Other peptides like PHM (peptide histidine methionine) and VIP (vasoactive intestinal peptide), which have no effect in normal subjects, occasionally trigger a convincing ACTH release in patients with Cushing's disease (151, 152). These data have been shown in well performed controlled studies. They strongly support the concept that abnormal or illegitimate hormone receptors may contribute to the pathophysiology of corticotroph adenomas.

Partial resistance to glucocorticoid suppression

Glucocorticoids exert a negative feedback on pituitary ACTH (153). In patients with primary adrenal deficiency, basal and stimulated ACTH are increased. On the other hand, excess glucocorticoid administration or secretion by a primary adrenocortical tumor inhibit basal and stimulated ACTH. Glucocorticoids inhibit hypothalamic CRF production (154) and also act directly at the corticotroph cell as demonstrated in various animal models.

They inhibit basal and stimulated ACTH release (155, 156), as well as POMC gene transcription in a dose dependent manner (157). Interestingly this inhibition is not complete and a small proportion of POMC transcription is not suppressed even by very high amounts of glucocorticoids (64, 158).

The hallmark of ACTH oversecretion in Cushing's disease is its partial resistance to the normal suppressive effect of glucocorticoids (159, 160). The dose-response curve between administered dexamethasone and plasma ACTH or endogenous cortisol production is shifted to the right (Fig. 3). Because ACTH secretion by the pituitary tumor is not normally restrained ACTH is overproduced with subsequent chronic hypercortisolism. Since peripheral tissues have retained their normal sensitivity to the action of cortisol (161, 162) they appropriately develop the features of Cushing's disease.

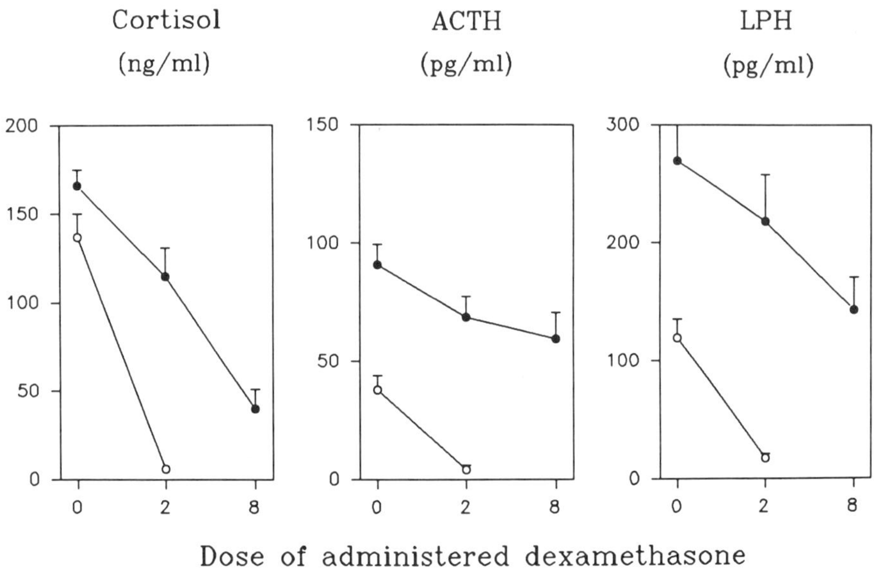

FIGURE 3. Variations (mean ± SEM of plasma Cortisol, ACTH and LPH in response to increasing daily doses of dexamethasone administered for two days in 15 normal subjects (o) and in 16 patients with Cushing's disease (•).

In vitro studies have confirmed that pituitary corticotroph adenomas are not autonomous and have indeed retained some sensitivity to the suppressive effect of glucocorticoids which invariably decrease basal and/or stimulated ACTH release (163-168). A direct comparison between the responses of normal and tumoral cells in vitro is lacking most of the time. A single study measured the effects of 2 doses of dexamethasone (1 and 10 ug/dl) on both ACTH release and POMCmRNA content in cultured cells obtained either from corticotroph adenomas or from their -presumably normal- periadenomatous tissues: whereas dexamethasone efficiently reduced both parameters in the periadenomatous cells, its suppressive effect was reduced in the tumoral cells (169).

Molecular pathogenesis

Most adenomas are monoclonal

The clonal origin of various human endocrine tumours has been recognised, based on the study of genetic markers born by the X chromosome in female heterozygous patients. Recent techniques using DNA probes directed at various genetic markers (hypoxanthine phosphoribosyl transferase and/or phosphoglycerate kinase) studied the X-inactivation pattern in peripheral and tumoral tissues through the combined DNA digestion with a methylation -specific enzyme and the restriction enzyme giving rise to a restriction fragment length polymorphism. They have shown the monoclonal nature of all non-functioning pituitary tumors (170, 171). Studies performed on functioning tumors (172) showed a monoclonal pattern in 3 of 3 GH secreting adenomas, 4 of 4 PRL secreting adenomas, and 3 of 4 corticotroph adenomas; the fourth corticotroph adenoma was substantially contaminated by interspersed normal adrenohypophyseal tissue which may have induced an apparent polyclonal pattern. More recent studies using another anonymus marker of the X chromosome, the M27beta probe, have confirmed these results in pituitary corticotroph adenomas (173-175).

Somatic defects are occasionally found

Somatic defects, (mutation, deletion or overexpression of specific genes) have been found in corticotroph adenomas. Some of them could explain the partial resistance to glucocorticoid feedback, the exagerated response to stimuli, or the abnormal proliferative potential of these tumors.

Glucocorticoid receptor(GR) defects:

Corticotroph adenomas are characterized by a relative resistance to the negative feedback action of cortisol on ACTH secretion. Corticosteroids act via activation of the glucocorticoid receptor (176). Two isoforms, GR alpha and GR beta are produced via an alternative splicing. GR beta could act as a negative regulator of the classical GR alpha by forming heterodimers (177-179). Hence, somatic disruption of the GR gene in a corticotroph cell, could explain an abnormal multiplication of this cell and the relative glucocorticoid resistance of the resulting clone.

Germinal point mutations or deletion of GR, impairing GR alpha function, are involved in syndromes of general glucocorticoid resistance (178-182).

In fact, a frameshift mutation in the GR gene has been found in the tumoral tissue of one patient with a corticotroph adenoma whose developement was exagerated after bilateral adrenalectomy (Nelson's syndrome). The mutation was expected to create a truncated GR protein, and was hypothesized to explain the corticotroph proliferation and hypercortisolism in this patient (183).

However, the search for similar defects in the different parts of the GR gene failed to identify mutations in a series of 19 ACTH secreting tumors, including16 corticotroph adenomas (184). Thus, point mutations in the GR gene are not frequently involved in the development of these tumors.

Inactivation of one allele of the GR gene can also be achieved by deletion of GR gene. Indeed, monoallelic deletions in the region of the GR gene have been described in corticotroph adenomas, 6/18 showing loss of heterozygoty near the GR gene. By comparison, LOH at the GR locus was present only once in 28 pituitary tumors of other origin. No somatic mutations were found in the remaining allele or in tumors without LOH. In the absence of functional studies allowing to determine whether haploinsufficiency does occur in these tumors, the significance of these deletions remains unknown (185).

In vitro studies suggested that besides deletions and mutations of the GR gene, an altered ratio of the two isoforms alpha and beta of the GR, could result in glucocorticoid resistance (182) as GR beta, the non ligand binding form of GR, exerts a dominant negative effect over the ligand binding form, GR alpha. Quantitative evaluation of the mRNAs encoding the two isoforms of GR in 17 corticotroph adenomas revealed the two isoforms to be expressed at levels similar to those seen in normal pituitary. Moreover, while variable expression of the beta subform was observed among the tumors, no correlation was found with the degree of glucocorticoid resistance, evaluated by the response to the high dose dexamethasone suppression test (184). Thus, although the relative expression of the two isoforms have been studied only at the mRNA level and not at the protein level, the results of this study does not support the hypothesis of an abnormal splicing of the GR at the origin of glucocorticoid resistance in corticotroph adenomas.

11 beta hydroxysteroid dehydrogenase overexpression:

As no gross abnormalities of the GR was found in corticotroph adenomas, an alternative mechanism was proposed to explain the relative insensitivity of these tumors to glucocorticoids. Two isoenzymes of the 11beta hydroxysteroid deshydrogenase (HSD) play an important prereceptor role in regulating corticosteroid hormone action, either by converting cortisone to cortisol (type1), or conversely by converting cortisol to cortisone (type 2). Renal 11βHSD type 2 is responsible for the peripheral inactivation of cortisol to cortisone in mineralocorticoid target tissues, thus preventing its binding to the mineralocorticoid receptor. Immunohistochemical studies revealed a marked induction of 11βHSD type 2 in ACTH secreting pituitary tumors (186). Although these data have to be confirmed, overexpression of type 2 11βHSD, by converting active cortisol reaching the corticotroph cells into inactive cortisone, could explain the relative resistance to glucocorticoid feedback in Cushing's disease.

CRH receptor:

CRH (187) is the main regulator of HPA axis and plays an important role in integrating the endocrine, autonomic and behavioural response to stress and immunological challenges.

CRH directly stimulates POMC gene expression and ACTH secretion through specific G protein coupled transmembrane receptors present on the corticotroph cell, which activate the adenylate cyclase and increase intracellular cAMP formation (188, 189).

There are at least two forms of CRH receptor, sharing significant sequence homology, but differing in their pharmacology and tissue distribution. CRH-R type 1 is predominantly present on pituitary cells (190), whereas type 2 is more widely expressed throughout central nervous system and peripheral tissues such as the heart (191). CRH-R1 is responsible for CRH stimulated release of ACTH from the pituitary. Indeed, the pituitary adrenal phenotype of CRH-R1 KO mice shows a striking resemblance to CRH KO mice. CRH-R1 KO mice exhibit pronounced glucocorticoid deficiency and atrophy of the adrenal gland. ACTH plasma levels were in the normal range but inappropriate to the very low corticosterone concentrations. CRH-R1 KO mice also show reduced anxiety related behaviours and impaired ACTH and corticosterone response to stress (192, 193). CRH is essential for the stress induce activation of the HPA axis. Together, CRH-R1 and CRH-R2 co-ordinate stress response in a complementary manner: CRH-R1 mediates the ACTH stimulation characteristic of the initial stress induced response whereas CRH-R2 mediates the stress coping responses during the recovery phase of stress including a late negative feed back on HPA axis (194, 195).

As gain of function mutations involving seven transmembrane domain receptor type have been found in pituitary tumors, such as somatotroph adenomas (196) search for such mutations was performed in corticotroph adenomas. No abnormalities in the coding sequence of the CRH receptor 1 were found in corticotroph tumors, but an increased expression of the gene with high mRNA levels . However, it is not known whether the up-regulation of the transcript corresponds to an increased receptor number at the cell surface or an adaptative mechanism to a processing defect or abnormal function of the protein in tumoral cells (197, 198). If these mRNA are efficiently translated in functional CRH-R 1, these finding could explain, in part, the exagerated response of ACTH to CRH in patients with Cushing 's disease.

It is noteworthy that CRH gene expression itself was demonstrated in corticotroph adenomas, whereas it is absent from normal pituitary gland. This expression seems to be correlated with the proliferative potential of the tumor *(199)*. So, CRH mediated autocrine or paracrine stimulation may be a mechanism associated with tumor proliferative potential.

Vasopressin receptor (V3):

In addition to its role on water conservation, vasopressin regulates pituitary ACTH secretion by potentiating the stimulatory effects of CRH.

The actions of vasopressin are mediated by plasma membrane receptors belonging to the guanyl nucleotide binding protein (G-protein) family. Three major receptor subtypes encoded by different genes have been identified and cloned; kidney V2 receptors, which are coupled to the G-protein, Gs, and adenylate cyclase; V1a receptors, present in smooth muscle, vascular muscle and liver, coupled to phospholipase C (PLC); and the pituitary V1b, also called V3 receptor, linked to PLC and responsible for vasopressin effects on HPA axis. (200-205.).

The action of AVP potentiates that of CRF (206) by further increasing cAMP formation (207-209): cross-talk between the 2 transducing systems provides the synergystic action which promotes the maximal ACTH response (210-212). This phenomenon, thoroughly studied in vitro on animal models, is also observed in man (213-216). In animal models, vasopressin participates in the adaptation of the HPA axis during stress, becoming the predominant regulator during chronic stress, through a relative refractoriness of its effect to glucocorticoid feedback (217).

As CRH 1 receptors, V3 receptor mRNA was qualitatively and quantitatively analyzed in corticotroph tumors. The V3 mRNA structure and sequence was identical in normal and tumoral pituitary, but marked over expression of V3R was present in corticotroph adenomas (141).

If the number of V3R increases at the cell surface, this would provide an explanation to the well known secretory hyperresponsiveness of corticotroph tumors to vasopressin (216).

Wether the increase inV3 mRNA levels is a consequence of chronic hypercortisolism or are induced independantly remains unknown.

However, it is tempting to speculate that in corticotroph adenomas, the V3 R overexpression may stimulate the level of second messengers, either contitutively or under stimulation of endogenous vasopressin *(218-219)* and act synergistically with CRH-R1 to stimulate POMC expression and corticotroph proliferation.

LIF, LIF receptor and other cytokines:

A neuro-immuno-endocrine loop is well documented, involving different cytokines and pituitary secretions (220). Stimulation of ACTH secretion is a characteristic of several cytokines: LIF strongly stimulates ACTH secretion and POMC gene expression, in striking synergy with CRH (221, 222). LIF signaling is mediated through the LIF receptor - gp130 coreceptor complex, the Jak-STAT signaling cascade and the STAT3 phosphorylation (223-226). It is controlled in a negative feed back loop by SOCS-3 (suppressor of cytokine signalling) (227). Interleukin 1 , IL 11 and -6 also exert a stimulatory action on ACTH release (228-231). IL1 is thought to activate corticotroph function predominantly through stimulation of hypothalamic CRH (232) and pituitary LIF expression (233), whereas IL6 is a CRH independent stimulator of the HPA axis (234).

LIF (235) LIF receptor and the gp 130 receptor subunit (236) are expressed in many pituitary cells, mainly corticotroph and folliculostellate cells where LIF gene expression is up regulated by lipopolysaccharide (237), IL1 (233) and down regulated by dexamethasone (238). Interleukin 1 and interleukin 6 are also normally present in the rat anterior pituitary (239-240) thus raising the possibility that all these cytokines act as autocrine or paracrine factors. In a variety of acute or chronic inflammatory conditions, LIF expression is up regulated in different tissues and serum levels of LIF increase, especially in sepsis (241). In keeping with all these results, LIF-/-KO mice mounted an attenuated basal, post-stress and post-IL1 ACTH response, in comparison with LIF+/+ mice (242).

LIF and LIF receptor, IL6 and IL6 receptor, IL11 and IL11 receptor transcripts were detected in ACTH non secreting pituitary tumors, corticotroph adenomas and in other subtypes of secreting adenomas. (243-245). However, their contribution to pituitary tumorigenesis remains hypothetical.

G Protein:

The G protein complex is a key pathway of receptor signaling in corticotroph cells.

In striking contrast with the situation observed in somatotropinomas (196) , where activating mutations in the stimulatory a subunit of the G protein complex are a frequent finding, mutations of genes encoding for the different subunits of the G protein complex (Gsα, Giα, Gqα), do not seem to be involved in the pathogenesis of corticotroph adenomas (246-247).

Although activation of the CRH receptor leads to corticotroph proliferation and POMC gene stimulation through a G protein, no corticotroph adenoma is observed in the Mac Cune Albright syndrome, where Gs alpha is constitutively activated.

Transcription/differenciation factors:

Ptx 1 appears like an upstream regulator required for pituitary development (14, 19, 20) and in adults, recruited for cell specific expression of the POMC gene (21).

The expression of **Prop 1** gene is restricted to the developping pituitary gland, and involved in the proliferation and differentation of somatotroph, lactotroph, thyreotroph and gonadotroph but not corticotroph cells. In mouse and in humal, loss of function or null mutations are responsible for a combined pituitary deficiency, involving GH, TSH, PRL, FSH and LH secreting cells (248-251*)* Impaired ACTH function has been described once (252). Moreover, transgenic mice constitutively expressing Prop1 after the end of pitutary differentiation and during adulthood exhibit misregulation of pituitary cell proliferation, including adenomatous hyperplasia with the formation of Rathke's cleft cysts and tumors (253).

Prop 1 was detected in 14 pituitary adenomas including 1 corticotroph umors, without mutations in the coding sequence (254). The expression of Ptx1 and Prop 1 was studied in a serie of 34 pituitary adenomas including 6 corticotroh. mRNA for both factors were detected in all 34 adenomas. Expression of Ptx1, but not that of Prop 1 was reduced in corticotroph adenomas (255). However, the involvment of these genes in tumor formation remains unknown.

Thus, the important role in embryological pituitary development of Ptx1 and other pituitary homeobox genes makes these genes natural candidates for involvment in abnormal pituitary growth. However, their potential role in pituitary tumorigenesis has not yet been elucidated

Nurr 77 was identified as a mediator of CRH stimulation of POMC gene transcription. It has also been demonstrated that Nur 77 could antagonise negative the signal exerted by the GR (61). Thus overexpression of Nur77 could lead to glucocortcoid resistance, as observed in corticotroph adenomas. However, Nur77 deficient mouse did not develop pituitary abnormalities (256) and the role of Nur 77 in corticotroph tumors has not yet been fully explored.

Regulators of the cell cycle:

The **Rb1** gene product plays an essential role in regulating the progression of the cell cycle. Progression of the cell cycle occurs when Rb1 is phosphorylated by complexes formed by cyclin and cyclin dependant kinase (257).

The activity of cyclin -CDK complexes is regulated by two families of CDK1 proteins that inhibit cell cycle progression. The INK4 group includes p16, p18, p15, and p19 (258). The kip family comprises p21, **p27**, and p57, all capable of binding most cyclin/CDK complexes. Thus, these agents are negative regulators of the cell cycle (259-261).

Rb1 and P27/Kip1 genes were examined because of the phenotype of KO mice, which develop intermediate lobe pituitary tumors(93-99).

No mutation in the Rb gene was found in corticotroph tumors, at the nucleotide level (262-265), although frequent loss of heterozygosity at the Rb locus was demonstrated. (264-265)

No mutation in the p27kip1 gene and no reduction of the p27mRNA expression have been detected in human corticotroph tumors (266-268). However, a diminution of p27/Kip 1 protein levels was observed in corticotroph tumors, and not in other types of pituitary adenomas. Malignant transformation leads to complete loss of p27 immunoreactuvity (269).

Like P27 and Rb KO mice, **p18** deficient mice develop melanotroph pituitary tumors. Mice lacking both p18 and p27 die from aggressive pituitary tumors (100). However the role of p18 in pituitary tumorsremains to be determined.

P53 has a tumor suppressor activity by blocking the progression of the cell cycle from G1, principally by activating P21, which in turn inhibits certain cyclin dependent kinases. TheP53 gene is considered as the most frequently mutated gene in human malignancy (270).

In one study investigating **p53** status in corticotroph adenomas, an abnormal p53 immunostaining was observed in more than 50% of both invasive and non invasive corticotroph tumors (271). However, p53 was normal at the nucleotide level (272-273), suggesting that the positive p53 immuno staining could be due to an up regulation of the wild type p53.

Overexpression ok the **cyclins** has been shown in many different tumour types The expression of these proteins has also been investigated in pituitary tumors, including corticotroph adenomas.

All tumors studied contained cells that immunostainedpositively for cyclin A, B, D, E whereas cyclin D is not normally expressed inanterior pituitary. The cyclin expression was positively correlated with tumor size. Cyclin E was specifically increased in corticotroph tumors from patients with Cushing's disease, may be in relation with the decrease expression of p27 protein in these tumors (274-275). Allelic inbalance was suggested for cyclin D, but with no correlation with cyclin D expression (276). Thus, although nuclear cyclin expression constitutes a good marker of tumor aggressivity, the signal leading to cyclin expression remains to be identified.

Other tumor suppressor genes:

hZAC. This zinc finger protein with antiproliferative properties has been cloned from pituitary cDNA libraries (277-278). This gene is the human homolog of a transcript previously isolated from AtT20.Its DNA binding activity is compatible with a transcription factor. Functional studies revealedthat antiproliferative properties of hZAC are related to its ability to induce apoptosis and cell cycle arrest in G1, like P53 (277). ZAC expression is lost in ovarian cell lines (278). A decrease expression of ZAC was observed in functionnal pituitary tumors, including corticotroph adenomas, whereasprotein expression was lost in nonfunctioning pituitary adenomas 279) Loss of expression was not associated with a mutation of the ZAC gene.

MEN 1. Corticotroph adenomas are most often presenting as sporadic tumors. However, they can occasionally occur as part of multiple endocrine neoplasia type 1 syndrome (280). The susceptibility gene for MEN 1 has been mapped on chromosome 11q13 (281). This gene product is named menin. Although 10 to 30% of pituitary tumors have been found to have LOH at the MEN 1 locus (282-284), incidence of mutations of the MEN 1 gene in sporadic pituitary adenomas is very low, whatever the cell lineage (285-287). The expression of MEN1 mRNA also appears normal il all tumors, including corticotropinomas (287-288).

Protooncogenes:

Most of protooncogene tested in corticotroph tumors (RET, cMYC, cMYB, cERB2, PKC, RAS, cFOS) have shown no abnormalities, except in the most aggressive macroadenomas, without specificyty for the corticotroph lineage. (cf § 4.2.3) (289-294)

Retinoic acid:

In a recent and rather provocative paper, Stalla's group in Munich (295), show the role of retinoic acid in the control of POMC gene expression and AtT20 cell growth: in this mouse pituitary derived cell line, retinoic acid inhibits both POMC gene transcription in vitro and tumor progression in most mice transplanted with AtT20 cells. This effect is inhibited when the orphan nuclear factor COUP/TF1 is expressed by transfection in the tumor cells. Quite interestingly, these authors also studied the action of retinoic acid in vitro on ACTH secreting pituitary adenomas from Cushing's disease: retinoic acid induces a clear decrease of ACTH secretion. This effect is not observed in normal pituitary cells. This correlates with the absence of COUP-TFI in adenomatous cells, whereas this transcription factor is present in normal cells.

Corticotroph macroadenomas

Although most corticotroph adenomas are small and often difficult to visualize by imaging techniques, a small number are macroadenomas defined as tumor diameter greater than 10 mm (Fig 4) Some of them are large tumors, particular by their aggressive course, frequent neurological complications and relapse after therapeutical attempts. These tumors are characterized by alteration in POMC maturation process, and some particular biochemical and molecular findings.

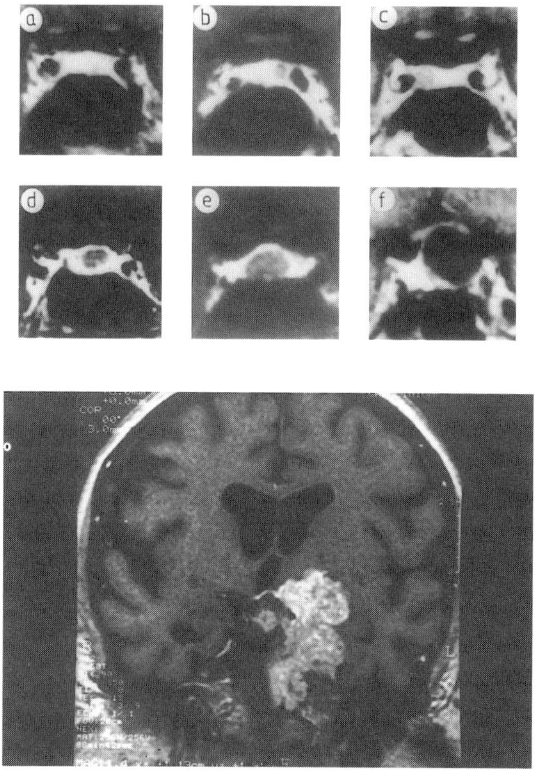

FIGURE 4. Coronal MRI views of the sellar region in different cases of corticotroph adenomas: different cases of microadenomas (upper pannel: a,b,c,d) and macroadenomas (upper pannel: e,f and lower pannel) are shown.

Biochemical features of corticotroph macroadenomas

ACTH and cortisol production tends to be greater in macroadenomas than in microadenomas although some of them are "silent" tumors, devoid of ACTH secretion. Some studies have shown less relative nycthemeral variation of ACTH concentrations, and less suppression of cortisol excretion by dexamethasone. Other pituitary hormonal secretion were more frequently impaired (296, 297). Also, The ACTH and cortisol hyperresponsiveness to hexarelin present in Cushing's disease with microadenomas is not observed in patients with macroadenomas (298).

Thus, in comparison with microadenomas, corticotroph macroadenomas are characterized by a greater and more autonomous ACTH secretion.

ACTH production is quantitatively variable and qualitatively altered

POMC transcription

Corticotroph micro and macroadenomas contain the 1200nt long-pituitary messenger, although at a lower concentrations than in normal corticotroph cells. No abnormality has been observed that would alter the quality of the POMC message, even in non-secreting ("silent") corticotroph pituitary adenomas (299)

Abnormal POMC maturation

POMC processing is often defective in corticotroph macroadenomas. Production of high molecular weight forms of ACTHimmunoreactivity by large pituitary tumors has been reported, sometimes associated with the secretion of small peptides such as CLIP and alpha-MSH, which are absent in the normal pituitary (300-302). More recently, systematic studies evaluated POMC secretion in ACTH secreting tumors, by a direct precursor- (303) or POMC assay (304) in patient plasma. Production of intact POMC was observed in 21 out of 25 corticotroph macroadenomas, whereas it was absent from the plasma of the 60 patients with corticotroph microadenomas (304 and personal datas) (Fig 5).

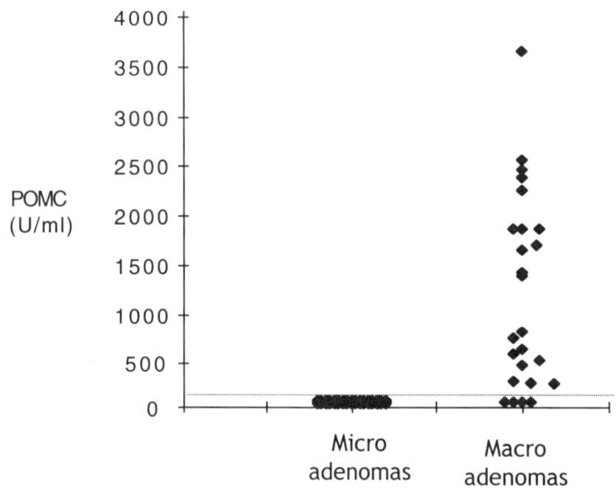

FIGURE 5. Proopiomelanocortin (POMC) plasma levels in patients with corticotroph micro (n=80) and macro (n=25) adenomas. The dashed line indicates the upper limit of the normal range.

It is noteworthy that defective POMC processing is also a hallmark of malignant non-pituitary tumors responsible from the ectopic ACTH syndrome, whereas highly differentiated- ACTH producing bronchial carcinoids do not secrete intact prescursor. Thus a defect in the normal process of maturation in ACTH secreting tumors, is associated with an aggressive course, reflecting the loss of neuroendocrine differentiation of ACTH secreting cells in both pituitary or non pituitary tumors (304).

Molecular defects: special features of aggressive adenomas

The mechanism of cellular proliferation leading to macroadenomas when most corticotroph tumors remain small, is still poorly understood. Patients with a macroadenoma had a significantly higher value of Ki67 index than patients with microadenoma, whereas the apoptotic index was not significantly different (305).

The CRH gene has been shown to be expressed in mouse corticotropic tumor cell and in human corticotroph adenomas. Interestingly, CRH mRNA signal intensity in pituitary corticotroph adenoma cells was linearly correlated with Ki 67 tumor growth fractions. In macroadenomas and locally invasive adenoma cells, it was significantly higher than in microadenoma cells. (199)

Abnormalities of some protooncogenes such as RAS, cMYC, PKC have been associated with aggressive corticotroph tumors, although those findings don't seem to be specifically related to the corticotroph subtype (306-309)

Brn3 belongs to the family of POU transcription factor. It has been found to stimulate POMC transcription and also to exert mitogenic effects. High expression of this factor was associated with aggressive forms of ACTH secreting tumors, both of pituitary and non-pituitary origin (310).

Angiogenesis essential for tumor growth and has been correlated with tumor behavior in a variety of different tumor types. In marked contrast with other tumors, pituitary adenomas (including ACTH secreting adenomas) are less vascular than the normal pituitary gland (311.). However, angiogenesis seems to have a pronostic value as surgical cure was more likely in ACTH secreting tumors with lower microvessel density (312).

Pituitary cancer

Although corticotroph adenomas often show invasive features and sometimes a high mitotic index, the existence of primary pituitary carcinomas is still a subject of debate. It is generally admitted that extra-neural metastases are required for this diagnosis. Several such patients have been reported (313-316) including cases where immunocytochemistry of distant metastases (liver, bone) proved the presence of POMC peptides (316). This situation which is more frequently -but not exclusively- associated with Nelson's syndrome, occurs equally in males and females, and the mean age is relatively low. Caution should be exerted to eliminate the possibility that an occult non-pituitary tumor secreting ACTH has metastasised to different sites, including the pituitary. Because the growth aggressiveness of the pituitary tumor may be the ultimate prognostic factor (317), ways to predict it via the mitotic index, nuclear aneuploidy, or other means are certainly to be searched.

HYPOTHALAMUS-DEPENDENT CUSHING'S DISEASES?

Although it is becoming more and more clear that a vast majority of the patients with Cushing's disease harbor a corticotroph adenoma in their pituitary there are still some cases -including patients from Cushing's monograph of 1932 (318) - where such lesions were apparently lacking. Therefore 2 questions remain to be solved:

- is there a subset of patients with Cushing's disease where pituitary ACTH oversecretion is caused by a primary hypothalamic dysfunction creating simple diffuse corticotroph hyperplasia ?

- in the vast majority of patients who harbor a pituitary corticotroph adenoma, does the lesion result from prior hyperplasia caused by a primary hypothalamic dysfunction ?

A subset of patients with Cushing's disease due to primary hypothalamic dysfunction?

Abnormalities in the quantitative and qualitative aspects of various CNS and anterior pituitary functions have been interpreted as evidence for a primary (and general?) hypothalamic dysfunction that would include CRF overproduction with subsequent ACTH oversecretion. Classically cited are the loss of normal ACTH circadian rhythm, the decreased suppressive effect of glucocorticoids, altered secretory patterns of GH, TSH, PRL and gonadotropins, and modified sleep EEG patterns (319-321). A detailed statistical analysis- performed retrospectively on already published studies- showed that the distribution of parameters of plasma cortisol fluctuations was compatible with the existence of 2 populations of patients with Cushing's disease : it was hypothesized that highly fluctuating cortisol concentrations were of hypothalamic origin, low fluctuations of primary pituitary origin (322). Also in favor of a primary hypothalamic disorder were the apparent- and sometimes persistent- cure of some patients treated with drugs acting directly at the CNS such as the serotonin antagonist cyproheptadine and the gabaergic sodium valproate (320,321).

All these evidences can be seriously challenged. Many of them are simply attributable to the hypercortisolic state: altered EEG patterns, abnormal GH, PRL, gonadotropin, and TSH secretory patterns all return to normal when the source of hormone excess is removed (320,323-325). The beneficial effet of CNS directed drug is debated particularly because it was based on uncontrolled studies: subsequent elegant studies have shown that many Cushing's diseases have spontaneous fluctuations with long lasting periods of apparent total remission... whether or not the patients are treated with CNS directed or dopaminergic drugs (326).

Corticotroph cell hyperplasia has been reported in pituitaries of patients with Cushing's disease (327-331). Because CRF exerts a growth stimulatory action on corticotroph cells in animals (332, 333) and because corticotroph cell hyperplasia has been observed in pituitaries obtained from patients with Cushing's syndrome resulting from chronic CRF oversecretion by hypothalamic (334) or ectopic (335, 336) tumors, this histological finding is generally taken as a strong argument for CRF involvement (337). Establishing this histological diagnosis is of utmost difficulty (338). It should not be accepted when obtained on limited surgical materials: some adenoma may escape surgical exploration because of an ectopic location, and the difficult histological diagnosis of corticotroph hyperplasia certainly requires that the whole gland be thoroughly examined since the corticotroph cells are not randomly scattered in the gland, and rather show clusters of densely aggregated cells (339, 340). If this histological diagnosis can be demonstrated under the necessary scrutiny in a patient with the clinical and biological features of Cushing's disease then it probably offers the best evidence of a CRF dependence. Yet it would still need to be proven that CRF is of hypothalamic origin, -and not from an occult ectopic source- and if so, that it is not merely a functional and transient hypothalamic CRF dysfunction associated with disorders like depression, chronic stress, general resistance to glucocorticoïds...

Even the finding of corticotroph cell hyperplasia is not a definitive proof that it is CRF-mediated. Evidence has been provided for the existence of primary multinodular corticotroph hyperplasia (341). Thus the ultimate proof still requires the unambiguous demonstration that hypothalamic CRF is indeed overproduced. Plasma CRF is normal or low in patients with Cushing's disease, although the significance of these peripheral plasma values may not be absolutely relevant (342,343). Corticotrophin-releasing factor rather appears to be low in the CSF of patients with Cushing's disease (344), in contrast with depressed patients (345, 346). Although CRF analogs which exert an antagonist activity have been used in the rat (347), such a molecule remains to be built for human studies. It certainly would constitute a unique heuristic tool and the sole way to definitively establish a pathophysiological role for CRF in Cushing's disease.

Corticotroph cell hyperplasia preceeding the adenoma?

That hyperplasia often precedes the formation of a clonal tumor has long been recognized in animal models and in man, and brilliantly demonstrated by the recent use of powerful transgenic animal models (348). Site directed expression of the SV40T antigen in the mouse pancreatic ß cell first induces hyperplasia of the endocrine cells (349). As a result of a second and spontaneous event, an occasional clonal ß cell tumor develops. It is thought that this experimental model -somehow- reproduces the long time proposed 2 steps theory of Knudson for spontaneous oncogenesis (350). In man, the familial form of medullary thyroid cancer (Sipple syndrome) provides the best example of a naturally occuring endocrine cancer where tumoral transformation is preceded by a state of general hyperplasia of C-cells (351). The molecular mechanisms responsible for the different steps have been unravelled in privileged situations in man, such as the retinoblastoma. In most other cases they are still unknown. The local expression of angiogenic compounds may be the tumor promoting factor (352). On these grounds it is interesting to remember that a recently characterized angiogenic factor, vascular endothelial growth factor or vasculotropin, was isolated from pituitary folliculo-stellate cells (353, 354) and the mouse corticotroph cell line AtT-20 (355).

More and more the pituitary is becoming the target of site-directed oncogenesis in various transgenic animals. Transgenic mice expressing the human GRF gene develop selective hyperplasia of the GH cells (356, 357). It will be of major interest to observe if authentic tumors eventually develop with time. Various genes have been targeted to pituitary corticotroph cells in transgenic mice using transgenes placed under the promoter of the rat POMC gene (358, 359). Expression of SV40T antigen provoked marked developement of both anterior pituitary corticotrophs and intermediate-lobe melanotrophs (358); whether true clonal tumors secondarily develop is not yet known.

In man, the evidence that hyperplasia precedes the formation of a pituitary adenoma is still scant: although pregnancy is a condition with prolactin cell hyperplasia, no association is clearly found to suggest that it is a risk factor for subsequent development of a prolactin secreting tumor. The ectopic GRF syndrome is almost always associated with GH cell hyperplasia when pituitary examination is performed (360). Similarly pituitary enlargement occurring in states of long lasting peripheral hormone deprivation (hypothyroidism, hypogonadism) is, as a rule, associated with hyperplasia and reverses under adequate treatment. Whether true tumors have really developed under such conditions remains to be unequivocally established.

The odds that primary CRF hyperactivity may be responsible for corticotroph cell hyperplasia secondarily initiating the formation of an adenoma hold on theoretical, histological and clinical grounds: 1) CRF exerts a growth stimulatory action on corticotroph cells: 2) associations of a corticotroph adenoma and corticotroph hyperplasia have been found (361-364): 3) an increased number of stressful events have been found in patients with Cushing's disease (365). It suggested that prolonged CRF overactivity eventually led to the formation of the adenoma. Clinical experience does not indicate that depressed patients are particularly at risk for Cushing's disease. More interesting are the patients whose disease recur after "successful" removal of a pituitary adenoma (366, 367). It is speculated that the original hypothalamic CRF overactivity had been transiently silenced by the hypercortisolism, and recovers its activity after removal of the pituitary adenoma. In fact, there is simply no way to be sure that the recurence is not due to the regrowth of a small amount of tumoral cells which had escaped the surgeon's skill and which can be removed by a second pituitary surgery (368, 369).

Thus, although such a mechanism remains a possibility (Fig.6), there is certainly at the present time no definitive proof that corticotroph cell hyperplasia is a prerequisite condition leading to the formation of a pituitary adenoma. Progress in this field might come from new experimental approaches generating chronic pituitary stimulation by CRF in animals (370) and whenever an anti-CRF analog will be available for clinical studies.

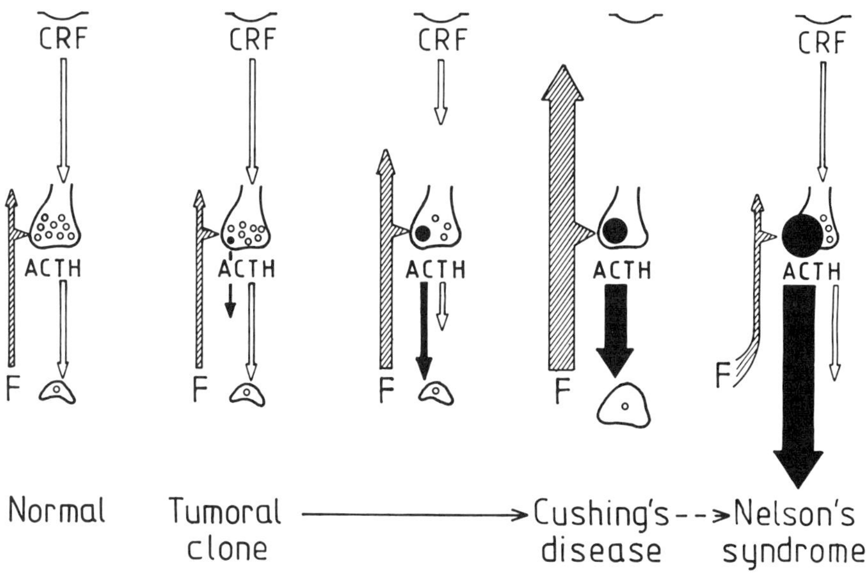

FIGURE 6. Pathogenesis of Cushing's disease. Schematic view of a tentative pituitary hypothesis. ACTH: adrenocorticotropic hormone, CRF: corticotropin releasing factor, F: cortisol.

CONCLUSIONS

In spite of numerous studies, the pathogenesis of corticotroph adenomas remains puzzling. Animal models are invaluable tools for characterization of the pathways allowing physiological HPA development and regulation, but do not provide a good model for human Cushing's disease. Many molecular studies have investigated the most obvious candidate genes of the corticotroph specific regulatory pathway or pituitary development. However, if molecular abnormalities have been described in a small number of tumors, there is no clear evidence of a common defect. Moreover, these abnormalities could be the consequences of corticotroph proliferation or hypercortisolism rather than the cause.

Some points should however be underlined:

- the monoclonal origin of corticotroph adenomas suggests the role of an initial molecular event.

- the relative resistance to glucocorticoid negative feedback is the hallmark of corticotroph tumors.

Thus a mutational abnormality occuring on glucocorticoid inhibitory pathway might give rise to a cell resistant to glucocorticoid feedback, with a growth advantage over the surrounding cells. This first hit could be eventually favoured by preexisting proliferative signals.

Additional – variable – hits might generate more rapidly growing and/or more invasive subclones.

Among the different hypersecretory adenomas that develop in the pituitary, a molecular explanation has been found in a minority of GH secreting adenomas bearing activating mutations of Gsalpha, and in the rare MEN1 adenomas. For all other adenomas, including Cushing's disease, the mystery still remains.

REFERENCES

Pathophysiology, epidemiology

1. Huff TA. Clinical syndromes related to disorders of adrenocorticotrophic hormone. In: Allen,M.B., and Makesh, V.B. eds. The Pituitary : A current review. New York: Academic Press, 1977.p.153-168.

2. Baxter JD, Tyrrell JB. The adrenal cortex. In: Felig P, Baxter JD, Broadus AE, Frohman LA, eds. Endocrinology and Metabolism. New York: McGraw Hill Book Company, 1981.p.385-510.
3. Plotz CM, Knowlton AL,Ragan C. The natural history of Cushing's syndrome. Am J Med 1952;13:597-614.
4. Neville AM, O'Hare MJ. Aspects of structure, function and pathology in James VHT ed. The Adrenal Gland. In: New York, Raven Press. 1979.p.1-66.

Corticotroph cell differentiation

5. Japon MA, Rubinstein M, Low MJ. 1994 In situ hybridization analysis of anterior pituitary hormone gene expression during fetal mouse development. J Histochem Cytochem.;42:1117-25.
6. Sornson M.W., Wu W., Dasen J.S., Flynn S.E., Norman D.J., O'Connell S.M., Gukovsky I., Carriere C., Ryan A.K. and Miller A.P. et al. 1996 Pituitary lineage determination by the Prophet of Pit-1 homeodomain factor defective in Ames dwarfism. Nature, 384:327-333.
7. Bodner M., Castrillo J.L., Theill L.E., Deerinck T., Ellisman M. and Karin M. 1988 The pituitary-specific transcription factor GHF-1 is a homeobox-containing protein. Cell, 55:505-518.
8. Ingraham H.A., Chen R., Mangalam H.J., Elsholtz H.P., Flynn S.E., Lin C.R., Simmons D.M., Swanson L. and Rosenfeld M.G. (1988) A tissue-specific transcription factor containing a homeodomain specifies a pituitary phenotype. Cell, 55:519-529.
9. Ingraham H.A., Lala D.S., Ikeda Y., Luo X., Shen W.H., Nachtigal M.W., Abbud R., Nilson J.H. and Parker K.L. (1994) The nuclear receptor steroidogenic factor 1 acts at multiple levels of the reproductive axis. Genes Dev., 8:2302-2312.
10. Ikeda Y., Luo X., Abbud R., Nilson J.H. and Parker K.L. (1995) The nuclear receptor steroidogenic factor 1 is essential for the formation of the ventromedial hypothalamic nucleus. Mol. Endocrinol., 9:478-486.
11. Kioussi C., Carriere C. and Rosenfeld M.G. (1999) A model for the development of the hypothalamic-pituitary axis: transcribing the hypophysis. Mech. Dev., 81:23-35.
12. Sheng H.Z. and Westphal H. (1999) Early steps in pituitary organogenesis. Trends Genet., 15:236-240.
13. Sheng H.Z., Moriyama K., Yamashita T., Li H., Potter S.S., Mahon K.A. and Westphal H. (1997) Multistep control of pituitary organogenesis. Science, 278:1809-1812.
14. Gage P.J. and Camper S.A. (1997) Pituitary homeobox 2, a novel member of the bicoid-related family of homeobox genes, is a potential regulator of anterior structure formation. Hum. Mol. Genet., 6:457-464.
15. Gage P.J., Suh H. and Camper S.A. (1999) The bicoid-related Pitx gene family in development. Mamm. Genome, 10:197-200.
16. I Netchine, ML Sobrier, H Krude, D Schnabe, M Maghnie, E Marcos, B Duriez, V Cacheux, A v. Moers, M Goossens, A Grüters, S Amselem (2000), Mutations in LHX3 result in a new syndrome revealed by combined pituitary hormone deficiency. Nat Genet, 25 , 182 - 186.
17. Akita S, Webster J, Ren SG, Takino H, Said J, Zand O, Melmed S 1995 Human and murine pituitary expression of leukemia inhibitory factor. Novel intrapituitary regulation of adrenocorticotropin hormone synthesis and secretion. J Clin Invest 95:1288-1298.
18. Lanctôt C., Gauthier Y. and Drouin J. (1999) Pituitary homeobox 1 (Ptx1) is differentially expressed during pituitary development. Endocrinology, 140:1416-1422.
19. Lanctôt C., Moreau A., Chamberland M., Tremblay M.L. and Drouin J. (1999) Hindlimb patterning and mandible development require the Ptx1 gene. Development, 126:1805-1810.

20. Tremblay J.J., Lanctôt C. and Drouin J. (1998) The pan-pituitary activator of transcription, Ptx-1 (pituitary homeobox1), acts in synergy with SF-1 and Pit1 and is an upstream regulator of the Lim-homeodomain gene Lim3/Lhx3. Mol. Endocrinol., 12:428-441.
21. Lamonerie T, Tremblay JJ, Lanctot C, Therrien M, Gauthier Y, Drouin J 1996 Ptx1, a bicoid-related homeo box transcription factor involved in transcription of the pro-opiomelanocortin gene. Genes Dev.10:1284-1295.
22. Ericson J, Norlin S, Jessel TM, Edlund T. 1998. Intregrated FGF and BMP signaling controls the progression of progenitor cell differenciation and the emergence of pattern in the embryonic anterior pituitary. Development. 125, 1005-1015.
23. Treier M, Gleiberman AS, O'Connell SM, Szeto DP, McMahon JA, McMahon AP, Rosenfeld MG. 1998, Multistep signaling requirements for pituitary organogenesis in vivo. Gene & Dev. 10: 1284-1295.
24. Yano H, Readhead C, Nakashima M, Ren SG, Melmed S 1998. Pituitary-directed leukemia inhibitory factor transgene causes Cushing's syndrome: neuro-immune-endocrine modulation of pituitary development. Mol Endocrinol 12:1708-1720.
25. Lee J.E., Hollenberg S.M., Snider L., Turner D.L., Lipnick N. and Weintraub H. (1995) Conversion of Xenopus ectoderm into neurons by NeuroD, a basic helix-loop-helix protein. Science, 268:836-844.
26. Poulin G., Turgeon B. and Drouin J. (1997) NeuroD1/BETA2 contributes to cell-specific transcription of the POMC gene. Mol. Cell. Biol.,17:6673-6682.
27. B Lamolet , AM Pulichino, T Lamonerie, Y Gauthier, T Brue, A Enjalbert, J Drouin, 2001, A Pituitary Cell-Restricted T Box Factor, Tpit, Activates POMC Transcription in Cooperation with Pitx Homeoproteins Cell, 104, 849-859.

ACTH biosynthesis

28. Smith AI,Funder JW. Proopiomelanocortin processing in the pituitary, central nervous system and peripheral tissues. Endocr Rev 1988;9:159-179.
29. Seidah N, Rochemont J, Hamelin J, Lis M,Chretien M. Primary structure of the major human pituitary proopiomelanocortin NH2-terminal glycopeptide. Evidence for an aldosterone-stimulating activity. J Biol Chem 1981;256:7977-7984.
30. Estivariz E, Hope J, McLean C,Lowry PJ. Purification and characterization of a gamma-melanotropin precursor from frozen human pituitary glands. Biochem J 1980;191:125-132.
31. Seidah NG, Rochemont J, Hamelin J, Benjannet S,Chretien M. The missing fragment of the pro-sequence of human pro-opiomelanocortin: sequence and evidence for C-terminal amidation. Biochem Biophys Res Commun 1981;102:710-716.
32. Bertagna X, Camus F, Lenne F, Girard F,Luton JP. Human Joining Peptide : a Proopiomelanocortin product secreted as a homodimer. Mol Endocrinol 1988;2:1108-1114.
33. Fenger M, Johnsen HA. Alpha-Amidated peptides derived from Proopiomelanocortin in normal human pituitary. Biochem J 1988;250:781-788.
34. Scott AP, Lowry PJ. Adrenocorticotrophic and melanocyte stimulating peptides in the human pituitary. Biochem J 1974;139:593-602.
35. Tanaka N, Abe K, Miyakawa S, Ohnami S, Tanaka M,Takeuchi T. Analysis of human pituitary and tumor adrenocorticotropin using isoelectric focusing. J Clin Endocrinol Metab 1979;48:559-565.
36. Allen RG, Orwoll E, Kendall JW,Herbert E. The distribution of forms of adrenocorticotropin and ß-endorphin in normal, tumorous and autopsy human anterior pituitary tissues ; virtual absence of 13K adrenocorticotropin. J Clin Endocrinol Metab 1980;51:376-380.
37. Tanaka K, Nicholson WE,Orth DN. The nature of immunoreactive lipotropins in human plasma and tissue extracts. J Clin Invest 1978;62:94-104.

38. Orth DN, Nicholson WE. High molecular weight forms of human ACTH are glycoproteins. J Clin Endocrinol Metab 1977;44:214-217.
39. Bradbury AF, Finnie MDA, Smyth DG. Mechanism of C-terminal amide formation by pituitary enzymes. Nature 1982;298:686-688.
40. Eipper BA, Park L, Keutmann HT,Mains R. Amidation of Joining Peptide, a major Pro-ACTH/Endorphin derived product peptide. J Biol Chem 1986;261:8686-8694.
41. Bennet HP, Browne CA,Solomon S. Biosynthesis of phosphorylated forms of corticotropin related peptides. Proc Natl Acad Sci USA 1981;78:4713-4717.
42. Eipper BA, Mains RE. Phosphorylation of proadrenocorticotropin/endorphin derived peptides. J Biol Chem 1982;257:4907-4915.
43. Scott AP, Lowry PJ, Ratcliffe JG, Rees LH,Landon J. Corticotropin-like peptides in the rat pituitary. J Endocrinol 1974;61:355-367.
44. Krieger DT. The multiple faces of proopiomelanocortin, a prototype precursor molecule. Clin Res 1983;31:342-353.
45. Silman RE, Chard T, Lowry PJ, Smith I,Young IM. Human foetal pituitary peptides and parturition. Nature 1976;260:716-718.
46. Chen CL, Mather JP, Morris PL,Bardin CW. Expression of pro-opiomelanocortin-like gene in the testis and epididymis. Proc Natl Acad Sci USA 1984;81:5672-5675.
47. Pintar JE, Shacter BS, Herman AB, Durgerian S,Krieger DT. Characterization and localization of pro-opiomelanocortin messenger RNA in the adult rat testis. Science (Wash DC) 1984;225:632-634.
48. Jeannotte L, Burbach JPH, Drouin J. Unusual proopiomelanocortin ribonucleic acids in extrapituitary tissues; intronless transcripts in testes and long poly (A)+ tails in hypothalamus. Mol Endocrinol 1987;1:749-757.
49. Lacaze-Masmonteil T, de Keyzer Y, Luton JP, Kahn A,Bertagna X. Characterization of Proopiomelanocortin transcripts in human non-pituitary tissues. Proc Natl Acad Sci USA 1987;84:7261-7265.
50. Jingami H, Nakanishi S, Imura H, Numa S. Tissue distribution of messenger RNAs coding for opioid peptide precursors and related RNA. Eur J Biochem 1984;142:441-447.
51. De Bold CR, Nicholson WE, Orth DN. Immunoreactive proopiomelanocortin (POMC) peptides and POMC-like messenger ribonucleic acid are present in many rat nonpituitary tissues. Endocrinology 1988;122:2648-2657.
52. Clark AJL, Lavender PM, Coates P, Johnson MR,Rees LH. In Vitro and in Vivo analysis of the processing and fate of the peptide products of the short proopiomelanocortin mRNA. Mol Endocrinol 1990;4:1737-1743.
53. Buzetti R, Mc Loughlin L, Lavender PM, Clark AJL,Rees LH. Expression of pro-opiomelanocortin gene and quantification of adrenocorticotrophic hormone-like immunoreactivity in human normal peripheral mononuclear cells and lymphoid and myeloid malignancies. J Clin Invest 1989;83:733-737.
54. De Keyzer Y, Bertagna X, Lenne F, Girard F, Luton JP,Kahn A. Altered proopiomelanocortin gene expression in ACTH producing non-pituitary tumors. J Clin Invest 1985;76:1892-1898.
55. Vieau D, Massias JF, Girard F, Luton JP, Bertagna X. Corticotrophin-like intermediary lobe peptide as a marker of alternate pro-opiomelanocortin processing in ACTH-producing non-pituitary tumours. Clin Endocrinol 1989;31:691-700.
56. Ratter SJ, Gillies G, Hope H, et al. Pro-opiocortin related peptides in human pituitary and ectopic ACTH secreting tumors. Clin Endocrinol 1983;18:211-218.

Nuclear factors involved in POMC transcription

57. Poulin, G, Turgeon, B, Drouin, J. (1997) NeuroD1/beta2 contributes to cell-specific transcription of the proopiomelanocortin gene. *Mol Cell Biol* 17, 6673-6682.

58. Lamonerie, T., Tremblay, J. J., Lanctôt, C., Therrien, M., Gauthier, Y., and Drouin, J. (1996) Ptx1, a bicoid-related homeo box transcription factor involved in the transcription of the pro-opiomelanocortin gene. *Genes Dev* 10, 1284-1295.
59. Lamolet, B., Pulichino, A., Lamonerie, T., Gauthier, Y., Brue, T., Enjalbert, A., and Drouin, J. (2001) A pituitary cell-restricted T box factor, Tpit, activates POMC transcription in cooperation with Pitx homeoproteins. *Cell* 104, 849-859.
60. Poulin, G., Lebel, M., Chalmberland, M., Paradis, F., and Drouin, J. (2000) Specific protein protein interaction between basic helix-loop-helix transcription factors and homeoproteins of the Pitx family. *Mol. Cell. Biol* 20, 4826-4837.
61. Murphy, E. P., and Conneely, O. M. (1997) Neuroendocrine regulation of the hypothalamic pituitary axis by the nurrl/nurr77 subfamily of nuclear receptors. *Mol. Endocrinol* 16, 39-47.
62. Philips, A., Lesage, S., Gingras, R., Maira, M.-H., Gauthier, Y., Hugo, P., and Drouin, J. (1997a) Novel dimeric Nur77 signaling mechanism in endocrine and lymphoid cells. *Mol. Cell. Biol* 17, 5946-5951.
63. Eberwine, J. H., Jonassen, J. A., Evinger, M. J. Q., and Roberts, J. L. (1987) Complex transcriptional regulation by glucocorticoids and corticotropin-releasing hormone of proopiomelanocortin gene expression in rat pituitary cultures. *DNA* 6, 483-492.
64. Gagner, J., and Drouin, J. (1987) Tissue-specific regulation of pituitary proopiomelanocortin gene transcription by corticotropin-releasing hormone, 3',5' -cyclic adenosine monophosphate, and glucocorticoids. *Mol. Endocrinol* 1, 677-682.
65. Loeffler, J. P., Kley, N., Pittius, C. W., and Höllt, V. (1985) Corticotropin-releasing factor and forskolin increase proopiomelanocortin messenger RNA levels in rat anterior and intermediate cells in vitro. *Neurosci. Lett* 62, 383-387.
66. Maira, M., Martens, C., Philips, A., and Drouin, J. (1999) Heterodimerization between members of the Nur subfamily of orphan nuclear receptors as a novel mechanism for gene activation. *Mol. Cell. Biol* 19, 7549-7557.
67. Philips, A., Maira, M., Mullick, A., Chamberland, M., Lesage, S., Hugo, P., and Drouin, J. (1997b) Antagonism between Nur77 and glucocorticoid receptor for control of transcription. *Mol. Cell. Biol* 17, 5952-5959.
68. Drouin, J., Sun, Y. L., Chamberland, M., Gauthier, Y., De Léan, A., Nemer, M., and Schmidt, T. (1993) Novel glucocorticoid receptor complex with DNA element of the hormone-repressed POMC gene. *EMBO J* 12, 145-156.
69. Reichardt, H. M., Kaestner, K. H., Tuckermann, J., Kretz, O., Wessely, O., Bock, R., Gass, P., Schmid, W., Herrlich, P., Angel, P., and Schütz, G. (1998) DNA binding of the glucocorticoid receptor is not essential for survival. *Cell* 93, 531-541.
70. Ray, D. W., Ren, S. G., and Melmed, S. (1996) Leukemia inhibiting factor (LIF) stimulates proopiomelanocortin (POMC) expression in a corticotroph cell line. *J. Clin. Invest* 97, 1852-1859.
71. Bousquet C, Zatelli, M., and Melmed, S. (2000) Direct regulation of pituitary proopiomelanocortin by STAT3 provides a novel mechanism for immuno-neuroendocrine interfacing. *J. Clin. Invest* 106, 14171425.

Transgenic mice

72. Stenzel-Poore MP, Cameron VA, Vaughan J, Sawchenko PE, Vale W 1992 Development of Cushing's syndrome in corticotropin-releasing factor transgenic mice. Endocrinology 130:3378-3386.
73. Stenzel-Poore MP, Heinrichs SC, Rivest S, Koob GF, Vale WW 1994 Overproduction of corticotropin-releasing factor in transgenic mice: a genetic model of anxiogenic behavior. J Neurosci 14:2579-2584.
74. Cole TJ, Blendy JA, Monaghan AP, Krieglstein K, Schmid W, Aguzzi A, Fantuzzi G, Hummler E, Unsicker K, Schutz G. 1995 Targeted disruption of the glucocorticoid

receptor gene blocks adrenergic chromaffin cell development and severely retards lung maturation. Genes Dev;9:1608-21.
75. Reichardt HM, KH Kaestner, J Tuckermann, O Kretz, O, Wessely, R Bock, P Gass, P Angel, G Schutz. 1998. DNA binding of the glucocorticoid receptor is not essential for suvival. Cell. 93, 531-541.
76. Tronche F, Kellendonk C, Kretz O, Gass P, Anlag K, Orban PC, Bock R, Klein R, Schutz G. 1999. Disruption of the glucocorticoid receptor gene in the nervous system results in reduced anxiety. Nat Genet. 23, 99-103.
77. Saiardi A, Borrelli E. 1998. Absence of dopaminergic control on melanotrophs leads to Cushing's-like syndrome in mice. Mol Endocrinol;12:1133-9.
78. Kelly MA, Rubinstein M, Asa SL, Zhang G, Saez C, Bunzow JR, Allen RG, Hnasko R, Ben-Jonathan N, Grandy DK, Low MJ. 1997 Pituitary lactotroph hyperplasia and chronic hyperprolactinemia in dopamine D2 receptor-deficient mice. Neuron;19:103-13.
79. Braks and Martens, 1994, 7B2 is a neuroendocrine chaperone that transiently interacts with prohormone convertase PC2 in the secretory pathway. Cell 29;78:263-73.
80. Benjannet S, Savaria D, Chretien M, Seidah NG 1995 7B2 is a specific intracellular binding protein of the prohormone convertase PC2. J Neurochem;64(5):2303-11.
81. Zhu X, Lindberg I. 1995 7B2 facilitates the maturation of proPC2 in neuroendocrine cells and is required for the expression of enzymatic activity.J Cell Biol;129(6):1641-50.
82. Muller L, Zhu X, Lindberg .Mechanism of the facilitation of PC2 maturation by 7B2: involvement in ProPC2 transport and activation but not folding. J Cell Biol 1997 ;139:625-38.
83. Westphal CH, Muller L, Zhou A, Zhu X, Bonner-Weir S, Schambelan M, Steiner DF, Lindberg I, Leder P. 1999 The neuroendocrine protein 7B2 is required for peptide hormone processing in vivo and provides a novel mechanism for pituitary Cushing's disease. Cell 5;96(5):689-700.
84. Furuta M, Zhou A, Webb G, Carroll R, Ravazzola M, Orci L, Steiner DF. 2001. Severe defect in proglucagon processing in islet A-cells of prohormone convertase 2 null mice. Biol Chem. 276(29):27197-202.
85. Low MJ, Liu B, Hammer GD, Rubinstein M, Allen RG. 1993 Post-translational processing of proopiomelanocortin (POMC) in mouse pituitary melanotroph tumors induced by a POMC-simian virus 40 large T antigen transgene. J Biol Chem; 268:24967-75.
86. Elledge SJ 1996 Cell cycle checkpoints: preventing an identity crisis. Science 274:1664-672.
87. Kaufmann WK, Paules RS 1996 DNA damage and cell cycle checkpoints. FASEB J 10:238-247.
88. Murray AW 1995 The genetics of cell cycle checkpoints. Curr Opin Genet.Dev 5:5-11.
89. Kato JY, Matsuoka M, Polyak K, Massague J, Sherr CJ 1994 Cyclic AMP-induced G1 phase arrest mediated by an inhibitor (p27Kip1) of cyclin-dependent kinase 4 activation. Cell 79:487-496.
90. Polyak K, Lee MH, Erdjument-Bromage H, Koff A, Roberts JM, Tempst P, Massague J 1994 Cloning of p27Kip1, a cyclin-dependent kinase inhibitor and a potential mediator of extracellular antimitogenic signals. Cell.78:59-66.
91. Polyak K, Kato JY, Solomon MJ, Sherr CJ, Massague J, Roberts JM, Koff A. 1994 p27Kip1, a cyclin-Cdk inhibitor, links transforming growth factor-§ and contact inhibition to cell cycle arrest. Genes Dev. 8:9-22.
92. Dahia PLM, Aguiar RCT, Honegger J, Fahlbush R, Jordan S, Lowe DG, Lu X, Besser GM, Grossman AB 1998 Mutation and expression analysis of the p27/kip1 gene in corticotropin-secreting tumours. Oncogene. 16:69-76.
93. Fero ML, Rivkin M, Tasch M, Porter P, Carow CE, Firpo E, Polyak K, Tsai LH, Broudy V, Perlmutter RM, Kaushansky K, Roberts JM 1996 A syndrome of multiorgan

hyperplasia with features of gigantism, tumorigenesis, and female sterility in p27(Kip1)-deficient mice. Cell 85:733-744.
94. Kiyokawa H, Kineman RD, Manova-Todorova KO, Soares VC, Hoffman ES, Ono M, Khanam D, Hayday AC, Frohman LA, Koff A 1996 Enhanced growth of mice lacking the cyclin-dependent kinase inhibitor function of p27(Kip1). Cell 85:721-732.
95. Nakayama K, Ishida N, Shirane M, Inomata A, Inoue T, Shishido N, Horii I, Loh DY, Nakayama K 1996 Mice lacking p27(Kip1) display increase body size, multiple organ hyperplasia, retinal dysplasia, and pituitary tumors. Cell 85:707-720.
96. Jacks T, Fazeli A, Schmitt EM, Bronson RT, Goodell MA, Weinberg RA. 1992 Effects of an Rb mutation in the mouse. Nature;359(6393):295-300.
97. Hu N, Gutsmann A, Herbert DC, Bradley A, Lee WH, Lee EY. 1994 Heterozygous Rb-1 delta 20/+mice are predisposed to tumors of the pituitary gland with a nearly complete penetrance. Oncogene;9(4):1021-7.
98. Harvey M, Vogel H, Lee EY, Bradley A, Donehower LA. 1995 Mice deficient in both p53 and Rb develop tumors primarily of endocrine origin. Cancer Res;55(5):1146-51.
99. Park MS, Rosai J, Nguyen HT, Capodieci P, Cordon-Cardo C, Koff A. 1999. P27 and Rb are on overlapping pathways suppressing tumorigenesis in mice. PNAS. 96 : 6382-6387.
100. Franklin DS, Godfrey VL, Lee H, Kovalev GI, Schoonhoven R, Chen-Kiang S, Su L, Xiong Y 1998 CDK inhibitors p18 (INK4c) and p27 (Kip1) mediate two separate pathways to collaboratively suppress pituitary tumorigenesis. Genes Dev 12:2899-2911.
101. Helseth A, Siegal GP, Haug E, Bautch VL. 1992 Transgenic mice that develop pituitary tumors. A model for Cushing's disease. Am J Pathol;140(5):1071-80.
102. Helseth A, Haug E, Nesland JM, Siegal GP, Fodstad O, Bautch VL 1995 Endocrine and metabolic characteristics of polyoma large T transgenic mice that develop ACTH-producing pituitary tumors. J Neurosurg; 82:879-85.

ACTH secretion in Cushing's disease
103. Tyrrell JB, Brooks RM, Fitzgeral PA, Cofoid PB, Forsham PH,Wilson CW. Cushing's disease : selective transsphenoidal resection of pituitary adenomas. N Engl J Med 1978;298:753-758.
104. Salassa RM, Laws ER, Carpenter PC,Northcutt RC. Transsphenoidal removal of pituitary microadenoma in Cushing's disease. Mayo Clin Proc 1978;53:24-28.
105. Guilhaume B, Bertagna X, Thomsen M, et al. Transsphenoidal pituitary surgery for the treatment of Cushing's disease : results in 64 patients and long term follow-up studies. J Clin Endocrinol Metab 1988;66:1056-1064.
106. Mampalam TJ, Tyrrell JB,Wilson CB. Transsphenoidal microsurgery for Cushing disease: a report of 216 cases. Ann Intern Med 1988;109:487-493.
107. Carpenter PC. Cushing's syndrome : update of diagnosis and management. Mayo Clin Proc 1986;61:49-58.
108. Ludecke DK. Present status of surgical treatment of ACTH-secreting pituitary adenomas in Cushing's disease. In: Lamberts SWJ, Tilders FJH, Van Der Veen EA, Assies J, eds. Trends in diagnosis and treatment of pituitary adenomas. Amsterdam: Free University Press, 1984.p.315-323.
109. Hardy J. Cushing's disease - 50 years later. Can J Neurol Sci 1982;9:375-381.
110. Fahlbusch R, Bushfelder M,Muller OA. Transsphenoidal surgery for Cushing's disease. J R Soc Med 1986;79:262-269.
111. Nakane T, Kuwayama A, Watanabe M, et al. Long term results of transsphenoidal adenomectomy in patients with Cushing's disease. Neurosurgery 1987;21:218-222.
112. Krieger DT. Physiopathology of Cushing's disease. Endocr Rev 1983;4:22-43.
113. Robert F, Pelletier G,Hardy J. Pituitary adenomas in Cushing's disease. Arch Pathol Lab Med 1978;102:448-455.

114. Kovacs K,Horvath E. Pathology of pituitary tumors. Endocrinol Metab Clin North Am 1987;16:529-551.
115. Schteingart DE, Chandler WF, LLoyd RV,Ibarra-Perez G. Cushing's syndrome caused by an ectopic pituitary adenoma. Neurosurgery 1987;21:223-227.
116. Kammer J,George R. Cushing's disease in a patient with an ectopic pituitary adenoma. JAMA 1981;246:2722-2724.
117. Duoskova J,Putz Z. An intracranial tumor in an unusual location associated with Cushing's syndrome. Cesk Patol 1985;21:113-117.
118. Burch WM, Kramer RS, Kenan PD,Hammond CB. Cushing's disease caused by an ectopic pituitary adenoma within the sphenoid sinus. N Engl J Med 1985;312:587-588.
119. Axiotis CA, Lippes HA, Merino MJ, DeLanerolle NC, Stewart AF,Kinder B. Corticotroph cell pituitary adenoma within an ovarian teratoma. A new cause of cushing's syndrome. Am J Surg Pathol 1987;11:218-224.
120. Olivier L, Vila-Porcile , Peillon F, Racadot J. Etude en microscopie electronique des grains de secrétion basophiles dans les cellules hypophysaires tumorales de la maladie de Cushing. C R Soc Biol 1972;166:1591-1600.
121. Crooke AC. A change in the basophil cells of the pituitary gland common to conditions which exhibit the syndrome attributed to basophil adenoma. J Path Bact 1935;41:339-349.
122. Porcile E,Racadot J. Ultrastructure des cellules de Crooke observées dans l'hypophyse humaine au cours de la maladie de Cushing. C R Acad Sci (Paris) 1966;263:948-951.
123. Phifer RF, Spicer SS,Orth DN. Specific demonstrations of the human hypophyseal cells which produce adrenocorticotrophic hormone. J Clin Endocrinol Metab 1970;31:347-361.
124. Olivier L, Vila-Porcile E, Dubois MP, Racadot J, Peillon F. Histological and cytological aspects of pituitary adenomas in Cushing's disease. In: Derome PJ, Jedynak CP, Peillon F, eds. In Derome P.J., Jedynack C.P., Peillon F eds. : Pituitary adenomas, biology, physiopathology and treatment. France: Asclepios Publishers,, 1980.p.19-32.
125. Lamberts SWJ, Stefanko SZ, De Lange S, et al. Failure of clinical remission after transsphenoidal removal of a microadenoma in a patient with Cushing's disease ; multiple hyperplastic and adenomatous cell nests in surrounding pituitary tissue. J Clin Endocrinol Metab 1980;50:793-798.
126. Ludecke D, Kautzky R, Saeger W,Schrader D. Selective removal of hypersecreting pituitary adenomas. Acta Neurochir 1976;35:27-42.
127. Saeger W, Ludecke DL, Geisler F. The anterior lobe in Cushing's disease/syndrome. In: Ludecke DL, Chrousos GP, Tolis G, eds. ACTH, Cushing's syndrome, and other hypercortisolemic states. New York: Raven Press, 1990.p.147-156.
128. Cook DM,McCarthy PJ. Failure of hypophysectomy to correct pituitary-dependent Cushing's disease in two patients. Arch Intern Med 1988;148:2497-500.

ACTH production is excessive but qualitatively normal
129. De Keyzer Y, Bertagna X, Lenne F, Girard F, Luton JP,Kahn A. Altered proopiomelanocortin gene expression in ACTH producing non-pituitary tumors. J Clin Invest 1985;76:1892-1898.
130. Suda T, Tozawa F, Yamada M, et al. Effects of Corticotropin-releasing Hormone and Dexamethasone on Proopiomelanocortin Messenger RNA level in human corticotroph adenoma cells in vitro. J Clin Invest 1988;82:110-114.
131. Clark AJL, Lavender PM, Besser GM,Rees LH. Pro-opiomelanocortin mRNA size heterogeneity in ACTH-dependent Cushing's syndrome. J Mol Endocrinol 1988;2:3-9.
132. Nayaya T, Seo H, Kuwayama A, et al. Pro-opiomelanocortin gene expression in silent corticotroph-cell adenoma and Cushing's disease. J Neurosurg 1990;72:262-267.
133. De Keyzer Y, Bertagna X, Luton JP,Kahn A. Variable modes of Proopiomelanocortin gene transcription in human tumors. Mol Endocrinol 1989;3:215-223.

134. Bertagna X, Seurin D, Pique L, Luton JP, Bricaire H,Girard F. Peptides related to the NH2 terminal end of pro-opiocortin in man. J Clin Endocrinol Metab 1983;56:489-495. peptides in human pituitary and ectopic ACTH secreting tumors. Clin Endocrinol 1983;18:211-218.
135. Shibasaki T, Masui H, Sato G, Ling N,Guillemin R. Secretion pattern of pro-opiomelanocortin-derived peptides by a pituitary adenoma from a patient with Cushing's disease. J Clin Endocrinol Metab 1981;52:350.
136. Bertagna X, Luton JP, Binoux M, Bricaire H,Girard F. Characterization of lipotropin-, corticotropin-, and ß-endorphin-immunoreactive materials secreted in vitro by a human pituitary adenoma responsible for a case of Nelson's syndrome. J Clin Endocrinol Metab 1979;49:527-532.
137. Raff H,Findling JW. A new immunoradiometric assay for corticotropin evaluated in normal subjects and patients with Cushing's syndrome. Clin Chem 1989;35:596-600.

ACTH secretion is dysregulated but not autonomous.: Increased sensitivity to secretagogs

138. Orth DN, De Bold CR, De Cherney GS, et al. Pituitary microadenomas causing Cushing's disease respond to corticotrophin-releasing factor. J Clin Endocrinol Metab 1982;55:1017-1019.
139. Chrousos GP, Schulte JM, Oldfield EH, Gold PW, Cutler GBJr,Loriaux DL. The corticotropin-releasing factor stimulation test. An aid in the evaluation of patients with Cushing's syndrome. N Engl J Med 1984;310:622-626.
140. de Keyzer Y, Rene P, Beldjord C, Lenne F, Bertagna X. Overexpression of vasopressin (V3) and corticotrophin-releasing hormone receptor genes in corticotroph tumours. Clin Endocrinol (Oxf) 1998 Oct;49(4):475-82.
141. Raux MC, Binoux M, Luton JP, Pham Huu Trung MT,Girard F. Studies of ACTH secretion control in 116 cases of Cushing's syndrome. J Clin Endocrinol Metab 1975;40:186-197.
142. Malerbi DA, Fragoso MC, Vieira Filho AH, Brenlha EM, Mendonca BB. Cortisol and adrenocorticotropin response to desmopressin in women with Cushing's disease compared with depressive illness. J Clin Endocrinol Metab 1996 Jun;81(6):2233-7.
143. Rene P, Lenne F, Ventura MA, Bertagna X, de Keyzer Y. Nucleotide sequence and structural organization of the human vasopressin pituitary receptor (V3) gene. Gene 2000 Jan 4;241(1):57-64.
144. Ventura MA, Rene P, de Keyzer Y, Bertagna X, Clauser E. Gene and cDNA cloning and characterization of the mouse V3/V1b pituitary vasopressin receptor. J Mol Endocrinol 1999 Jun;22(3):251-60.
145. Saito M, Tahara A, Sugimoto T. 1-desamino-8-D-arginine vasopressin (DDAVP) as an agonist on V1b vasopressin receptor. Biochem Pharmacol 1997 Jun 1;53(11):1711-7.
146. Dahia PL, Ahmed-Shuaib A, Jacobs RA, Chew SL, Honegger J, Fahlbusch R, Besser GM, Grossman AB. Vasopressin receptor expression and mutation analysis in corticotropin-secreting tumors. J Clin Endocrinol Metab 1996 May;81(5):1768-71.
147. Korbonits M, Grossman AB. 1995 Growth hormone-releasing peptide and its analogues. Novel stimuli to growth hormone release. Trends Endocrinol Metab. 6:43–49.
148. Ghigo E, Arvat E, Ramunni J, Colao A, Gianotti L, Deghenghi R, Lombardi G, Camanni F. Adrenocorticotropin- and cortisol-releasing effect of hexarelin, a synthetic growth hormone-releasing peptide, in normal subjects and patients with Cushing's syndrome. J Clin Endocrinol Metab 1997 Aug;82(8):2439-44.
149. Korbonits M, Jacobs RA, Aylwin SJ, Burrin JM, Dahia PL, Monson JP, Honegger J, Fahlbush R, Trainer PJ, Chew SL, Besser GM, Grossman AB. Expression of the growth hormone secretagogue receptor in pituitary adenomas and other neuroendocrine tumors. J Clin Endocrinol Metab 1998 Oct;83(10):3624-30.

150. Watanobe H, Tamura T. Stimulation by peptide histidine methionine (PHM) of adrenocorticotropin secretion in patients with Cushing's disease: a comparison with the effect of vasoactive intestinal peptide (VIP) and a study on the effect of combined administration of corticotropin-releasing hormone with PHM or VIP. J Clin Endocrinol Metab 1994 Jun;78(6):1372-7.
151. Mazzocchi G, Rebuffat P, Gottardo L, Nussdorfer GG. Vasoactive intestinal peptide stimulates rat adrenal glucocorticoid secretion, through an ACTH receptor-dependent activation of the adenylate cyclase signaling pathway. Horm Metab Res 1998 May;30(5):241-3.

Partial resistance to glucocorticoid suppression

152. Liddle GW, Island D,Meador CK. Normal and abnormal regulation of corticotropin secretion in man. Rec Prog Horm Res 1962;18:125-166.
153. Jingami H, Matsukura S, Numa S,Imura H. Effects of adrenalectomy and dexamethasone administration on the level of prepro-corticotropin-releasing factor messenger ribonucleic acid (mRNA) in the hypothalamus and adrenocorticotropin/beta-lipotropin precursor mRNA in the pituitary in rats. Endocrinology 1985;117:1314-1320.
154. Vale W, Vaughan J, Smith M, Yamamoto G, Rivier J,Rivier C. Effects of synthetic ovine corticotropin-releasing factor, glucocorticoids, catecholamines, neuro-hypophysial peptides, and other substances on cultured corticotropic cells. Endocrinology 1983;113:1121-1131.
155. Bilezikjian LM,Vale WW. Glucocorticoids inhibit corticotropin-releasing factor-induced production of adenosine 3',5' -monophosphate in cultured anterior pituitary cells. Endocrinology 1983;113:657-662.
156. Roberts JL, Chen CLC, Eberwine JH, et al. Glucocorticoid regulation of proopiomelanocortin gene expression in rodent pituitary. Rec Prog Horm Res 1982;38:227-256.
157. Eberwine JH,Roberts JL. Glucocorticoid regulation of proopiomelanocortin gene transcription in the rat pituitary. J Biol Chem 1984;259:2166-2170.
158. Liddle GW. Tests of pituitary-adrenal suppressibility in the diagnosis of Cushing's syndrome. J Clin Endocrinol Metab 1960;20:1539-1560.
159. Wolfsen AR,Odell WD. The dose-response relationship of ACTH and cortisol in Cushing's disease. Clin Endocrinol 1980;12:557-568.
160. Pardes EM, de Yampey JW, Soto RJ, Moses DF,De Nicola AF. A correlative study between glucocorticoid receptor levels in human mononuclear leukocytes and biochemical data in Cushing's disease. Acta Endocrinol 1989;120:55-61.
161. Kontula K, Pelkonen R, Andersson L,Sivula A. Glucocorticoid receptors in adrenocorticoid disorders. J Clin Endocrinol Metab 1980;51:654-657.
162. Grino M, Boudouresque F, Conte-Devolx B, et al. In vitro corticotropin-releasing hormone (CRH) stimulation of adrenocorticotropin release from corticotroph adenoma cells : effect of prolonged exposure to CRH ans its interaction with cortisol. J Clin Endocrinol Metab 1988;66:770-775.
163. Oosterom R, Verleun T, Vitterlinden P, et al. ACTH and ß-endorphin secretion by three corticotropic adenomas in culture. Effect of culture time, dexamethasone, vasopressin and synthetic corticotropin-releasing factor. Acta Endocrinol 1984;106:21-29.
164. Suda T, Tomori N, Tozawa F, Demura H,Shizume K. Effects of corticotropin-releasing factor and other materials on adrenocorticotropin secretion from pituitary glands of patients with Cushing's disease in vitro. J Clin Endocrinol Metab 1984;59:840-845.
165. White MC, Adams EF, Loizou M, Mashiter K,Fahlbush R. Ovine corticotropin-releasing factor stimulates ACTH release from human corticotrophinoma cells in culture: interaction with hydrocortisone and arginine vasopressin. Clin Endocrinol 1985;23:295-302.

166. Mashiter K, Adams EF, Gillies G, Van Noorden S,Ratter S. Adrenocorticotropin and lipotropin secretion by dispersed cell cultures of a human corticotropic adenoma: effect of hypothalamic extract, arginine vasopressin, hydrocortisone and serotonin. J Clin Endocrinol Metab 1980;51:566.
167. Lamberts SWJ, Bons EG,Vitterlinden P. Studies on the glucocorticoid-receptor blocking action of RU 38486 in cultured ACTH-secreting human pituitary tumour cells and normal rat pituitary cells. Acta Endocrinol 1985;109:64-69.
168. Suda T, Tozawa F, Yamada M, Ushiyama T, Tomori N, Sumitomo T, Nakagami Y, Demura H, Shizume K. Effects of corticotropin-releasing hormone and dexamethasone on proopiomelanocortin messenger RNA level in human corticotroph adenoma cells in vitro. J Clin Invest 1988 Jul;82(1):110-4.

Molecular pathogenesis

169. Klibanski A. Editorial: further evidence for a somatic mutation theory in the pathogenesis of human pituitary tumors. J Clin Endocrinol Metab 1990;71:1415A-1415C.
170. Alexander JM, Biller BMK, Bikkal H, Arnold A,Klibanski A. Clinically non-functioning pituitary tumors are monoclonal in origin. J Clin Invest 1990;86:336-340.
171. Herman V, Fagin J, Gonsky R, Kovacs K,Melmed S. Clonal origin of pituitary adenomas. J Clin Endocrinol Metab 1990;71:1427-1433.
172. Schulte HM, Oldfield EH, Allolio B, Katz DA, Berkman RA, Ali IU. 1991Clonal composition of pituitary adenomas in patients with Cushing's disease: determination by X-chromosome inactivation analysis. Clin Endocrinol Metab;73(6):1302-8.
173. Gicquel C, Le Bouc Y, Luton JP, Girard F, Bertagna X. 1992 Monoclonality of corticotroph macroadenomas in Cushing's disease. J Clin Endocrinol Metab Aug;75(2):472-5.
174. Biller BM, Alexander JM, Zervas NT, Hedley-Whyte ET, Arnold A, Klibanski A. 1992. Clonal origins of adrenocorticotropin-secreting pituitary tissue in Cushing's disease. J Clin Endocrinol Metab;75(5):1303-9.

GR defects

175. Hollenberg SM, Weinberger C, Ong ES, Cerelli G, Oro A, Lebo R, Thompson EB, Rosenfeld MG, Evans RM 1985 Primary structure and expression of a functional human glucocorticoid receptor cDNA. Nature; 318:635-41
176. Oakley RH, Jewell CM, Yudt MR, Bofetiado DM, Cidlowski JA. 1999. The dominant negative activity of the human glucocorticoid receptor beta isoform. Specificity and mechanisms of action. J Biol Chem.;274:27857-66.
177. Oakley RH, Webster JC, Sar M, Parker CR, Cidlowski JA. 1997 Expression and subcellular distribution of the beta-isoform of the human glucocorticoid receptor. Endocrinology; 138:5028-38.
178. de Castro M, Elliot S, Kino T, Bamberger C, Karl M, Webster E, Chrousos GP. 1996 The non-ligand binding beta-isoform of the human glucocorticoid receptor (hGR beta): tissue levels, mechanism of action, and potential physiologic role. Mol Med.;2:597-607.
179. Lamberts SW 1996 The glucocorticoid insensitivity syndrome. Horm Res. 45:2-4.
180. Stratakis CA, Karl M, Schulte HM, Chrousos GP 1994 Glucocorticosteroid resistance in humans. Elucidation of the molecular mechanisms and implications for pathophysiology. Ann NY Acad Sci 746:362-3.
181. de Castro M, Chrousos GP 1997 Glucocorticoid resistance. Curr Ther Endocrinol Metab 6:188-189.
182. Karl M, Von Wichert G, Kempter E, Katz DA, Reincke M, Monig H, Ali IU, Stratakis CA, Oldfield EH, Chrousos GP, Schulte HM 1996 Nelson's syndrome associated with a somatic frame shift mutation in the glucocorticoid receptor gene. J Clin Endocrinol Metab 81:124-129.

183. Dahia PL, Honegger J, Reincke M, Jacobs RA, Mirtella A, Fahlbusch R, Besser GM, Chew SL, Grossman AB 1997 Expression of glucocorticoid receptor gene isoforms in corticotropin-secreting tumors. J Clin Endocrinol Metab 82:1088-1093.
184. Huizenga NA, de Lange P, Koper JW, Clayton RN, Farrell WE, van der Lely AJ, Brinkmann AO, de Jong FH, Lamberts SW 1998 Human adrenocorticotropin-secreting pituitary adenomas show frequent loss of heterozygosity at the glucocorticoid receptor gene locus. J Clin Endocrinol Metab 83:917-921.

11 b HSD

185. Korbonits M, Bujalska I, Shimojo M, Nobes J, Jordan S, Grossman AB, Stewart PM. 2001 Expression of 11 beta-hydroxysteroid dehydrogenase isoenzymes in the human pituitary: Induction of the type 2 enzyme in corticotropinomas and other pituitary tumors. J Clin Endocrinol Metab;86(6):2728-33*CRH R.*
186. Vale W, Spiess J, Rivier C,Rivier J. Characterization of a 41-residue ovine hypothalamic peptide that stimulates secretion of corticotropin and ß-endorphin. Science 1981;213:1394-1397.
187. Labrie F, Veilleux R, Lefevre G, Coy DH, Sueiras-Diaz J,Schally AV. Corticotropin-releasing factor stimulates accumulation of adenosine 3', 5', -monophosphate in rat pituitary corticotrophs. Science 1982;216:1007-1008.
188. Aguilera G, Harwood JP, Wilson JX, Morell J, Brown JH,Catt KJ. Mechanisms of action of corticotropin-releasing factor and other regulators of corticotropin release in the rat pituitary cells. J Biol Chem 1983;258:8039-8045.
189. Chen R, Lewis KA, Perrin MH, Vale WW. 1993 Expression cloning of a human corticotropin-releasing-factor receptor. Proc Natl Acad Sci U S A;90:8967-71.
190. Liaw CW, Lovenberg TW, Barry G, Oltersdorf T, Grigoriadis DE, de Souza EB 1996 Cloning and characterization of the human corticotropin-releasing factor-2 receptor complementary deoxyribonucleic acid. Endocrinology 137:72-7.
191. G.W. Smith et al., Corticotropin releasing factor receptor 1-deficient mice display decreased anxiety, impaired stress response, and aberrant neuroendocrine development. Neuron 20 6 (1998), pp. 1093-1102.
192. P. Timpl et al., Impaired stress response and reduced anxiety in mice lacking a functional corticotropin-releasing hormone receptor. Nature Genetics 192 (1998), pp. 162-166.
193. Coste SC, Murray SE, Stenzel-Poore MP. 2001 Animal models of CRH excess and CRH receptor deficiency display altered adaptations to stress. Peptides; 22:733-41.
194. Hsu SY, Hsueh AJ. 2001. Human stresscopin and stresscopin-related peptide are selective ligands for the type 2 corticotropin-releasing hormone receptor. Nat Med;7:605-11.
195. Landis CA, Masters SB, Spada A, Pace AM, Bourne HR, Vallar L. GTPase inhibiting mutations activate the alpha chain of Gs and stimulate adenylyl cyclase in human pituitary tumours. Nature 1989;340(6236):692-6.
196. de Keyzer Y, Rene P, Beldjord C, Lenne F, Bertagna X. Overexpression of vasopressin (V3) and corticotrophin-releasing hormone receptor genes in corticotroph tumours. Clin Endocrinol (Oxf) 1998;49(4):475-82.
197. Dieterich KD, Gundelfinger ED, Ludecke DK, Lehnert H. 1998 Mutation and expression analysis of corticotropin-releasing factor 1 receptor in adrenocorticotropin-secreting pituitary adenomas. J Clin Endocrinol Metab;83(9):3327-31.
198. Xu B, Sano T, Yamada S, Li CC, Hirokawa M. Expression of corticotropin-releasing hormone messenger ribonucleic acid in human pituitary corticotroph adenomas associated with proliferative potential. J Clin Endocrinol Metab 2000 Mar;85(3):1220-5 *V3 R.*
199. Seibold, P. Brabet, W. Rosenthal and M. Birnbaumer, Structure and chromosomal localization of the human antidiuretic hormone receptor gene. Am J Hum Genet 51 (1992), pp. 1078-1083.

200. S.J. Lolait, A.M. O'Carroll, O.W. McBride, M. Konig, A. Morel and M.J. Brownstein, Cloning and characterization of a vasopressin V2 receptor and possible link to nephrogenic diabetes insipidus. Nature 357 (1992), 336-339.
201. Morel, A.M. O'Carroll, M.J. Brownstein and S.J. Lolait, Molecular cloning and expression of a rat V1a arginine vasopressin receptor. Nature 356 (1992), 523-526.
202. S.J. Lolait, E. Mezey, A.-M. O'Carroll, L.C. Mahan, C.C. Felder, O.C. Button, W.S. Young, III and M.J. Brownstein, Extrapituitary expression of the rat V1b vasopressin receptor gene. Proc Natl Acad Sci USA 92 (1994), 6783-6787.
203. T. Sumimoto, M. Saito, S. Mochizuki, Y. Wanatabe, S. Hashimoto and H. Kawashima, Molecular cloning and functional expression of a cDNA encoding the human V1b vasopressin receptor. J Biol Chem 269 (1994), 27088-27092.
204. de Keyzer Y, Auzan C, Lenne F, Beldjord C, Thibonnier M, Bertagna X, Clauser E. Cloning and characterization of the human V3 pituitary vasopressin receptor. FEBS Lett 1994 Dec 19;356(2-3):215-20.
205. Gillies GE, Linton EA,Lowry PJ. Corticotropin releasing activity of the new CRF is potentiated several times by vasopressin. Nature 1982;299:355-357.
206. Abou-Samra AB, Harwood JP, Manganiello VC, Catt KJ,Aguilera G. Phorbol 12-myristate 13-acetate and vasopressin potentiate the effect of corticotropin-releasing factor on cyclic AMP production in rat anterior pituitary cells. J Biol Chem 1987;262:1129-1136.
207. Oki Y, Nicholson WE,Orth DN. Role of protein kinase-C in the adrenocorticotropin secretory response to arginine vasopressin (AVP) and the synergistic response to AVP and corticotropin-releasing factor by perifused rat anterior pituitary cells. Endocrinology 1990;127:350-357.
208. Giguere V,Labrie F. Vasopressin potentiates cyclic AMP accumulation and ACTH release induced by corticotropin-releasing factor (CRF) in rat anterior pituitary cells in culture. Endocrinology 1982;111:1752-1754.
209. Affolter HU, Reisine T. Corticotropin releasing factor increases proopiomelanocortin messenger RNA in mouse anterior pituitary tumor cells. J Biol Chem 1985;260:15477-15481.
210. Gagner JP,Drouin J. Opposite regulation of pro-opiomelanocortin gene transcription by glucocorticoids and CRH. Mol Cell Endocrinol 1985;40:25-32.
211. Bruhn TO, Sutton RE, Rivier C,Vale W. Corticotropin releasing factor regulates proopiomelanocortin messenger RNA levels in vivo. Neuroendocrinology 1984;39:170-175.
212. De Bold CR, Sheldon WR, De Cherney GS, et al. Arginine vasopressin potentiates adrenocorticotropin release induced by ovine corticotropin releasing factor. J Clin Invest 1984;73:533-538.
213. Lamberts SWJ, Verleun T, Oosterom R, de Jong F,Hackeng WHL. Corticotrophin releasing factor (ovine) and vasopressin exert a synergistic effect on adrenocorticotropin release in man. J Clin Endocrinol Metab 1984;58:298-303.
214. Liu JH, Muse K, Contrepas P, et al. Augmentation of ACTH-releasing activity of synthetic corticotropin releasing factor by vasopressin in women. J Clin Endocrinol Metab 1983;57:1087-1089.
215. Raux MC, Binoux M, Luton JP, Gourmelen M, Girard F.J Studies of ACTH secretion control in 116 cases of Cushing's syndrome. Clin Endocrinol Metab 1975;40(2):186-97.
216. G Aguilera C Rabadan-Diehl. 2000 Vasopressinergic regulation of the hypothalamic-pituitary-adrenal axis: implications for stress adaptation Regulatory Peptides 96, 23-29.
217. Colao A, Ferone D, Di Sarno A, Tripodi FS, Cerbone G, Marzullo P, Boudouresque F, Oliver C, Merola B, Lombardi G. 1996 Vasopressin levels in Cushing's disease: inferior petrosal sinus assay, response to corticotrophin-releasing hormone and comparison with patients without Cushing's disease. Clin Endocrinol (Oxf);45(2):157-66.

218. Friedman TC, Yanovski JA, Nieman LK, Doppman JL, Cutler GB Jr, Oldfield EH, Gold PM, Chrousos GP, Kalogeras KT. Inferior petrosal sinus arginine vasopressin concentrations in normal volunteers and patients with Cushing's disease. J Clin Endocrinol Metab 1996;81(8):3068-72.

Lif, LIF R, autres cytokines

219. Bateman A, Singh A, Kral T,Solomon S. The immune-hypothalamic-pituitary-adrenal axis. Endocr Rev 1989;10:92-112.
220. Chesnokova V, Auernhammer CJ, Melmed S 1998 Murine leukemia inhibitory factor gene disruption attenuates the hypothalamo-pituitary-adrenal axis stress response. Endocrinology 139:2209-2216.
221. Akita S, Conn PM, Melmed S 1996 Leukemia inhibitory factor (LIF) induces acute adrenocorticotrophic hormone (ACTH) secretion in fetal rhesus macaque primates: a novel dynamic test of pituitary function. J Clin Endocrinol Metab 81:4170Ð4178.
222. Ray DW, Ren SG, Melmed S 1996 Leukemia inhibitory factor (LIF)stimulates proopiomelanocortin (POMC) expression in a corticotroph cell line. Role of STAT pathway. J Clin Invest 97:1852-1859.
223. Ray DW, Ren SG, Melmed S 1998 Leukemia inhibitory factor regulates proopiomelanocortin transcription. Ann NY Acad Sci 840:162-173.
224. Bousquet C, Melmed S 1999 Critical role for STAT3 in murine pituitary ACTH leukemia inhibitory factor (LIF) signaling. J Biol Chem 274:10723-10730.
225. Auernhammer CJ, Bousquet C, Melmed S 1999 Autoregulation of pituitary corticotroph SOCS-3 expression: characterization of the murine SOCS-3 promoter. Proc Natl Acad Sci USA 96:6964-6969.
226. Auernhammer CJ, Chesnokova V, Bousquet C, Melmed S 1998 Pituitary corticotroph SOCS-3: novel intracellular regulation of leukemia-inhibitory factor-mediated proopiomelanocortin gene expression and adrenocorticotropin secretion. Mol Endocrinol 12:954-961.
227. Berkenbosch F, Van Oers J, Del Rey A, Tilders F,Besedovsky H. Corticotropin releasing factor producing neurons in the rat activated by interleukin 1. Science 1987;238:524-526.
228. Naitoh Y, Fukata J, Tominaga T, et al. Interleukin-6 stimulates the secretion of adrenocorticotropic hormone in conscious freely moving rats. Biochem Biophys Res Commun 1988;155:1459-1463.
229. Spangelo BL, Judd AM, Isakson PC,MacLeod RM. Interleukin-6 stimulates anterior pituitary hormone release in vitro. Endocrinology 1989;125:575-577.
230. Auernhammer CJ, Melmed S 1999 Interleukin-11 stimulates POMC gene expression and ACTH secretion in corticotroph cells: evidence for a redundant cytokine network in the HPA axis. Endocrinology 140:1559-1566.
231. Chrousos GP 1995 The hypothalamic-pituitary-adrenal axis and immune-mediated inflammation. N Engl J Med 332:1351-1362.
232. Auernhammer CJ, Chesnokova V, Melmed S 1998 Leukemia inhibitory factor modulates interleukin-1§-induced activation of the hypothalamo-pituitary-adrenal axis. Endocrinology. 139:2201-2208.
233. Bethin KE, Vogt SK, Muglia LJ. 2000 Interleukin-6 is an essential, corticotropin-releasing hormone-independent stimulator of the adrenal axis during immune system activation. Proc Natl Acad Sci U S A Aug 1;97(16):9317-22.
234. Akita S, Webster J, Ren SG, Takino H, Said J, Zand O, Melmed S 1995 Human and murine pituitary expression of leukemia inhibitory factor. Novel intrapituitary regulation of adrenocorticotropin hormone synthesis and secretion. J Clin Invest 95:1288-1298.
235. Shimon I, Yan X, Ray DW, Melmed S 1997 Cytokine-dependent gp130 receptor subunit regulates human fetal pituitary adrenocorticotropin hormone and growth hormone secretion. J Clin Invest 100:357-363.

236. Wang Z, Ren SG, Melmed S 1996 Hypothalamic and pituitary leukemia inhibitory factor gene expression in vivo: a novel endotoxin-inducible neuro-endocrine interface. Endocrinology 137:2947-2953.
237. Carter DA 1995 Leukaemia inhibitory factor expression in cultured rat anterior pituitary is regulated by glucocorticoids. J Neuroendocrinol 7:623-628.
238. Koenig JI, Snow K, Clark BD, et al. Intrinsic pituitary interleukin-1 ß is induced by bacterial lipopolysaccharide. Endocrinology 1990;126:3053-3058.
239. Vankelecom H, Carmeliet P, Van Damme J, Billiau A,Denef C. Production of interleukin-6 by folliculo-stellate cells in the anterior pituitary gland in a hystiotypic cell aggregate culture system. Neuroendocrinology 1989;49:102-106.
240. Ren SG, Seliktar J, Li X, Braunstein GD, Melmed S 1998 Measurement of leukemia inhibitory factor in biological fluids by radioimmunoassay. J Clin Endocrinol Metab 83:1275-1283.
241. Chesnokova V, Auernhammer CJ, Melmed S 1998 Murine leukemia inhibitory factor gene disruption attenuates the hypothalamo-pituitary-adrenal axis stress response. Endocrinology 139:2209-2216.
242. Hanisch A, Dieterich KD, Dietzmann K, Ludecke K, Buchfelder M, Fahlbusch R, Lehnert H. Expression of members of the interleukin-6 family of cytokines and their receptors in human pituitary and pituitary adenomas. J Clin Endocrinol Metab 2000 Nov;85(11):4411-4.
243. Kurotani R, Yasuda M, Oyama K, Egashira N, Sugaya M, Teramoto A, Osamura RY. 2001 Expression of interleukin-6, interleukin-6 receptor (gp80), and the receptor's signal-transducing subunit (gp130) in human normal pituitary glands and pituitary adenomas. Mod Pathol Aug;14(8):791-7.
244. Kontogeorgos G, Patralexis H, Tran A, Kovacs K, Melmed S. 2000 Expression of leukemia inhibitory factor in human pituitary adenomas: a morphologic and immunocytochemical study. Pituitary; 2(4):245-51.

G proteins

245. Williamson EA, Ince PG, Harrison D, Kendall-Taylor P, Harris PE 1995 G-protein mutations in human pituitary adrenocorticotrophic hormone-secreting adenomas. Eur J Clin Invest 25:128-131.
246. Oyesiku NM, Evans CO, Brown MR, Blevins LS, Tindall GT, Parks JS 1997 Pituitary adenomas: screening for G alpha q mutations. J Clin Endocrinol Metab 82:4184‹4188.

Transcription, differentiation factors

247. Sornson, M.W., Wu, W., Dasen, J.S., Flynn, S.E., Norman, D.J., O'Connell, S.M., Gukovsky, I., Carrière, C., Ryan, A.K., Miller, A.P. et al. (1996)
248. Pituitary lineage determination by the Prophet of Pit-1 homeodomain factor defective in Ames dwarfism. Nature, 384, 327–333.
249. Cogan JD, Wu W, Phillips III JA, et al. 1998 The PROP1 2-base pair deletion is a common cause of combined pituitary hormone deficiency. J Clin Endocrinol Metab. 83:3346–3349.
250. Wu W, Cogan, JD, Pfaffle RW, et al. 1999 Mutations in PROP1 cause familial combined pituitary hormone deficiency. Nat Genet 18:147–149.
251. Fluck C, Deladoey J, Rutishauser K, et al. 1998 Phenotypic variability in familial combined pituitary hormone deficiency caused by a Prop-1 gene mutation resulting in the substitution of ArgCys mutation at codon 120 (R120C). J Clin Endocrinol Metab. 83:3727–3734.
252. Pernasetti, F., Toledo, S.P., Vasilyev, V.V., Hayashida, C.Y., Cogan, J.D., Ferrari, C., Lourenco, D.M. and Mellon, P.L. (2000) Impaired adrenocorticotropin-adrenal axis in

combined pituitary hormone deficiency caused by a two-base pair deletion (301–302delAG) in the prophet of Pit-1 gene. J. Clin. Endocrinol. Metab., 85, 390–397.
253. Lisa J. Cushmanl, Dawn E. Watkins-Chowl, Michelle L. Brinkmeierl, Lori T. Raetzmanl, Amy L. Radakl, Ricardo V. Lloyd and Sally A. Camper. 2001. Persistent Propl expression delays gonadotrope differentiation and enhances pituitary tumor susceptibility. Human Molecular Genetics, 10, 1141-1153.
254. Yoshio Nakamura, Takeshi Usui, Haruo Mizuta, Hiroyuki Murabe, Seiji Muro, Michio Suda, Kiyoshi Tanaka, Issei Tanaka, Akira Shimatsu and Kazuwa Nakao Characterization of Prophet of Pit-1 Gene Expression in Normal Pituitary and Pituitary Adenomas in Humans J. Clinical Endocrinol Metab. 84, No. 4 1414-1419.
255. Skelly RH, Korbonits M, Grossman A, Besser GM, Monson JP, Geddes JF, Burrin JM. 2000 Expression of the pituitary transcription factor Ptx-1, but not that of the trans-activating factor prop-1, is reduced in human corticotroph adenomas and is associated with decreased alpha-subunit secretion. J Clin Endocrinol Metab;85(7):2537-42.
256. Crawford PA, Sadovsky Y, Woodson K, Lee SL, Milbrandt J 1995 Adrenocortical function and regulation of the steroid 21-hydroxylase gene in NGFI-B-deficient mice. Mol Cell Biol 15:4331-4316.

Tumor suppressor genes
Rbl
257. Jacks T, Fazeli A, Schmitt EM, Bronson RT, Goodell MA, Weinberg RA. 1992 Effects of an Rb mutation in the mouse. Nature. 55:1206–1210.
258. Hirama T, Koeffler HP. 1995 Role of cyclin-dependent kinase inhibitors in the development of cancer. Blood. 86:841–845.
259. Toyoshima H, Hunter T. 1994 p27, a novel inhibitor of G1 cyclin-Cdk protein kinase activity, is related to p21. Cell. 78:67–74.
260. Lee MH, Reynisdottir I, Massague J. 1995 Cloning of p57KIP2, a cyclin-dependent kinase inhibitor with unique domain structure and tissue distribution. Genes Dev. 9:639–649.
261. Polyak K, Kato JY, Solomon MJ, et al. 1994 p27Kipl, a cyclin-Cdk inhibitor, links transforming growth factor-ß and contact inhibition to cell cycle arrest. Genes Dev. 8:9–22.
262. Cryns VL, Alexander JM, Klibanski A, Arnold A 1993 The retinoblastoma gene in human pituitary tumors. J Clin Endocrinol Metab 77:644‹646.
263. Melmed S 1994 Pituitary neoplasia. Endocrinol Metab Clin North Am 23:81-92.
264. Woloschak M, Roberts JL, Post KD 1994 Loss of heterozygosity at the retinoblastoma locus in human pituitary tumors. Cancer 74:693‹696.
265. Pei L, Melmed S, Scheithauer B, Kovacs K, Benedict WF, Prager D 1995 Frequent loss of heterozygosity at the retinoblastoma susceptibility gene (RB) locus in aggressive pituitary tumors: evidence for a chromosome 13 tumor suppressor gene other than RB. Cancer Res 55:1613‹1616.

P27
266. Dahia PL, Aguiar RC, Honegger J, et al. 1998 Mutation and expression analysis of the p27/kipl gene in corticotrophin-secreting tumours. Oncogene. 16:69–76.
267. Ikeda H, Yoshimoto T, Shida N. 1997 Molecular analysis of p21 and p27 genes in human pituitary adenomas. Br J Cancer. 76:1119–1123.
268. Takeuchi S, Koeffler HP, Hinton DR, Miyoshi I, Melmed S, Shimon I. 1998 Mutation and expression analysis of the cyclin-dependent kinase inhibitor gene p27/Kipl in pituitary tumors. J Endocrinol. 157:337–341.
269. Kulbir Lidhar, Márta Korbonits, Suzanne Jordan, Zamira Khalimova, Gregory Kaltsas, Xin Lu, Richard N. Clayton, Paul J. Jenkins, John P. Monson, G. Michael Besser, David G. Lowe and Ashley B. Grossman 1999. Low Expression of the Cell Cycle Inhibitor

p27Kip1 in Normal Corticotroph Cells, Corticotroph Tumors, and Malignant Pituitary Tumors J Clin Endocrinol Metab. 84, No. 10 3823-3830.

P53

270. Hollstein M, Sidransky D, Vogelstein B, Harris CC 1991 p53 mutations in human cancers. Science 253:49-53.
271. Buckley N, Bates AS, Broome JC, Strange RC, Perrett CW, Burke CW, Clayton RN 1995 p53 Protein accumulates in Cushings adenomas and invasive nonfunctional adenomas. J Clin Endocrinol Metab 80:692-696.
272. Herman V, Drazin NZ, Gonsky R, Melmed S 1993 Molecular screening of pituitary adenomas for gene mutations and rearrangements. J Clin Endocrinol Metab 77:50-55
273. Levy A, Hall L, Yeudall WA, Lightman SL 1994 p53 Gene mutations in pituitary adenomas: rare events. Clin Endocrinol (Oxf) 41:809-814.

Cyclin

274. Turner HE, Nagy Z, Sullivan N, Esiri MM, Wass JA. 2000 Expression analysis of cyclins in pituitary adenomas and the normal pituitary gland Clin Endocrinol.;53(3):337-44.
275. Jordan S, Lidhar K, Korbonits M, Lowe DG, Grossman AB. Cyclin D and cyclin E expression in normal and adenomatous pituitary. Eur J Endocrinol 2000 Jul;143(1):R1-6.
276. Hibberts NA, Simpson DJ, Bicknell JE, Broome JC, Hoban PR, Clayton RN, Farrell WE. Analysis of cyclin D1 (CCND1) allelic imbalance and overexpression in sporadic human pituitary tumors. Clin Cancer Res 1999; 5 (8):2133-9.

Other tumor suppressor genes:

ZAC

277. Varrault A, Ciani E, Apiou F, Bilanges B, Hoffmann A, Pantaloni C, Bockaert J, Spengler D, Journot L 1998 hZAC encodes a zinc finger protein with antiproliferative properties and maps to a chromosomal region frequently lost in cancer. Proc Natl Acad Sci USA 95:8835-8840.
278. Abdollahi A, Roberts D, Godwin AK, Schultz DC, Sonoda G, Testa JR, Hamilton TC 1997 Identification of a zinc-finger gene at 6q25: a chromosomal region implicated in development of many solid tumors. Oncogene 14:1973-1979.
279. Pagotto U, Arzberger T, Theodoropoulou M, Grubler Y, Pantaloni C, Saeger W, Losa M, Journot L, Stalla GK, Spengler D. The expression of the antiproliferative gene ZAC is lost or highly reduced in nonfunctioning pituitary adenomas. Cancer Res 2000 Dec 15;60(24):6794-9.

MEN 1

280. Burgess JR, Shepherd JJ, Parameswaran V, Hoffman L, Greenaway TM 1996 Spectrum of pituitary disease in multiple endocrine neoplasia type 1 (MEN 1): clinical, biochemical, and radiological features of pituitary disease in a large MEN 1 kindred. J Clin Endocrinol Metab 81:2642-2646.
281. Larsson C, Skosgeid B, Oberg K, Nakamura Y, Nordenskjsld M 1988 Multiple endocrine neoplasia type 1 gene maps to chromosome 11 and is lost in insulinoma. Nature 332:85-87.
282. Clayton RN, Boggild M, Bates AS, Bicknell J, Simpson D, Farrell W 1997 Tumour suppressor genes in the pathogenesis of human pituitary tumours. Horm Res 47:185-193.
283. Bystrom C, Larsson C, Blomberg C, Sandelin K, Falkmer U, Skogseid B, Oberg K, Werner S, Nordenskjold M 1990 Localization of the MEN1 gene to a small region within chromosome 11q13 by deletion mapping in tumors.Proc Natl Acad Sci USA 87:1968-1972.

284. Thakker RV 1993 The molecular genetics of the multiple endocrine neoplasia syndromes. Clin Endocrinol (Oxf) 38:1-14.
285. Tanaka C, Yoshimoto K, Yamada S, Nishioka H, Ii S, Moritani M, Yamaoka T, Itakura M 1998 Absence of germ-line mutations of the multiple endocrine neoplasia type 1 (MEN1) gene in familial pituitary adenoma in contrast to MEN1 in Japanese. J Clin Endocrinol Metab 83:960-965.
286. Zhuang Z, Ezzat SZ, Vortmeyer AO, Weil R, Oldfield EH, Park WS, Pack S, Huang S, Agarwal SK, Guru SC, Manickam P, Debelenko LV, Kester MB, Olufemi SE, Heppner C, Crabtree JS, Burns AL, Spiegel AM, Marx SJ, Chandrasekharappa SC, Collins FS, Emmert-Buck MR, Liotta LA, Asa SL, Lubensky IA 1997 Mutations of the MEN1 tumor suppressor gene in pituitary tumors. Cancer Res 57:5446-5451.
287. Prezant TR, Levine J, Melmed S 1998 Molecular characterization of the men1 tumor suppressor gene in sporadic pituitary tumors. J Clin Endocrinol Metab 83:1388-1391.
288. Satta MA, Korbonits M, Jacobs RA, Bolden-Dwinfour DA, Kaltsas GA, Vangeli V, Adams E, Fahlbusch R, Grossman AB. 1999. Expression of menin gene mRNA in pituitary tumours. Eur J Endocrinol;140(4):358-61.

Oncogenes

289. Alvaro V, Levy L, Dubray C, Roche A, Peillon F, Querat B, Joubert D 1993 Invasive human pituitary tumors express a point-mutated alpha-protein kinase-C. J Clin Endocrinol Metab 77:1125-1129.
290. Ezzat S, Zheng L, Smyth HS, Asa SL 1997 The c-erbB-2/neu proto-oncogene in human pituitary tumours. Clin Endocrinol (Oxf) 46:599-606
291. Pei L, Melmed S, Scheithauer B, Kovacs K, Prager D 1994 H-ras mutations in human pituitary carcinoma metastases. J Clin Endocrinol Metab 78:842-846.
292. Ikeda H, Yoshimoto T 1992 The relationship between c-myc protein expression, the bromodeoxyuridine labeling index and the biological behavior of pituitary adenomas. Acta Neuropathol (Berl) 83:361-364.
293. Komminoth P, Roth J, Muletta-Feurer S, Saremaslani P, Seelentag WK, Heitz PU 1996 RET proto-oncogene point mutations in sporadic neuroendocrine tumors. J Clin Endocrinol Metab 81:2041-2046.
294. Boggild MD, Jenkinson S, Pistorello M, Boscaro M, Scanarini M, McTernan P, Perrett CW, Thakker RV, Clayton RN 1994 Molecular genetic studies of sporadic pituitary tumors. J Clin Endocrinol Metab 78:387-392.

Ac retinoic

295. Paez-Pereda M, Kovalovsky D, Hopfner U, Theodoropoulou M, Pagotto U, Uhl E, Losa M, Stalla J, Grubler Y, Missale C, Arzt E, Stalla GK. Retinoic acid prevents experimental Cushing syndrome. J Clin Invest 2001 Oct;108(8):1123-31.

Macroadenomas

296. Selvais P, Donckier J, Buysschaert M, Maiter D. 1998. Cushing's disease: a comparison of pituitary corticotroph microadenomas and macroadenomas. Eur J Endocrinol. 138; 153-9.
297. Katznelson L, Bogan JS, Trob JR, Schoenfeld DA, Hedley-Whyte ET, Hsu DW, Zervas NT, Swearingen B, Sleeper M, Klibanski A. Biochemical assessment of Cushing's disease in patients with corticotroph macroadenomas. J Clin Endocrinol Metab 1998 May;83(5):1619-23.
298. Arvat E, Giordano R, Ramunni J, Arnaldi G, Colao A, Deghenghi R, Lombardi G, Mantero F, Camanni F, Ghigo E. Adrenocorticotropin and cortisol hyperresponsiveness to hexarelin in patients with Cushing's disease bearing a pituitary microadenoma, but not in those with macroadenoma. J Clin Endocrinol Metab 1998 Dec;83(12):4207-11.

299. Nayaya T, Seo H, Kuwayama A, et al. Pro-opiomelanocortin gene expression in silent corticotroph-cell adenoma and Cushing's disease. J Neurosurg 1990;72:262-267.
300. Fuller PJ, Lim ATW, Barlow JW, et al. A pituitary tumor producing high molecular weight adrenocorticotropin-related peptides: clinical and cell culture studies. J Clin Endocrinol Metab 1984;58:134-142.
301. Reincke M, Allolio B, Saeger W, Kaulen D,Winkelmann W. A pituitary adenoma secreting high molecular weight adrenocorticotropin without evidence of Cushing's disease. J Clin Endocrinol Metab 1987;65:1296-1300.
302. Hale AC, Millar JB, Ratter SJ, Pickard JD, Doniach I,Rees LH. A case of pituitary dependent Cushing's disease with clinical and biochemical features of the ectopic ACTH syndrome. Clin Endocrinol 1985;22:479-488.
303. Gibson S, Ray DW, Crosby SR et al. 1996, Impaired processing of proopiomelanocortin in corticotroph macroadenomas. J Clin Endocrinol Metab. 81; 497-502.
304. Raffin-Sanson ML, Massias JF, Dumont C, Raux-Demay MC, Proeschel MF, Luton JP, Bertagna X. High plasma proopiomelanocortin in aggressive adrenocorticotropin – secreting tumors. J Clin Endocrinol Metab. 1996. 81, 4272-4277.
305. Losa M, Barzaghi RL, Mortini P, Franzin A, Mangili F, Terreni MR, Giovanelli M. Determination of the proliferation and apoptotic index in adrenocorticotropin-secreting pituitary tumors: comparison between micro- and macroadenomas. Am J Pathol 2000 Jan;156(1):245-51.
306. Alvaro V, Levy L, Dubray C, Roche A, Peillon F, Querat B, Joubert D 1993 Invasive human pituitary tumors express a point-mutated {alpha}-protein kinase-C. J Clin Endocrinol Metab 77:1125-1129.
307. Ezzat S, Zheng L, Smyth HS, Asa SL 1997 The c-erbB-2/neu proto-oncogene in human pituitary tumours. Clin Endocrinol (Oxf) 46:599-606.
308. Pei L, Melmed S, Scheithauer B, Kovacs K, Prager D 1994 H-ras mutations in human pituitary carcinoma metastases. J Clin Endocrinol Metab 78:842-846.
309. Pei L, Melmed S, Scheithauer B, Kovacs K, Benedict WF, Prager D. Frequent loss of heterozygosity at the retinoblastoma susceptibility gene (RB) locus in aggressive pituitary tumors: evidence for a chromosome 13 tumor suppressor gene other than RB. Cancer Res 1995 Apr 15;55(8):1613-6.
310. Leblond-Francillard M, Picon A, Bertagna X, De Keyzer Y. 1997 High expression of the POU factor Brn3a in aggressive neuroendocrine tumors. J Clin Endocrinol Metab 82:89-94.
311. J Turner HE, Nagy Z, Gatter KC, Esiri MM, Harris AL, Wass JA. Angiogenesis in pituitary adenomas - relationship to endocrine function, treatment and outcome. Endocrinol 2000 May;165(2):475-81.
312. Turner HE, Nagy Z, Gatter KC, Esiri MM, Harris AL, Wass JA. Angiogenesis in pituitary adenomas and the normal pituitary gland. J Clin Endocrinol Metab 2000 Mar;85(3):1159-62.

Pituitary cancer
313. Gabrilove, J. L., P. J. Anderson, and N. S. Halmi. 1986. Pituitary proopiomelanocortin-cell carcinoma occurring in conjunction with a gliobastoma in a patient with Cushing's disease and subsequent Nelson's syndrome. *Clin. Endocrinol.* 25:117-126.
314. Forbes, W. 1947. Carcinoma of the pituitary gland with metastases to the liver in a case of Cushing's syndrome. *J. Path. Bact.* 59:137-144.
315. Salassa, R. M., T. P. Kearns, J. W. Kernohan, R. G. Sprague, and C. S. Mac Carty. 1959. Pituitary tumors in patients with Cushing's syndrome. *J. Clin. Endocrinol. Metab.* 19:1523-1539.
316. Casson, I. F., B. A. Walker, L. J. Hipkin, J. C. Davis, P. H. Buxton, and R. V. Jeffreys. 1986. An intrasellar pituitary tumour producing metastases in liver, bone and lymph

glands and demonstration of ACTH in the metastatic deposits. *Acta Endocrinol.* 111:300-304.
317. Welbourn, R. B. 1985. Survival and causes of death after adrenalectomy for Cushing's disease. *Surgery* 97:16-20.

Hypothalamus-dependent Cushing's diseases?

318. Cushing, H. 1932. The basophil adenomas of the pituitary body and their clinical manifestations. *Bull. Johns Hopkins Hosp.* 50:137-195.
319. Krieger, D. T. 1983. Physiopathology of Cushing's disease. *Endocr. Rev.* 4:22-43.
320. Baxter, J. D. and J. B. Tyrrell. 1981. The adrenal cortex. In Endocrinology and Metabolism. P. Felig, J. D. Baxter, A. E. Broadus, and L. A. Frohman, editors. New York, McGraw Hill Book Company. 385-510.
321. Krieger, D. T. and G. P. Gewirtz. 1974. Recovery of hypothalamic-pituitary-adrenal function, growth homrone responsiveness, and sleep EEG pattern in a patient following removal of an adrenal cortical adenoma. *J. Clin. Endocrinol. Metab.* 38:1075-1082.
322. Van Cauter, E. and S. Refetoff. 1985. Evidence for two subtypes of Cushing's disease based on the analysis of episodic cortisol secretion. *N. Engl. J. Med.* 312:1343-1349.
323. Kuwayama, A., N. Kageyama, T. Nakane, M. Watanabe, and N. KAnie. 1981. Anterior pituitary function after transsphenoidal selective adenomectomy in patients with Cushing's disease. *J. Clin. Endocrinol. Metab.* 53:165-172.
324. Schnall, A. M., J. S. Brodkey, B. Kaufman, and O. H. Pearson. 1978. Pituitary function after removal of pituitary microadenomas in Cushing's disease. *J. Clin. Endocrinol. Metab.* 47:410-417.
325. Fitzgerald, P. A., D. C. Aron, J. W. Findling, R. M. Brooks, C. B. Wilson, P. H. Forsham, and J. B. Tyrell. 1982. Cushing's disease : transient secondary adrenal insufficiency after selective removal of pituitary microadenomas : evidence for a pituitary origin. *J. Clin. Endocrinol. Metab.* 54:413-422.
326. Atkinson, A. B., A. L. Kennedy, D. J. Carson, D. R. Hadden, J. A. Weaver, and B. Sheridan. 1985. Five cases of cyclical Cushing's syndrome. *Br. Med. J.* 291:1453-1457.
327. Saeger, W. and D. K. Ludecke. 1983. Pituitary hyperplasia. Definition, light and electron microscopical structures and significance in surgical specimens. *Virchows Arch.* 399:277-287.
328. Schnall, A. M., K. Kovacs, J. S. Brodkey, and O. H. Pearson. 1980. Pituitary Cushing's disease without adenoma. *Acta Endocrinol.* 94:297-303.
329. Molitch, M. E. 1987. Pathogenesis of pituitary tumors. *Endocrinol. Metab. Clin. North Am.* 16:503-527.
330. Martin, R., Y. Cetin, G. L. Fehm, R. Fahlbusch, and K. H. Voigt. 1982. Multiple cellular forms of corticotrophs in surgically removed pituitary adenomas and periadenomatous tissue in Cushing's disease. *Am. J. Pathol.* 106:332-341.
331. Mc Keever, P. E., M. C. Koppelman, D. Metcalf, E. Quindlen, P. L. Kourblith, and C. A. Stroot. 1982. Refractory Cushing's disease caused by multinodular ACTH cell hyperplasia. *J. Neuropathol. Exp. Neurol.* 41:490-499.
332. Gertz, B. J., L. N. Contreras, D. J. McComb, K. Kovacs, J. B. Tyrrell, and M. F. Dallman. 1987. Chronic administration of corticotropin releasing factor increases pituitary corticotroph number. *Endocrinology* 120:381-388.
333. Horacek, M. J., G. T. Campbell, and C. A. Blake. 1989. Effects of corticotrophin-releasing hormone on corticotrophs in anterior pituitary gland allografts in hypophysectomized, orchidectomized hamsters. *Cell Tissue Res.* 258:65-68.
334. Asa, S. L., K. Kovacs, G. T. Tindall, D. L. Barrow, E. Howath, and P. Versei. 1984. Cushing's disease associated with an intrasellar gangliocytoma producing corticotrophin-releasing factor. *Ann. Intern. Med.* 101:789-793.

335. Belsky, J. L., B. Cuello, L. W. Swanson, D. M. Simmons, R. M. Sarrett, and F. Braza. 1985. Cushing's syndrome due to ectopic production of corticotropin-releasing factor. *J. Clin. Endocrinol. Metab.* 60:496-500.
336. Carey, R., S. Varma, C. Drake, M. Thorner, K. Kovacs, J. Rivier, and W. Vale. 1984. Ectopic secretion of corticotropin-releasing factor as a cause of Cushing's syndrome. *N. Engl. J. Med.* 311:13-20.
337. McNicol, A. M. 1985. Current topics in neuropathology. Cushing's disease. *Neuropathol. Appl. Neurobiol.* 11:485-494.
338. Kovacs, K. and E. Horvath. 1987. Pathology of pituitary tumors. *Endocrinol. Metab. Clin. North Am.* 16:529-551.
339. Phifer, R. F., S. S. Spicer, and D. N. Orth. 1970. Specific demonstrations of the human hypophyseal cells which produce adrenocorticotrophic hormone. *J. Clin. Endocrinol. Metab.* 31:347-361.
340. Phifer, R. F., D. N. Orth, and S. S. Spicer. 1974. Specific demonstration of the human hypophyseal adrenocorticomelanotropic (ACTH/MSH) cell. *J. Clin. Endocrinol. Metab.* 39:684-692.
341. Young, W. F., B. W. Scheithauer, H. Gharib, E. R. J. Laws, and P. C. Carpenter. 1988. Cushing's syndrome due to primary multinodular corticotrope hyperplasia. *Mayo Clin. Proc.* 63:256-262.
342. Suda, T., N. Tomori, F. Yajima, T. Sumitomi, Y. Nakagami, T. Yshiyama, H. Demura, and K. Shizume. 1985. Immunoreactive corticotropin-releasing factor in human plasma. *J. Clin. Invest.* 76:2026-2029.
343. Linton, E. A. and P. J. Lowry. 1989. Corticotrophin releasing factor in man and its measurements : a review. *Clin. Endocrinol.* 31:225-249.
344. Altemus, M. and P. W. Gold. 1990. Neuroendocrinology and Psychiatric Illness. *Front. Neuroendocrinol.* 11:3-238.
345. Nemeroff, C. B., E. Widerlov, G. Bissette, H. Wallens, I. Karlsson, K. Eklund, C. D. Kilts, P. T. Loosen, and W. Vale. 1984. Elevated concentrations of corticotropin-releasing-factor like immunoreactivity in depressed patients. *Science* 226:1342-1344.
346. Roy, A., D. Pickar, S. Paul, A. Doran, G. P. Chrousos, and P. W. Gold. 1987. CSF corticotropin releasing hormone in depressed patients and normal control subjects. *Am. J. Psychiatry* 144:641-644.
347. Rivier, C., J. Rivier, and W. Vale. 1984. Synthetic competitive antagonists of corticotropin-releasing factor : effect on ACTH secretion in the rat. *Science* 224:889-891.
348. Hanahan, D. 1988. Dissecting multistep tumorigenesis in transgenic mice. *Annu. Rev. Genet.* 22:479-519.
349. Hanahan, D. 1985. Heritable formation of pancreatic □-cell tumors in transgenic mice expressing recombinant insulin/simian virus 40 oncogenes. *Nature* 315:115-122.
350. Knudson, A. G. 1986. Genetics of human cancer. *Annu. Rev. Genet.* 20:231-251.
351. Wolfe, H. J., K. E. W. Melvin, S. J. Cervio-Skinner, A. A. Saadi, J. F. Juliar, C. E. Jackson, and A. N. Tashjian. 1973. C-Cell hyperplasia preceding medullary thyroid carcinoma. *N. Engl. J. Med.* 289:437-441.
352. Folkman, J., K. Watson, D. Ingber, and D. Hanahan. 1989. Induction of angiogenesis during the transition from hyperplasia to neoplasia. *Nature* 339:58-61.
353. Gospodarowicz, D., J. A. Abraham, and J. Schilling. 1989. Isolation and characterization of a vascular endothelial cell mitogen produced by pituitary-derived folliculo stellate cells. *Biochem. J.* 86:7311-7315.
354. Ferrara, N. and W. J. Henzel. 1989. Pituitary follicular cells secrete a novel heparin-binding growth factor specific for vascular endothelial cells. *Biochem. Biophys. Res. Commun.* 161:851-856.

355. Plouet, J., J. Schilling, and D. Gospodarowicz. 1989. Isolation and characterization of a newly identified endothelial cell mitogene produced by pituitary AtT-20 cells. *EMBO J.* 8:3801-3806.
356. Mayo, K. E., R. L. Hammer, L. W. Swanson, R. L. Brinster, M. G. Rosenfeld, and R. M. Evans. 1988. Dramatic pituitary hyperplasia in transgenic mice expressing a human growth hormone-releasing factor gene. *Mol. Endocrinol.* 2:606-612.
357. Stefaneanu, L., K. Kovacs, E. Horvath, S. L. Asa, N. E. Losinski, N. Billestrup, J. Price, and W. Vale. 1989. Adenohypophysial changes in mice transgenic for human growth hormone-releasing factor : a histological, immunocytochemical, and electron microscopic investigation. *Endocrinology* 125:2710-2718.
358. Hammer, G. D., V. Fairchild-Huntress, and M. J. Low. 1990. Pituitary-specific and hormonally regulated gene expression directed by the rat proopiomelanocortin promoter in transgenic mice. *Mol. Endocrinol.* 4:1689-1697.
359. Tremblay, Y., I. Tretjakoff, A. Pterson, T. Antakly, C. Xian Zhang, and J. Drouin. 1988. Pituitary-specific expression and glucocorticoid regulation of a pro-opiomelanocortin fusion gene in transgenic mice. *Proc. Natl. Acad. Sci. USA* 85:8890-8894.
360. Sano, T., S. L. Asa, and K. Kovacs. 1988. Growth homrone-releasing hormone-producing tumors : clinical, biochemical, and morphological manifestations. *Endocr. Rev.* 9:357-373.
361. Lamberts, S. W. J., S. Z. Stefanko, S. De Lange, M. Fermin, J. C. M. Van der Vijver, R. F. A. Weber, and F. H. de Jong. 1980. Failure of clinical remission after transsphenoidal removal of a microadenoma in a patient with Cushing's disease ; multiple hyperplastic and adenomatous cell nests in surrounding pituitary tissue. *J. Clin. Endocrinol. Metab.* 50:793-798.
362. Ludecke, D., R. Kautzky, W. Saeger, and D. Schrader. 1976. Selective removal of hypersecreting pituitary adenomas. *Acta Neurochir.* 35:27-42.
363. Saeger, W., D. L. Ludecke, and F. Geisler. 1990. The anterior lobe in Cushing's disease/syndrome. In ACTH, Cushing's syndrome, and other hypercortisolemic states. D. L. Ludecke, G. P. Chrousos, and G. Tolis, editors. New York, Raven Press. 147-156.
364. Cook, D. M. and P. J. McCarthy. 1988. Failure of hypophysectomy to correct pituitary-dependent Cushing's disease in two patients. *Arch. Intern. Med.* 148:2497-2500.
365. Sonino, N., G. A. Fava, S. Grandi, F. Mantero, and M. Boscaro. 1988. Stressful life events in the pathogenesis of Cushing's syndrome. *Clin. Endocrinol.* 29:617-623.
366. Guilhaume, B., X. Bertagna, M. Thomsen, C. Bricaire, E. Vila-Porcile, L. Olivier, J. Racadot, P. Derome, M. H. Laudat, F. Girard, H. Bricaire, and J.-P. Luton. 1988. Transsphenoidal pituitary surgery for the treatment of Cushing's disease : results in 64 patients and long term follow-up studies. *J. Clin. Endocrinol. Metab.* 66:1056-1064.
367. Lamberts, S. W. J., J. G. M. Klijn, and F. H. de Jong. 1987. The definition of true recurrence of pituitary-dependent Cushing's syndrome after transsphenoïdal operation. *Clin. Endocrinol.* 26:707-712.
368. Kuwayama, A. 1990. Long-term results of pituitary surgery in Cushing's disease. In ACTH, Cushing's syndrome, and other hypercortisolemic states. D. L. Ludecke, G. P. Chrousos, and G. Tolis, editors. New York, Raven Press. 289-295.
369. Friedman, R. B., E. H. Oldfield, L. K. Nieman, G. P. Chrousos, J. L. Doppman, G. B. Cutler, and D. L. Loriaux. 1989. Repeat transsphenoïdal surgery for Cushing's disease. *J. Neurosurg.* 71:520-527.
370. Hammer, G. D. and M. J. Low. 1989. Prolonged ectopic secretion of corticotropin-releasing hormone by a transplantable cell line in rats. *Endocrine Society 71st Annual Meeting,Seattle,Abstract* 956:261(Abstr.)

8
SYNDROME OF ECTOPIC ACTH HYPERSECRETION

William J. Kovacs, M.D.
Division of Endocrinology, Diabetes and Metabolism, Department of Medicine, Vanderbilt School of Medicine, Nashville, TN, USA

Although the development of bilateral adrenocortical hyperfunction in the setting of a nonendocrine tumor had been reported in the medical literature on numerous occasions in the first half of the twentieth century it was not until 1962 that evidence for direct secretion of adrenocorticotropin (ACTH) from such tumors was documented (1). Ectopic (that is, nonpituitary) ACTH production accounts for about 10-15% of cases of ACTH-dependent Cushing's syndrome. A variety of tumor types have been reported; most are tumors of the lung, pancreas and thymus. The lung tumors include both malignancies (chiefly small cell lung cancer, which has been reported to account for up to 50% of cases of ectopic ACTH syndrome in some centers (2)) as well as benign bronchial carcinoid tumors, which account for an increasing proportion of cases (3). Among patients with small cell lung cancer, only about 2-5% have clinically apparent hypercortisolism from ectopic ACTH-producion. Thymic and gastrointestinal tract carcinoid tumors, pancreatic islet cell tumors, medullary carcinomas of the thyroid, and pheochromocytomas also each account for a small percentage of the recognized cases of nonpituitary ACTH production (3, 4).

CLINICAL PRESENTATIONS OF PATIENTS WITH ECTOPIC ACTH SECRETION

Some patients with ACTH-producing nonpituitary tumors may present to medical attention with a clinical picture dominated by features of the neoplastic disease itself rather than the attendant hormonal abnormalities. The clinical presentation of patients with ectopic ACTH production in early reports was remarkably devoid of the usual physical stigmata of chronic glucocorticoid excess. Most cases occurred in men and symptomatology was generally of brief duration, usually less than six months. The characteristic central obesity, facial plethora, abdominal striae, and supraclavicular and dorsocervical fat deposition were usually absent. Hyperpigmentation could be quite marked and glucose intolerance, hypokalemia, proximal muscle weakness, and frank weight loss were generally among the predominant features that brought the patient to clinical attention. In more recently reported series of ectopic ACTH cases, which include many more indolent tumors as the source of ACTH, most patients show classical features of Cushing's syndrome that are difficult to distinguish from those of patients with pituitary corticotroph tumors (5).

The biochemical findings in patients with ectopic ACTH secreting tumors do not clearly distinguish these patients from those with pitutitary source of excess ACTH in every case, but hypokalemia, and alkalosis are much more common in ectopic ACTH syndrome. ACTH and cortisol levels tend to overlap with the values observed in patients with pituitary corticotroph tumors when ectopic ACTH cases are analysed as a single group. However, some investigators have shown that patients with clearly evident tumors have much higher ACTH and cortisol levels than those with "occult" nonpituitary tumors or those with corticotroph adenomas as the source of excess ACTH (5) .

CLINICAL PATHOPHYSIOLOGY

ACTH is the product of the proopiomelanocortin gene that encodes a number of peptides including ACTH, the lipotropins, the melanocyte stimulating hormones and β-endorphin. The human proopiomelanocortin gene, located on chromosome 2, is comprised of three exons that encode a number of different possible mRNA transcripts by usage of at least three different promoters. While the pituitary expresses a single transcript of 1200 bases, the nonpituitary tissues contain a shorter transcript of about 800 bases in length. The ovary and testis express larger transcripts of up to 1500 bases in length. The pituitary-specific transcript is translated into a large precursor peptide of 31 kDa called proopiomelanocortin. Post translational processing of this precursor by a specific prohormone convertase yields peptide products including ACTH, the lipotropins, endorphins, and melanocyte stimulating hormones. In pituitary corticotroph cells the processing yields ACTH, the principal regulator of adrenal steroidogenesis. In nonpituitary tissues the mRNA transcripts from the POMC gene have been estimated to be only 0.008 to 0.08% as abundant as those in the pituitary. Translation of these nonpitutitary mRNAs into the POMC molecule is very inefficient; only minute amounts of POMC peptides are detectable in extrapituitary sites (7). ACTH production by nonpituitary tumors appears to require expression of the 1200 base "pituitary-type" mRNA (7, 8).

ACTH regulation is highly disordered in nonpituitary tumours that produce Cushing's syndrome. ACTH secretion from the normal pituitary corticotroph is regulated by an intrinsic circadian rhythm, by the positive stimulatory effects of corticotropin releasing hormone (CRH) and arginine vasopressin (AVP) and by the negative feedback effects of circulating glucocorticoids. In general, each of these regulatory mechanisms is disturbed in ACTH-producing nonpituitary tumors. The diurnal rhythm of ACTH secretion is lost, the stimulatory effects of CRH and AVP are blunted or absent (9) and ACTH (and cortisol) production are not suppressible by the administration of large doses of exogenous glucocorticoids (1,10).

Precursor peptides that give rise to ACTH by post-translational processing circulate in excess in patients with ectopic ACTH production (Figure 1). In the normal pituitary corticotroph the POMC messenger RNA is translated into a large precursor protein that is cleaved into smaller products by the action of two serine proteases called prohormone convertase 1 and 2 (PC1 and PC2). Normal action of these prohormone convertases in the anterior pituitary yields the 39 amino acid ACTH

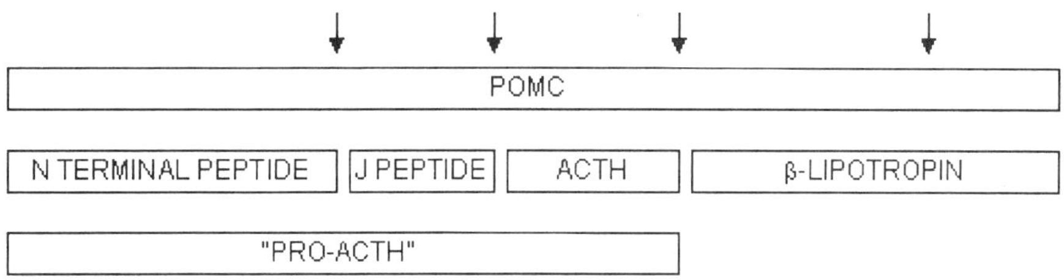

Figure 1. ACTH is cleaved from the proopiomelancortin molecule. The intact POMC precursor peptide (top) is cleaved at pairs of dibasic amino acids (arrows) to yield several smaller molecules in the normal anterior pituitary corticotroph (middle). In non-pituitary ACTH-producing tumors this post-translational processing is defective, so that the ACTH precursors POMC (top) and pro-ACTH (bottom) are produced in excess.

peptide. Nonpituitary tumors exhibit defects in this processing of ACTH precursor molecules (intact proopiomelanocortin and pro-ACTH). Levels of these precursors are greatly elevated in the plasma of patients with such tumors in comparison to normal persons or patients with ACTH-producing pituitary corticotroph tumors (11). The ratio of ACTH precursors to circulating authentic ACTH has been found to be elevated more than ten-fold in the plasma of patients with ectopic ACTH production compared to those with Cushing's syndrome (11).

Acquisition of features of the corticotroph phenotype

The nature of the regulatory signals driving POMC gene expression in nonpituitary ACTH-producing tumours is an area of active inquiry. Acquisition of this principal feature of the pituitary corticotroph phenotype by cells outside the pituitary could result from aberrant expression of the transcription factors that normally dictate corticotroph differentiation. Such developmentally regulated transcription factors have been found to be of importance in the differentiation of all the other hormone-producing cells of the anterior pituitary (12). Until recently the developmental program for corticotroph differentiation remained obscure. Evidence from animal models and human mutations has recently revealed a corticotroph-specific transcription factor designated Tpit, that directs differentiation of pituitary corticotrophs and expression of the POMC gene (13). Whether Tpit is involved in the ectopic expression of the POMC gene in nonpituitary tumors is unknown.

The DNA sequences within the 300-400 proximal bases of the proopiomelanocortin gene promoter have been found sufficient to direct transgene expression to the anterior pituitary and a number of other transcription factors that bind to DNA sequences in this promoter region have been identified. These include Nur77 (an orphan nuclear receptor whose expression is induced in corticotrophs by the action of CRH) (14-16), NeuroD1 (a helix-loop-helix protein) (17, 18), PTX1 (a homeobox protein) (19) and Brn3.0 (a POU-class DNA-binding protein) (20) . None of these transcription factors has been identified as responsible for driving ectopic POMC gene expression in any tumour, although Brn 3.0 has been found to be expressed at a high level in an aggressive corticotroph tumor as well as in a number of small cell lung cancers. These high levels of Brn 3.0 did not, however, correlate with POMC gene expression (21).

The activity of mammalian genes can also be modulated by the methylation of cytosine residues. The methylation-related silencing is thought to be due to recruitment of histone deacetylases that modify histones in a way that stabilizes chromatin and suppresses gene expression. It has been recently recognized that the POMC promoter, embedded within a CpG island that appears to be methylated in normal tissues that do not express POMC, is unmethylated in POMC expressing tissues and tumors (22).

Mutations in the POMC promoter itself or expression of transcription factors not normally operative at the POMC promoter might also be capable of driving POMC gene expression. No mutations in the POMC promoter have been found to date in ACTH-producing tumors of any kind (23) but one ACTH-producing cell line derived from a small cell lung cancer has been found to express a transcription factor of the E2F family that is capable of binding to a DNA element in the POMC promoter and driving POMC gene transcription by a mechanism that appears to be distinct from that operative in the normal corticotroph (24).

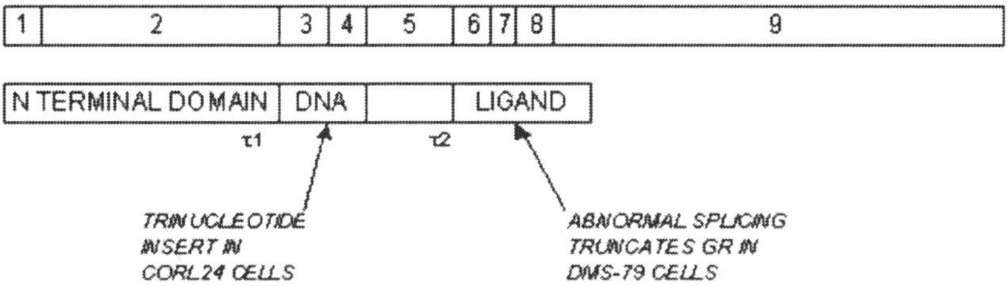

Figure 2. Disruption of glucocorticoid signalling may be responsible for the complete insensitivity of most non-pituitary ACTH-producing tumors to negative feedback inhibition of POMC gene expression. Glucocorticoids normally exert these effects through the glucocorticoid receptor. GR mRNA (top) is transcribed and spliced from the 9 exons of the GR gene. The resultant protein is a ligand–activated transcription factor with discrete DNA binding domain (DNA) ligand binding domain (LIGAND) and transcriptional activation domains (τ1 and τ2). Alterations in GR in ACTH-secreting tumor cell lines include a trinucleotide insert in the COR L24 cell GR that reduces its maximal activity and an abnormally spliced GR mRNA in DMS-79 cells that introduces a stop codon that truncates the ligand binding domain and inactivates the protein.

Insensitivity to feedback regulation by glucocorticoids

Clinical evidence suggests that nonpituitary ACTH producing tumors are, in general, profoundly insensitive to the negative feedback effects of glucocorticoids on POMC gene expression and ACTH secretion. A number of laboratories have sought to define the molecular and cellular alterations responsible for this apparent glucocorticoid resistance using cell culture models derived from ACTH-producing tumours. These cell lines are also resistant to the negative feedback effects of glucocorticoids (25, 26). In one such small cell lung cancer line (DMS-79) splicing of the glucocorticoid receptor mRNA is apparently defective, resulting in expression of a truncated GR protein that lacks a ligand binding domain and fails to activate positively regulated target genes or to exert suppressive effects on the POMC promoter (Figure 2)

(27-29). Another cell line derived from a small cell lung carcinoma (CORL24) has compound heterozygous mutations in its GR genes. One allele harbours an Asp to Ser mutation at residue 363 and appears to function normally. The other allele has a trinucleotide insert that positions an extra arginine between the two zinc fingers of the DNA binding domain (Figure 2). This mutant GR functions at a little less than half of the maximum of the wild-type receptor (30).

Hypokalemia

Hypokalemic alkalosis is a prominent clinical feature of the syndrome of ectopic ACTH secretion. It is clearly not mediated by aldosterone, since levels of this hormone are suppressed. Corticosterone and deoxycorticosterone are only variably elevated in patients with nonpituitary ACTH-producing tumours, and the levels do not correlate with accompaniment or degree of hypokalemia. Under normal physiologic conditions cortisol does not exert effects through the mineralocorticoid receptor, although MR's affinity for cortisol is equal to that for aldosterone (31). A specific enzyme called 11β-hydroxysteroid dehydrogenase, type II is expressed in the same cells as MR and serves too inactivate cortisol signalling by conversion of cortisol to cortisone, which has no activity at MR. This local "protection" of MR from circulating cortisol has been found to be defective in patients with ectopic ACTH syndrome (32, 33). It is not known whether ACTH itself or circulating steroid hormones down-regulate expression of the enzyme or inhibit the enzyme's activity (33). Alternatively, it has been proposed that saturation of the enzyme's capacity to inactivate cortisol may occur in the renal tubule in the setting of ectopic ACTH syndrome (34).

DIAGNOSTIC CONSIDERATIONS

Once it has been established that a given patient suffers from ACTH-dependent hypercortisolism it can still be significant diagnostic challenge to distinguish the infrequent case of ectopic ACTH secretion from the more frequent cause of ACTH-dependent Cushing's Syndrome, a pituitary corticotroph neoplasm. For over thirty-five years the apparent glucocorticoid resistance of nonpituitary ACTH-producing tumours has been exploited in the use of the high dose dexamethasone suppression test to distinguish ectopic ACTH secretion from Cushing's syndrome (35). Several variations of the high dose test have been used, including the standard (2 mg dexamethasone every 6 hours for two days), a single 8 mg dose at 11pm-12 mn (36), and a seven hour infusion (37). Each of these tests correctly distinguishes between the suppressibility of Cushing's syndrome and the non-suppressibility of ACTH-dependent tumours in the majority of cases. However, because a small proportion of corticotroph tumours fail to suppress with high dose dexamethasone administration, and a small proportion of nonpituitary ACTH-producing tumors do exhibit suppressibility, not every patient can be correctly classified by even the modified tests. Such patients require additional procedures to localize ACTH secretion to the pituitary (by inferior petrosal sinus sampling (38)) or to localize a nonpituitary neuroendocrine tumor (by somatostatin receptor scintigraphy (39)).

Somatostatin receptor scintigraphy exploits the expression of somatostatin recptors by neuroendocrine tumours as targets for radiopharmaceuticals suitable for image generation (Figure 3).

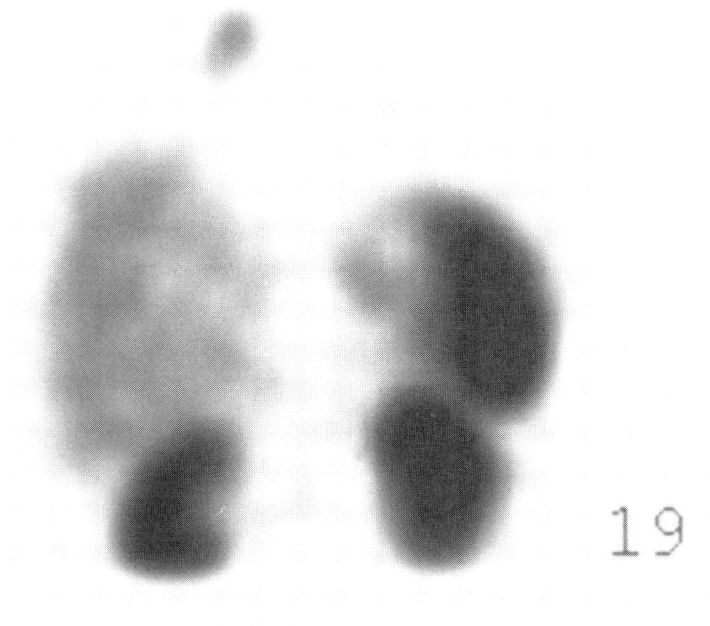

Figure 3. Octreotide scan of a patient with an ACTH-secreting bronchial carcinoid tumor located in the right upper lung.

^{111}Indium labelled pentetreotide, which binds to two of the five subclasses of somatostatin receptors, has been used successfully to image a wide variety of tumors including carcinoids, gastrinomas, insulinomas, and glucagonomas. Numerous case reports of their successful application in identification of tumours responsible for ectopic ACTH syndrome have been reported. However, the general superiority of somatostatin receptor scintigraphy over conventional CT and MR imaging has not been evident in larger series of patients with nonpituitary ACTH producing tumors (40,

41). False positive scans may be fairly frequent and it appears that, once a negative SRS image is obtained, repeat studies are not useful in detecting occult ACTH-producing neoplasms. Most authorities consider somatostatin receptor scintigraphy as an adjunct to conventional imaging techniques.

DIFFERENTIAL DIAGNOSIS

Bronchial Carcinoid

ACTH-secreting bronchial carcinoid tumors are usually located in the mid lung zones (A). They are small tumors, often < 1 cm in diameter, and two-thirds of them are occult in that the ultimate diagnosis is delayed by more than six months following recognition of the presence of ectopic ACTH hypersecretion (A). Patients typically present with clinical features indistinguishable from those seen in pituitary-dependent Cushing's syndrome. Hypokalemia is, however, more frequent and is seen in more than two-thirds of affected patients (B). Respiratory symptoms are unusual. Only 15% of patients have abnormal 24h Urine 5-HIAA levels (B). More than one-half of patients exhibit some degree of suppression following the administration of high-dose dexamethasone and nearly two-thirds respond to metyrapone (B). Chest radiographs are abnormal in 36% of patients whereas computed tomography of the chest is abnormal in 89% of patients (B). Bronchial carcinoids have a high signal intensity on T2-weighted and STIR sequences on magnetic resonance imaging (C). These characteristics may permit the localization of central lesions and facilitate their distinction from pulmonary vasculature when CT is equivocal (Figure 4). Bronchoscopy and lavage usually fails to localize these tumors (D,E). When possible, lobectomy is the treatment of choice. Approximately two-fifths of bronchial carcinoids are malignant (B). In many cases, the metastases appear years after the initial diagnosis and treatment.

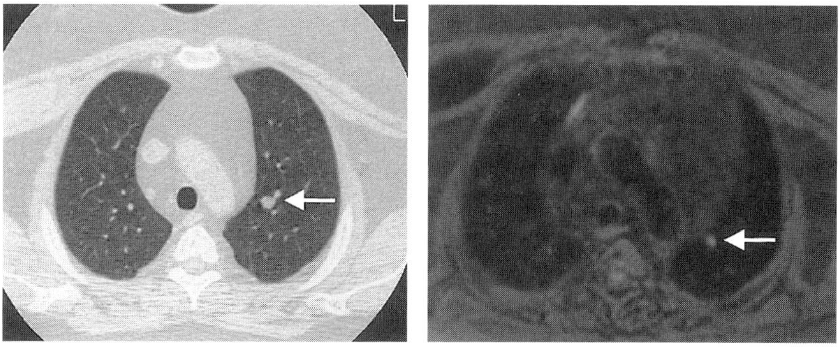

Figure 4. CT and MRI of the chest in a patient with an ACTH-secreting bronchial carcinoid tumor located in the left upper lung. The lesion was initially overlooked on the CT images as the radiographic abnormality was felt to represent a dilated blood vessel. The lesion was ultimately recognized due to its hyperintense character on T-2 weighted MRI sequences.

Thymic Carcinoid

Carcinoid tumors of the thymus gland are rare. Most present as large mediastinal masses with local symptoms (F). They are usually apparent of CT examintions of the chest and mediastinum but microadenomas have been reported. These tumors commonly invade local structures. Complications may include the superior vena cava syndrome, supravalvular pulmonary artery stenosis, and pericardial tamponade. Distant metastases occur in 20-30% of patients. One-half of these tumors are associated with an endocrinopathy. The ten-year mortality in patients with ACTH-secreting thymic carcinoids is 65% (G).

Islet-Cell Carcinoid

ACTH-secreting islet-cell carcinoid tumors are large malignant lesions (H). Most have metastasized to the liver and regional lymph nodes by the time of presentation of Cushing's syndrome. In a review by Doppman et al, the tumors ranged in size from 2.5 to 6 cm (H). Eighty per-cent of patients also had the Zollinger-Ellison syndrome do to co-secretion of gastrin. The Zollinger-Ellison syndrome preceded a diagnosis of Cushing's syndrome by one to three years in 63% of patients with both disorders.

Medullary Thyroid Carcinoma

Cushing's syndrome may herald a diagnosis of medullary thyroid carcinoma. A thyroid mass is often palpable and most tumors are metastatic at the time of diagnosis. In one series of 184 patients with medullary thyroid carcinoma, fifty (33%) developed distant metastases; four of these developed the syndrome of ectopic ACTH hypersecretion (I).

Pheochromocytoma

ACTH-secreting pheochromocytomas are nearly always benign solitary unilateral tumors. In a series by Chen et al, the mean tumor size was 2.8 cm (J). A review of the literature disclosed tumors ranging in size form 2.3 to 11 cm. Patients may present with clinical features similar to those seen in pituitary-dependent disease or with features consistent with overt ectopic ACTH hypersecretion. Urinary excretion of catecholamines and their metabolites is usually elevated. Contralateral ACTH-induced adrenal hyperplasia may suggest bilateral disease. MRI is useful in determining which adrenal gland harbors the pheochromocytoma as these tumors are bright or hyperintense on T2-weighted images (Figure 5).

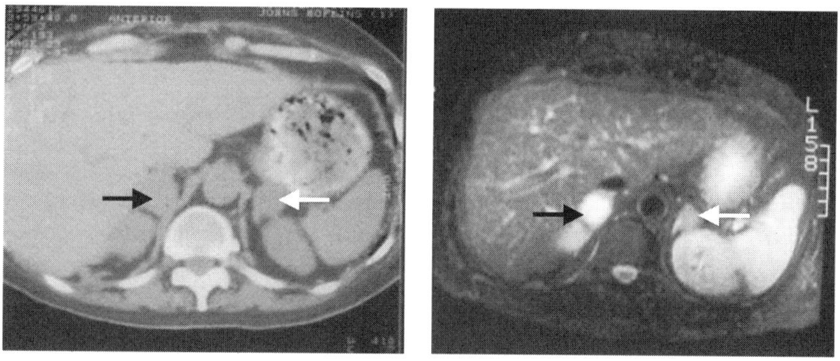

Figure 5. CT and MRI of the abdomen in a patient with an ACTH-secreting pheochromocytoma. The black arrows denote the pheochromocytoma that was hyperintense on T2-weighted MRI. The white arrows denote the contralateral hyperplastic adrenal gland. Unilateral right adrenalectomy resulted in cure of the hypercortisolism and the catecholamine excess.

Small Cell Carcinoma of the Lung

Small cell carcinoma of the bronchus accounts for 20% of patients with the syndrome of ectopic ACTH hypersecretion. While the syndrome may herald the diagnosis of metastatic carcinoma, most patients have a known lung mass and distant metastases. Ectopic ACTH hypersecretion should be suspected in male cigarette smokers who present with weight loss, hypertension, diabetes mellitus, proximal myopathy, and hypokalemia.

Other Tumors

Pancreatic cystadenomas, ovarian adenocarcinoma, colon carcinoma, intestinal carcinoid tumors and adenocarcinoma of the prostate have each been reported in association with ectopic ACTH hypersecretion.

THERAPEUTIC CONSIDERATIONS

The first line of therapy for non-pituitary ACTH-producing tumors is, of course, resection. However, most of these tumors are not resectable and further therapy is required to control hypercortisolism. Metastatic disease, if localized to the liver, may be treatable by resection or ablative procedures such as hepatic arterial embolization (3) and liver transplantation has been used (42).

For most patients with ectopic ACTH hypersecretion the therapeutic goal is control of hypercortisolism. A variety of agents, including somatostatin receptor agonists, inhibitors of adrenal steroidogenesis, and adrenolytic agents have been used. Suppression of tumor ACTH production has been achieved in some patients with somatostatin receptor-positive tumors using octreotide, a long-acting somatostatin analog as either pre-operative or palliative therapy (43-46) aminoglutethimide, and ketoconazole have all been used for control of the hypercortisolism resulting from ectopic ACTH secretion (47-53). Steroidogenic inhibition is believed to be effective in ectopic ACTH syndrome because there is no compensatory rise in ACTH levels as occurs in ACTH-dependent Cushing's syndrome caused by pituitary corticotroph adenomas. Combinations of these steroidogenic inhibitors are effective, and may allow the use of lower doses with reduction in resultant side effects.

CONCLUSION

Cortisol excess driven by ACTH production from nonpituitary tumors remains a less common cause of Cushing's syndrome, but a considerable diagnostic and therapeutic challenge. Insight into the molecular pathophysiology of ectopic ACTH secretion may provide new strategies for the diagnostic distinction of this entity from Cushing's syndrome and may reveal new possibilities for the therapeutic control of

ACTH production from occult tumors that cannot be localized and surgically resected.

REFERENCES

1. Meador CK, Liddle GW, Island DP et al. Cause of Cushing's syndrome in patients with tumors arising from "nonendocrine" tissue. J Clin Endocrinol Metab 1962; 22:693-703.
2. Liddle GW, Nicholson WE, Island DP, Orth DN, Abe K, Lowder SC. Clinical and laboratory studies of ectopic humoral syndromes. Recent Prog Horm Res 1969; 25:283-314.
3. Wajchenberg BL, Mendonca BB, Liberman B et al. Ectopic adrenocorticotropic hormone syndrome. Endocr Rev 1994; 15(6):752-787.
4. Orth DN. Ectopic hormone production. In: Felig P, Baxter JD, Broadus AE, Frohman LA, editors. Endocrinology and Metabolism. New York: McGraw-Hill, 1981: 1191-1217.
5. Howlett TA, Drury PL, Perry L, Doniach I, Rees LH, Besser GM. Diagnosis and management of ACTH-dependent Cushing's syndrome: comparison of the features in ectopic and pituitary ACTH production. Clin Endocrinol (Oxf) 1986; 24:699-713.
6. Howlett TA, Drury PL, Perry L, Doniach I, Rees LH, Besser GM. Diagnosis and management of ACTH-dependent Cushing's syndrome: comparison of the features in ectopic and pituitary ACTH production. Clin Endocrinol (Oxf) 1986; 24:699-713.
7. DeBold CR, Menefee JK, Nicholson WE, Orth DN. Proopiomelanocortin gene is expressed in many normal human tissues and in tumors not associated with ectopic adrenocorticotropin syndrome. Mol Endocrinol 1988; 2:862-870.
8. DeBold CR, Schworer ME, Connor TB, Bird RE, Orth DN. Ectopic pro-opiolipomelanocortin: sequence of cDNA coding for beta- melanocyte-stimulating hormone and beta-endorphin. Science 1983; 220:721-723.
9. Chrousos GP, Schulte HM, Oldfield EH, Gold PW, Cutler GB, Jr., Loriaux DL. The corticotropin-releasing factor stimulation test. An aid in the evaluation of patients with Cushing's syndrome. N Engl J Med 1984; 310:622-626.
10. Liddle GW. Tests of pituitary-adrenal suppressibility in the diagnosis of Cushing's syndrome. J Clin Endocrinol Metab 1960; 20:1539-1560.
11. Stewart PM, Gibson S, Crosby SR et al. ACTH precursors characterize the ectopic ACTH syndrome. Clin Endocrinol (Oxf) 1994; 40:199-204.
12. Rosenfeld MG, Bach I, Erkman L et al. Transcriptional control of cell phenotypes in the neuroendocrine system. Recent Prog Horm Res 1996; 51:217-238.
13. Lamolet B, Pulichino AM, Lamonerie T et al. A pituitary cell-restricted T box factor, Tpit, activates POMC transcription in cooperation with Pitx homeoproteins. Cell 2001; 104:849-859.
14. Drouin J, Maira M, Philips A. Novel mechanism of action for Nur77 and antagonism by glucocorticoids: a convergent mechanism for CRH activation and glucocorticoid repression of POMC gene transcription. J Steroid Biochem Mol Biol 1998; 65:59-63.
15. Philips A, Maira M, Mullick A et al. Antagonism between Nur77 and glucocorticoid receptor for control of transcription. Mol Cell Biol 1997; 17:5952-5959.
16. Murphy EP, Conneely OM. Neuroendocrine regulation of the hypothalamic pituitary adrenal axis by the nurr1/nur77 subfamily of nuclear receptors. Mol Endocrinol 1997; 11:39-47.
17. Poulin G, Turgeon B, Drouin J. NeuroD1/beta2 contributes to cell-specific transcription of the proopiomelanocortin gene. Mol Cell Biol 1997; 17:6673-6682.
18. Therrien M, Drouin J. Cell-specific helix-loop-helix factor required for pituitary expression of the pro-opiomelanocortin gene. Mol Cell Biol 1993; 13:2342-2353.
19. Lamonerie T, Tremblay JJ, Lanctot C, Therrien M, Gauthier Y, Drouin J. Ptx1, a bicoid-related homeo box transcription factor involved in transcription of the pro-opiomelanocortin gene. Genes Dev 1996; 10:1284-1295.

20. Gerrero MR, McEvilly RJ, Turner E et al. Brn-3.0: a POU-domain protein expressed in the sensory, immune, and endocrine systems that functions on elements distinct from known octamer motifs. Proc Natl Acad Sci U S A 1993; 90:10841-10845.
21. Leblond-Francillard M, Picon A, Bertagna X, De Keyzer Y. High expression of the POU factor Brn3a in aggressive neuroendocrine tumors. J Clin Endocrinol Metab 1997; 821:89-94.
22. Newell-Price J, King P, Clark AJ. The CpG island promoter of the human proopiomelanocortin gene is methylated in nonexpressing normal tissue and tumors and represses expression. Mol Endocrinol 2001; 15:338-348.
23. Monig H, Ali IU, Oldfield EH, Schulte HM. Structure of the POMC promoter region in pituitary and extrapituitary ACTH producing tumors. Exp Clin Endocrinol 1993; 101:36-38.
24. Picon A, Bertagna X, De Keyzer Y. Analysis of the human proopiomelanocortin gene promoter in a small cell lung carcinoma cell line reveals an unusual role for E2F transcription factors. Oncogene 1999; 18:2627-2633.
25. Clark AJ, Stewart MF, Lavender PM et al. Defective glucocorticoid regulation of proopiomelanocortin gene expression and peptide secretion in a small cell lung cancer cell line. J Clin Endocrinol Metab 1990; 70:485-490.
26. Farrell WE, Stewart MF, Clark AJ, Crosby SR, Davis JR, White A. Glucocorticoid inhibition of ACTH peptides: small cell lung cancer cell lines are more resistant than pituitary corticotroph adenoma cells. J Mol Endocrinol 1993; 10:25-32.
27. Gaitan D, DeBold CR, Turney MK, Zhou P, Orth DN, Kovacs WJ. Glucocorticoid receptor structure and function in an adrenocorticotropin-secreting small cell lung cancer. Mol Endocrinol 1995; 9:1193-1201.
28. Parks LL, Gaitan D, Turney MK, Detera-Wadleigh SD, Kovacs WJ. An ACTH-producing small cell lung cancer expresses aberrant glucocorticoid receptor transcripts from a normal gene. Mol.Cell Endocrinol 1998; 142:175-181.
29. Turney MK, Kovacs WJ. Function of a truncated glucocorticoid receptor form at a negative glucocorticoid response element in the proopiomelanocortin gene. J Mol Endocrinol 2001; 26:43-49.
30. Ray DW, Davis JR, White A, Clark AJ. Glucocorticoid receptor structure and function in glucocorticoid- resistant small cell lung carcinoma cells. Cancer Res 1996; 56:3276-3280.
31. Arriza JL, Weinberger C, Cerelli G et al. Cloning of human mineralocorticoid receptor complementary DNA: structural and functional kinship with the glucocorticoid receptor. Science 1987; 237:268-275.
32. Stewart PM, Walker BR, Holder G, O'Halloran D, Shackleton CH. 11 beta-Hydroxysteroid dehydrogenase activity in Cushing's syndrome: explaining the mineralocorticoid excess state of the ectopic adrenocorticotropin syndrome. J Clin Endocrinol Metab 1995; 80:3617-3620.
33. Walker BR, Campbell JC, Fraser R, Stewart PM, Edwards CR. Mineralocorticoid excess and inhibition of 11 beta- hydroxysteroid dehydrogenase in patients with ectopic ACTH syndrome Clin Endocrinol (Oxf) 1992; 37:483-492.
34. Ulick S, Wang JZ, Blumenfeld JD, Pickering TG. Cortisol inactivation overload: a mechanism of mineralocorticoid hypertension in the ectopic adrenocorticotropin syndrome. J Clin Endocrinol Metab 1992; 74:963-967.
35. Orth DN. Medical Progress: Cushing's syndrome. N Engl J Med 1995; 332:791-803.
36. Tyrrell JB, Findling JW, Aron DC, Fitzgerald PA, Forsham PH. An overnight high-dose dexamethasone suppression test for rapid differential diagnosis of Cushing's syndrome. Ann Intern Med 1986; 104:180-186.
37. van den Bogaert DP, de Herder WW, de Jong FH, Biemond P, van der Lely AJ, Lamberts SW. The continuous 7-hour intravenous dexamethasone suppression test in the differential diagnosis of ACTH-dependent Cushing's syndrome. Clin Endocrinol (Oxf) 1999; 51:193-198.
38. Oldfield EH, Doppman JL, Nieman LK et al. Petrosal sinus sampling with and without corticotropin-releasing hormone for the differential diagnosis of Cushing's syndrome. N Engl J Med 1991; 325:897-905.

39. de Herder WW, Lamberts SW. Tumor localization--the ectopic ACTH syndrome. J Clin Endocrinol Metab 1999; 84:1184-1185.
40. Tabarin A, Valli N, Chanson P et al. Usefulness of somatostatin receptor scintigraphy in patients with occult ectopic adrenocorticotropin syndrome. J Clin Endocrinol Metab 1999; 84:1193-1202.
41. Torpy DJ, Chen CC, Mullen N et al. Lack of utility of (111)In-pentetreotide scintigraphy in localizing ectopic ACTH producing tumors: follow-up of 18 patients. J Clin Endocrinol Metab 1999; 84:1186-1192.
42. Lehnert T. Liver transplantation for metastatic neuroendocrine carcinoma: an analysis of 103 patients. Transplantation 1998; 66:1307-1312.
43. de Herder WW, Lamberts SW. Octapeptide somatostatin-analogue therapy of Cushing's syndrome. Postgrad Med J 1999; 75:65-66.
44. Hearn PR, Reynolds CL, Johansen K, Woodhouse NJ. Lung carcinoid with Cushing's syndrome: control of serum ACTH and cortisol levels using SMS 201-995 (sandostatin). Clin Endocrinol (Oxf) 1988; 28:181-185.
45. Phlipponneau M, Nocaudie M, Epelbaum J et al. Somatostatin analogs for the localization and preoperative treatment of an adrenocorticotropin-secreting bronchial carcinoid tumor. J Clin Endocrinol Metab 1994; 78:20-24.
46. Bertagna X, Favrod-Coune C, Escourolle H et al. Suppression of ectopic adrenocorticotropin secretion by the long-acting somatostatin analog octreotide. J Clin Endocrinol Metab 1989; 68:988-991.
47. Winquist EW, Laskey J, Crump M, Khamsi F, Shepherd FA. Ketoconazole in the management of paraneoplastic Cushing's syndrome secondary to ectopic adrenocorticotropin production. J Clin Oncol 1995; 13:157-164.
48. Tabarin A, Navarranne A, Guerin J, Corcuff JB, Parneix M, Roger P. Use of ketoconazole in the treatment of Cushing's disease and ectopic ACTH syndrome. Clin Endocrinol (Oxf) 1991; 34:63-69.
49. Farwell AP, Devlin JT, Stewart JA. Total suppression of cortisol excretion by ketoconazole in the therapy of the ectopic adrenocorticotropic hormone syndrome. Am J Med 1988; 84:1063-1066.
50. Shepherd FA, Hoffert B, Evans WK, Emery G, Trachtenberg J. Ketoconazole. Use in the treatment of ectopic adrenocorticotropic hormone production and Cushing's syndrome in small-cell lung cancer. Arch Intern Med 1985; 145:863-864.
51. Comi RJ, Gorden P. Long-term medical treatment of ectopic ACTH syndrome. South Med J 1998; 91:1014-1018.
52. Gorden P, Becker CE, Levey GS, Roth J. Efficacy of amino-glutethimide in the ectopic ACTH syndrome. J Clin Endocrinol Metab 1968; 28:921-923.
53. .Coll R, Horner I, Kraiem Z, Gafni J. Successful metyrapone therapy of the ectopic ACTH syndrome. Arch Intern Med 1968; 121:549-553.
54. Doppman JL. The search for Ecult Ectopic ACTH-Producing Tumors. The Endocrinolgist 2:41-46, 1992.
55. Leinung MC, Young WF, Whitaker MD, Scheithauer BW, Trastek VF, Kvols LK. Diagnosis of Corticotrophin-Producing Bronchial Carcinoids Tumors Tausing Cushing's Syndrome. Mayo Clin Proc 65:1314-1321, 1990.
56. Doppman JL, Pass HI, Nieman LK, Findling JW, Dwyer AJ, Feuerstin IM, Ling A, Travis WD, Cutler GB, Chrousos GP, Lorio DL. Detection of ACTH-Producing Bronchial carcinoid tumors: MR imaging vs. CT. AJR 156:39-43, 1991.
57. Doppman JL, Pass HI, Nieman L, Cutler GB, Chrousos GP, Loriaux BL. Failure of Bronchial Levage to Detect Elevated Levels of Adrenacorticotrophin (ACTH) in Patients with ACTH-Producing Bronchial Carcinoids. J Clin Endocrinol Metab 69:1302-1304, 1989.
58. Limper AH, Carpender PC, Scheithauer B, Staats BA. Cushing's Syndrome Induced by Bronchial Carcinoid Tumors. Annals of Internal Medicine 117:209-214, 1992.
59. Economopoulos GC, Lewis JW, Lee MW, Silverman NA. Carcinoid Tumors of the Thymus. Ann Thorac Surg 50:58-61, 1990.
60. Gartner LA, Vorhees ML. During the Corticotrophic Hormone- Producing Thymic Carcinoid in a Teenager. Cancer 71:106-111, 1993.

61. Doppman JL, Nieman LP, Cutler GB, et al. Adrinocorticotrophic Hormone-Secreting Islet Cell Tumors: Are They Always Malignant? Radiology 190:59-64, 1994.
62. Mure A, Gicquel C, Abdelmoumen N, et al. Cushing's Syndrome in Medularitharic Carcinoma. J Endocrinol Invest 18:180-185, 1995.
63. Chen H, Doppman JL, Chrousos GP, Norton JA, Nieman LK, Udelsman R. Adrenocorticotrophic Hormone-Secreting Cartocycomas: The Exception to the Rule.

9
CUSHING'S SYNDROME DUE TO ADRENAL DISORDERS

Lewis S. Blevins, Jr., M.D.
Division of Endocrinology, Diabetes and Metabolism, Department of Medicine, Vanderbilt School of Medicine, Nashville, TN, USA

Hypercortisolism due to primary adrenal diseases is responsible for approximately one-quarter of all cases of endogenous Cushing's syndrome (1,2). Disorders of the adrenal glands that secrete cortisol in an autonomous fashion and lead to the development of Cushing's syndrome are listed in Table 1. Benign and malignant adrenocortical neoplasms account for a majority of the cases. Although rare, the various adrenal hyperplasias represent an interesting spectrum of disorders encountered by physicians evaluating patients with Cushing's syndrome. This chapter will focus on the pathophysiology, clinical presentation, evaluation and management of these disorders.

With the exception of adrenocortical carcinoma (ACC), diseases of the adrenal glands that cause Cushing's syndrome usually progress rather slowly. Early in the stages of these disorders, serum cortisol levels and 24-hour urine free cortisol (24-h UFC) excretion rates may be normal (1). Often, however, there is a demonstrable loss of the normal diurnal variation in cortisol secretion. Plasma adrenocorticotrophic hormone (ACTH) levels are often detectable because the mass of cortisol secreted per day is not yet sufficient to fully suppress pro-opiomelanocortin (POMC) gene transcription and ACTH secretion (Figure 1) (1). As a result, cortisol secretion may be partially suppressible, down to the level of autonomous cortisol secretion, by the administration of dexamethasone. In this setting, it can be rather difficult to distinguish primary adrenal disorders from ACTH-dependent causes of Cushing's syndrome. Administration of corticotropin-releasing hormone (CRH) may provoke ACTH secretion from a pituitary tumor and allow one to separate these patients from those with primary adrenal diseases (1). As the primary adrenal disorder progresses, serum cortisol levels and the 24-h UFC rise to above their respected normal ranges, plasma ACTH levels become fully suppressed, and cortisol secretion, as assessed by serum cortisol levels and 24-h UFC, fails to decline following the administration of dexamethasone.

Table 1. Adrenal Causes of Cushing's Syndrome

Cortisol-producing Adrenal Adenoma
Adrenocortical Carcinoma
Adrenal Hyperplasia
- Primary Pigmented Nodular Adrenocortical Disease
- McCune –Albright syndrome
- Ectopic Receptor Expression
  - Food-Dependent (GIP receptor)
  - Pregnancy/Menopause (LH receptor)
  - Adrenergic Stimulation (β-adrenergic receptor)
  - Volume –Dependent (AVP receptor)
  - ? (5-HT receptor)
- Unclear Causes
  - Micronodular changes
  - Macronodular changes
  - Massive Macronodular changes

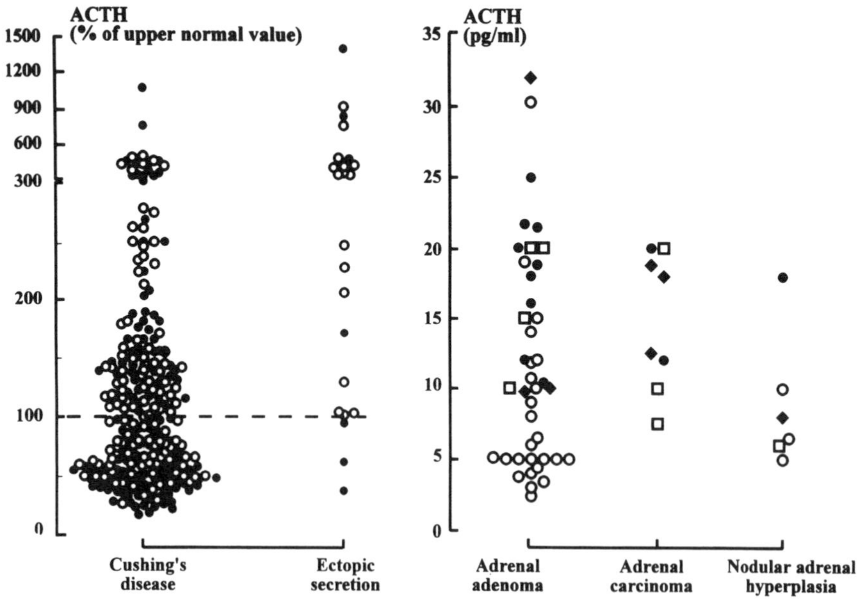

Figure 1. Plasma ACTH levels in patients with Cushing's syndrome. *Reproduced with permission from reference 1.*

Overt Cushing's syndrome may be preceded by a latent or subclinical phase characterized by one or more common clinical features such as obesity, hypertension, and noninsulin-dependent diabetes mellitus (3). Patients with ACC often have an explosive onset and rapid progression of their symptoms and signs (2). In general, the clinical features of patients with Cushing's syndrome due to primary adrenocortical disorders are similar to those seen in patients with ACTH-dependent Cushing's syndrome (Table 2) (2-4). There are, however, a few exceptions. Hyperpigmentation is usually not encountered in those patients with adrenal Cushing's since ACTH levels are usually low or undetectable. Patients with the Carney Complex may, however, have hyperpigmented macules about their centrofacial and upper torso regions (5). Patients with pure cortisol-secreting adrenal disorders often have hypertrichosis but almost never have hirsutism as adrenal androgen production is negligible in the setting of ACTH suppression. Dehydroepiandrosterone-sulfate (DHEA-S) levels are usually low in these patients (1,4,6). Patients with adrenocortical carcinoma may, however, have marked hirsutism due to autonomous hypersecretion of adrenal androgens. DHEA-S and 24-h urinary 17-ketosteroid (17-KS) levels are usually elevated in these patients (1,4,7). Hypokalemia is an uncommon finding except in those patients with adrenocortical carcinoma who also secrete potent mineralocorticoids (8-11). In these patients, plasma renin levels are usually suppressed. Aldosterone levels are often elevated but they may be undetectable in the setting of 11-deoxycorticosterone (11-DOC) hypersecretion (8,9). Other unusual historical and clinical features that are largely dependent upon the underlying disorder may be present in patients with adrenal Cushing's. These often serve as subtle clues to the underlying diagnosis and will be discussed in detail in the pertinent sections of this chapter.

Table 2. Clinical Features of Cushing's Syndrome Due to Primary Adrenal Diseases.

	%		
	Scott & Orth[2]	*Reincke*[3]	*Daitch*[4]
Central Obesity	93	88	86
Hypertension	93	63	89
Weakness/Myopathy	82		96
Hirsutism	79	88	96
Glucose intolerance/DM	79	13	47
Psychiatric Abnormalities	57	50	
Ecchymoses	57	63	83
Osteoporosis	54	38	
Edema	46	25	70
Striae	36	63	74
Abnormal Menses	75	63	94

CORTISOL-PRODUCING ADRENOCORTICAL ADENOMA

Cortisol-producing adrenocortical adenomas (CPA) are slow-growing, solitary, unilateral, benign adrenal gland tumors (2,4). Bilateral adenomas have been reported (12). Cushing's syndrome due to CPA has been reported in association with a synchronous contralateral nonfunctioning adenoma (13).

CPA's are more common in women than in men. The female to male ratio typically ranges from 5:1 to 20:1 (1,2,4). In most cases of overt Cushing's syndrome, symptoms and signs of hypercortisolism have been present for two to three years prior to diagnosis (1,14). In a study of 27 CPA's by Daitch et al, 26 of 27 adenomas occurred in women (4). The mean age of their patients was 39.6 ± 14.4 years. The mean adenoma size was 3.3 ± 1.0 cm with a range in size from 2 to 5 cm.

CPA's are well-circumscribed tumors that are usually golden yellow on cut section but some are black adenomas that exhibit a mottled or diffusely brown color due to lipofuscin deposition (15). Tumor cells are usually arranged in small nests but a solid growth pattern may be seen. Cells are usually clear to foamy due to the presence of accumulated sterols and lipids. Nuclei are generally small to medium-sized and pleomorphism is slight or absent. Necrosis is unusual and mitotic figures are infrequent. Adenoma capsules vary in thickness and may be imperceptible. Vascular and capsular invasion are not seen (16). Adjacent unaffected adrenal cortex is usually atrophic and this observation correlates well with suppressed baseline ACTH levels (3).

CPA's usually do not cause local symptoms due to mass effect but some have grown to massive proportions and have caused abdominal and flank pain. A subset of patients with CPA's are now recognized in a subclinical stage where they may have only one or more nonspecific signs such as obesity, hypertension, and diabetes (3). These patients often first present for evaluation following incidental detection of a unilateral solitary adrenal mass. McLeod et al identified six patients with subclinical Cushing's syndrome out of 122 patients with incidental asymptomatic adrenal adenomas (17). Important criteria for subclinical Cushing's syndrome included nonsuppression of cortisol following dexamethasone, suppressed ACTH levels, loss of diurnal variation in cortisol secretion, or unilateral uptake of iodocholesterol with suppression of the contralateral gland. Tumors in these patients ranged from 2 to 7 cm in diameter with a mean of 3.7 ± 1.9 cm. Five of six patients had suppressed ACTH levels, five of five had abnormal diurnal rhythms of cortisol secretion, and two of five had elevated 24-h UFC. Reincke et al studied 68 patients with asymptomatic adrenal tumors and found eight who had pathological cortisol production (3). The mean tumor size in their series was 3.7 ± 1.0 cm. All patients had abnormal morning serum cortisol levels following the administration of dexamethasone. ACTH levels were suppressed in five of the eight patients. Urinary cortisol levels were normal in three of four tested patients. Patients seemed to exhibit different stages of autonomous cortisol secretion suggesting a natural course of CPA's evolving from an asymptomatic stage and progressing to overt Cushing's syndrome. Rossi et al identified subclinical Cushing's syndrome in nearly one-quarter of 50 patients with consecutive incidentally detected adrenal adenomas (18). Clinical features in those patients who did not undergo adrenalectomy remained stable and no patient developed overt Cushing's syndrome. While it seems that relatively few, if any, of the patients with subclinical Cushing's syndrome will develop overt Cushing's syndrome, it seems prudent to individualize decisions regarding the need for surgery. Tumor size, the presence of potentially reversible cardiovascular risk factors such as diabetes, hypertension, and hyperlipidemia, and patient preference should all be taken into account once a diagnosis of subclinical Cushing's due to a CPA has been made.

Computed tomography (CT) of the adrenal glands usually demonstrates a well-defined unilateral adrenal mass that has smooth contours (Figure 2) (19). The adenomas are most often homogenous, may enhance following the administration of iodinated contrast, and on thin-section unenhanced scans often have an attenuation ≤ 10 Hounsfield Units (20). The contralateral unaffected adrenal gland may appear normal or even atrophic. Adrenal adenomas have a similar appearance on magnetic resonance imaging studies of the adrenal glands. Evaluation of signal intensity on gadolinium-enhanced fat-suppressed T1-weighted spin echo images can help differentiate benign adenomas from malignant lesions (21). Iodocholesterol uptake is increased in solitary unilateral CPA's and the uptake by the contralateral unaffected adrenal gland is typically absent (22,23).

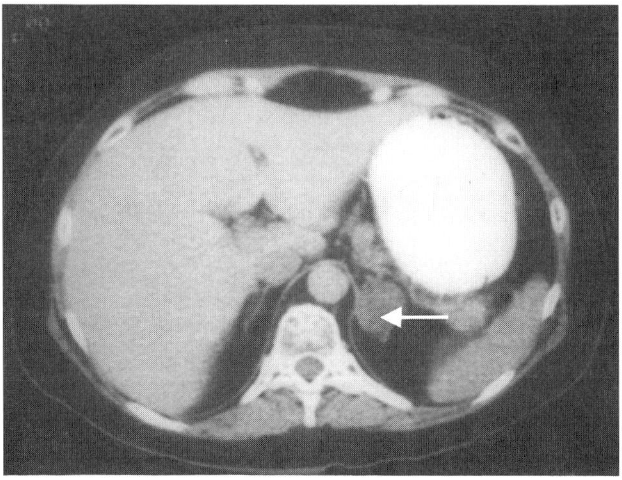

Figure 2. Computed tomography of the abdomen in a patient with a left adrenal cortisol-producing adrenocortical adenoma.

The treatment of choice for CPA is surgical excision. Most resections can be successfully accomplished employing a fiberoptic laprascope but some surgeons prefer the posterolateral flank approach. Operative morbidity and mortality is low and patients correctly diagnosed with CPA are usually rendered disease-free (12,4). Cushingoid features and related complications resolve in nearly all patients. Successfully treated patients may experience central adrenocortical insufficiency following surgery as the hypothalamic-pituitary-adrenal axis has usually been suppressed due to prolonged hypercortisolism. Scott et al reported that their CPA patients were successfully weaned off of steroids within three to six months (2). Daitch et al observed a mean length of postoperative exogenous steroid replacement therapy of 16.8 ± 9.1 months and noted that over one-half of their patients required steroid supplementation for more than two years (4). McLeod et al noted that four of five patients with subclinical Cushing's required steroid supplementation after unilateral adrenalectomy (17).

X-chromosome inactivation analysis has shown that nearly all CPA's are monoclonal tumors (24). Adenoma development and growth does not seem to be related to activating mutations of the ACTH receptor nor dependent on ACTH (25). Most tumors will, however, secrete cortisol in response to exogenous ACTH (26). Willenberg et al reported a case of an adenoma characterized by lymphocytic infiltration with CD-45 and CD-68 positive lymphocytes (27). Interleukin-1 receptor mRNA was detected in the adenoma cells but not in adjacent normal adrenal gland tissue. In culture, interleukin-1β resulted in cortisol secretion from the adenoma cells. De Herder et al described a 45-year old woman with aberrant expression of glucose-dependent insulinotrophic peptide (GIP) receptors in a unilateral adrenocortical adenoma (28). They hypothesized that a somatic mutation led to the abnormal expression of GIP receptors and that rises in GIP levels after food ingestion led to the development of Cushing's syndrome. Chabre et al reported a case of a 32-year old woman with an adrenocortical adenoma and food-dependent Cushing's syndrome (26). Cells from the adenoma expressed GIP receptors. Adenoma cells secreted cortisol in response to GIP. GIP also stimulated increases in mitogen-activated protein kinase activity suggesting that abnormal receptor expression may have played a role in proliferation of the tumor cells. Other studies have showed that vasopressin receptors are aberrantly expressed in some adrenocortical adenomas and that cortisol secretion from adenoma cells can be stimulated by receptor-ligand interaction (29,30). These observations suggest that somatic mutations may play a role in the development, progression, and the clinical presentation of CPA's.

ADRENOCORTICAL CARCINOMA

Adrenocortical carcinoma (ACC) is a lethal neoplasm that accounts for approximately 0.2% of all deaths due to cancer (31). The approximate annual incidence is 0.5 to 2 cases per million persons per year (31). ACC is more common in women and the reported female to male ratio ranges from 3:2 to 4:1(16,31,32). ACC most often presents in the 4th to 5th decade of life but these tumors can occur in all ages (31).

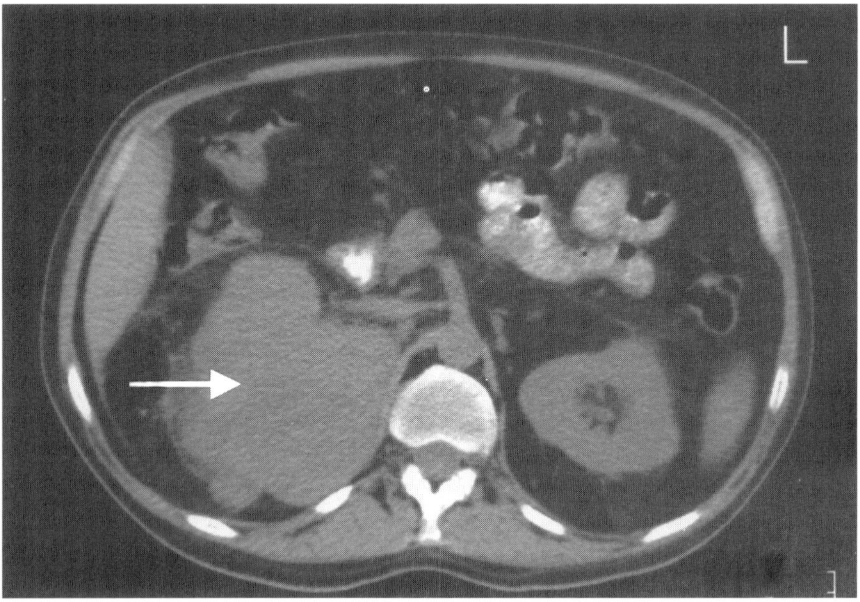

Figure 3. Computed tomography of the abdomen in a patient with a right-sided adrenocortical carcinoma.

Most ACC's are large solitary lobulated tumors (Figure 3). In a series of 54 patients by Evans and Vassilopoulou-Sellin, the median tumor size was 15 cm and the range was 5.5 to 25 cm (16). ACC's account for approximately 7% of incidentally detected adrenal tumors (33). Approximately 2% of ACC's are bilateral (31). Most tumors are either locally invasive or metastatic at the time of diagnosis. The TNM staging of ACC is shown in Table 3. The distribution of metastases in three large series is shown in Table 4 (15,16,31). Histopathologic findings of potential prognostic significance of ACC's include frequent mitotic figures, extensive tumor necrosis and intratumoral hemorrhage (15,16).

Table 3. TNM Staging of Adrenocortical Carcinoma (Adapted from reference 15).

Stage	*% of patients*	*TNM*	*Criteria*	
I	2.8	**T1N0M0**	T1	Tumor ≤ 5cm, no invasion
			T2	Tumor > 5cm, no invasion
II	29	T2N0M0	T3	Locally invasive tumor (soft tissues)
			T4	Locally invasive tumor (adjacent organs)
III	19	T1N1M0	N0	No regional nodes
		T2N1M0	N1	Regional nodes involved
		T3N0M0	M0	No distant metastases
IV	48.9	TxNxM1	M1	Distant metastases
		T3N1		
		T4		

Table 4. Distribution of Metastases in Adrenocortical Carcinoma.

		%	
Site	*Luton*[31]	*Evans et al*[16]	*Lack*[15]
Lung	49	72	78
Liver	49	69	92
Bone	15		18
Lymph Nodes		31	32
Other Adrenal	1	11	
Intestine	3	3	22
Brain	3	3	4
Pancreas	5	3	8
Diaphragm		3	12

Other sites: kidney, heart, spleen, thyroid, orbit, omentum, pleura, ovary, skin, breast.

Clinical symptoms and signs present at the time of diagnosis of ACC are depicted in Table 5 and Table 6. The mean time from onset of symptoms has been reported as varying between 8 months and 14 months (14,31).

Nonfunctional ACC's usually present with tumoral mass effects while functional tumors may present with mass effects and symptoms and signs of hormone excess. Functional or hormone-secreting ACC's account for 40 to 75% of the tumors (14,16,31,32). Cushing's syndrome is the most commonly encountered endocrine disorder and may be rapid and explosive in onset. Virilization and other features of hyperandrogenism are common and may occur either alone or in conjunction with other clinical features due to hypercortisolism (7). Clinical and biochemical features due to excess mineralocorticoid secretion are uncommon (8-11). Feminization is unusual but has been reported and ACC should be considered in the differential diagnosis of gynecomastia (34,35).

Table 5. Clinical Features of Adrenocortical Carcinoma.

	%	
Feature	***Luton***[31]	***Kasperlik-Zahuska***[32]
Endocrine disorder	68	69
Abdominal Pain	28	8
Weight loss/Constitutional	23	6
Abdominal mass	8	2
Metastases	6	

Table 6. Clinical Syndromes in Patients with Hormonal Disorders due to Adrenocortical Carcinoma.

	%	
Syndrome	***Luton***[31]	***Kasperlik-Zahuska***[32]
Cushing's	41	69
Cushings/Hyperandrogenism	42	22
Hyperandrogenism	10	6
Hyperaldosteronism	1	
Cushing's/Hyperaldosteronism	1	
Cushings/Hyperandrogenism/Feminization	1	
Feminization		3

Serum cortisol and 24h-UFC levels are usually elevated in patient's with Cushing's syndrome due to ACC (1,14). In some cases, levels are dramatically elevated and patients present with opportunistic infections due to corticosteroid-induced immunosuppression. Urinary 17-hydroxycorticosteroids are usually elevated and, like the 24h-UFC, fail to decline following the administration of dexamethasone (14). Urinary 17-ketosteroids, ketogenic steroids, and serum DHEA, DHEA-sulfate, and testosterone levels are occasionally elevated (7). Serum cortisol rises in only a small proportion of patients with ACC's following the exogenous administration of ACTH (14). Plasma ACTH levels are usually undetectable but may be subnormal or even in the lower part of the normal range (Figure 1). Deoxycorticosterone and aldosterone levels may be elevated in some patients in the setting of classical features of mineralocorticoid excess (8-11). Urinary and serum estrogens are usually elevated in men presenting with gynecomastia and other signs of feminization (34,35).

Five-year survival rates in patient's with ACC who have had treatment range from 20-30% (16, 31, 36). Luton et al observed an overall median survival of 14.5 months and a 5-year survival of 22% (Figure 4 and Figure 5) (31). They found that survival was lower in those patients with distant metastases at the time of initial presentation and in patient's over 40 years of age. Evans and Vassilopoulou-Sellin reported a median survival of 33 months in their series of 48 patients with ACC (16). They also observed a shorter survival in those patients who had distant metastases. They noted a trend towards shorter survival in patient's whose tumors had ≥ 20 mitoses per 10 high power fields. Lawrence et al found that the presence of tumors larger than 12 cm, ≥6 mitoses per 10 high power fields, and intratumoral hemorrhage were predictors of a shorter overall survival (37). Patients with none of these risk factors had an overall 5-year survival of 83%, those with one risk factor had a 5-year survival of 42%, those with two had a survival rate of 33%, and those with three risk factors had a survival rate of 0%. Jensen et al showed that a longer disease-free interval from the time of initial treatment to the time of recurrence was associated with a significantly improved overall survival (38).

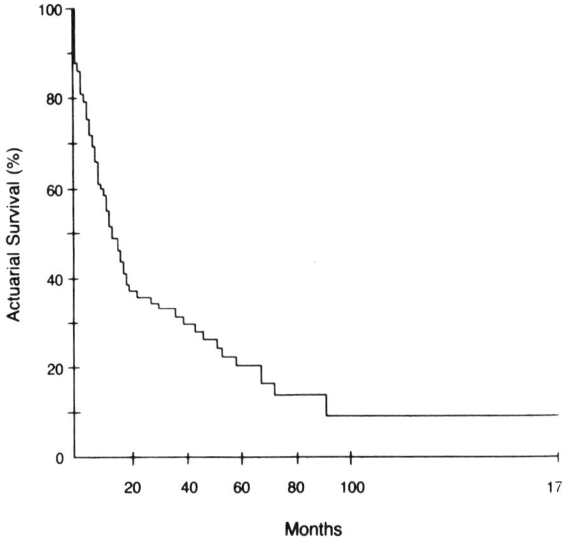

Figure 4. Survival from the time of diagnosis in 88 patients with adrenocortical carcinoma. *Reproduced with permission from reference 31.*

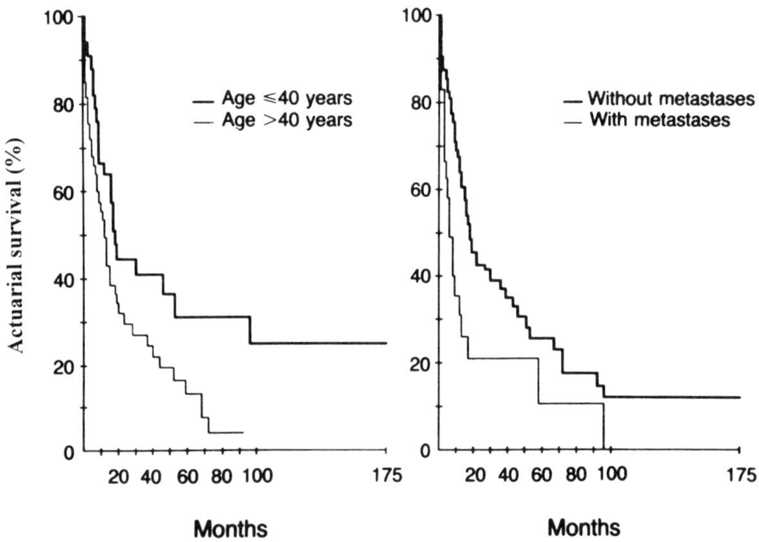

Figure 5. Survival from the time of diagnosis in 88 patients with adrenocortical carcinoma according to age and presence of metastases at diagnosis. *Reproduced with permission from reference 31.*

Surgical resection is the only treatment that offers patients with ACC, and in particular those with limited disease, a chance of long-term survival and cure. When complete resection is not possible, debulking procedures to remove as much tumor as possible may improve the hypercortisolism and relieve local mass effects. Luton et al reported a median disease free interval of 12.1 months in their patients with ACC who had undergone curative surgery (31). Lawrence et al studied 46 patients who underwent curative surgery for ACC (37). The 5-year survival in their series was 36%. These observations indicate that many patients with ACC have local unresected disease or distant metastases at the time of their initial surgery that went unrecognized at the time of the initial presentation. Some of their patients, however, were long-term survivors and, thus, aggressive initial surgery can be justified for most patients. When feasible, surgery is useful in the control of locally advanced recurrent and metastatic ACC (36,38). This approach may prolong survival and is often effective in the management of hypercortisolism but rarely leads to cure of patients with recurrent ACC. Bellantone et al reported results of follow-up of 20 patients with ACC who underwent surgical treatment for recurrent disease and another 32 patients who did not undergo reoperation in a multiinstitutional series (36). The mean survival in the group who had surgery was 15.9 ± 14.9 months. Mean survival in patients who did not undergo reoperation was 3.2 ± 2.9 months. Median survivals for the two groups were approximately 14 months and three months, respectively (Figure 6). Five-year survival rates from the time of initial surgery in the two groups were 49.7% and 8.3%, respectively. The authors do not, however, make it clear whether the group of patients selected for aggressive reoperation had limited disease at the time of recurrences when compared to the group with recurrences that did not undergo reoperation. Schulick and Brennan reported their results of a series of 113 patients who presented to Memorial Sloan-Kettering Cancer Center for treatment of ACC (39). Their results, summarized in Table 7, suggest improved survival in patients who present with early stage disease and are able to undergo complete resection. They also noted that patients who are able to have complete resections of locally recurrent and metastatic disease have improved survival.

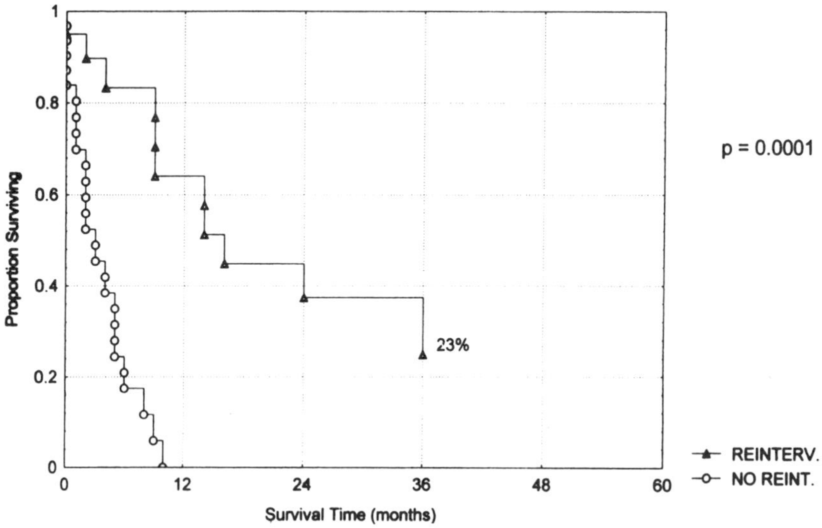

Figure 6. Survival from diagnosis of recurrence of adrenocortical carcinoma in 52 patients based on surgical reintervention. *Reproduced with permission from reference 36.*

Table 7. Survival in Adrenocortical Carcinoma is Dependent on Surgical Success (Data from Schulick and Brennan[39]).

Category	*Median Survival (mos)*	*5-year Survival (%)*
All Patients	38	37
Stage I or II	101	60
Stage III or IV	15	10
Complete Resection, 1st Procedure	74	55
Residual Disease, 1st Procedure	12	5
Complete Resection, 2nd Procedure	74	57
Residual disease, 2nd Procedure	16	0

Mitotane, o,p'-DDD, acts within adrenal mitochondria to inhibit the cholesterol side-chain cleavage and 11-β-hydroxylase enzymes (33). Mitotane also causes mitochondrial destruction and resultant adrenocortical cell injury and death. These actions have led to routine use of the drug in patient's with ACC. Wooten and King reviewed the literature and found 64 reports encompassing 551 patients with ACC who had been treated with mitotane (40). Approximately one-third of patients had been deemed responders in some fashion. There was insufficient data to determine if tumor responses correlated with improved survival. Luton et al administered mitotane to a majority of their patients (31). They found that mitotane slightly prolonged survival and induced tumor regression in some of their patients. Kasperlik-Zahuska et al also found improved survival in patient's with ACC who were treated with adjuvant mitotane (32). Their conclusion was based on the simple fact that there were more survivors in the group of patients who had been treated with mitotane. They did not perform a typical product-limit survival analysis of their results. Vassilopoulou-Sellin et al reviewed the M.D. Anderson Cancer Center experience with 19 patients with ACC who had limited or regional disease at the time of diagnosis (41). They had the leisure of following three groups of patients: a group of eight patients who received mitotane postoperatively and continued it until disease recurrence or the end of the follow-up period; a group of five patients who took mitotane for only a short period of time in the postoperative period; and, a group of six patients who did not take mitotane. There were no significant differences among the groups with respect to potentially confounding variables. Most patients were treated with 4-6 g of mitotane daily for a short period of time and then reduced their doses to1-2 g daily. Interestingly, the disease-free survival was shortest in the group of patients receiving mitotane. There were no apparent differences in overall survival. The authors concluded that their observations did not confirm that mitotane provided an outcome benefit to patients with local or regional ACC.

Various chemotherapeutic regimens have been employed in the treatment of patients with ACC (42-44). Bonacci et al demonstrated transient responses in 33% of patients with advanced ACC treated with etoposide and cisplatin (42). Williamson et al conducted a phase II evaluation of cisplatin and etoposide with mitotane employed at the time of disease progression in patients with advanced and metastatic ACC (43). They noted objective partial responses in only 11% of 45 patients. The Italian Group for the Study of Adrenal Cancer treated 28 patients with advanced ACC with mitotane, etoposide, doxorubicin and cisplatin (44). They noted complete responses in 2 patients and partial responses in 13 patients giving an overall response rate of 54%. Eight patients had stable disease while five were noted to have had progressive disease. Nine of 16 evaluable patients with hormone-secreting tumors enjoyed normalization of their biochemical indices of disease. Expression of the multidrug resistance 1 (mdr-1) gene in ACC cells leads to high levels of membrane expression of P-glycoprotein, a protein that increases efflux of natural chemotherapeutic agents from cells (43,44). Mitotane is able to reverse the mdr-1 expression and, thus, may lead to cellular retention of cytotoxic drugs and promote their efficacy. These observations are encouraging and require further study employing doses of mitotane adjusted on the basis of serum levels of the drug to ensure uniformity of efficacy and to limit toxicity.

To date, no long-term prospective, randomized, placebo-controlled studies have been performed in order to address the issues regarding the dose, duration, and efficacy of mitotane as adjuvant therapy in controlling residual disease or improving survival in patient's with ACC. Based on the available evidence and anectdotal reports, it seems prudent to at least consider the use of mitotane in patients with advanced locally unresectable or distant metastatic disease and especially in those with hypercortisolism. Expected outcomes of therapy and drug toxicities should be discussed with patients and they should be given the opportunity to choose whether to undergo treatment. Drug toxicity often limits dose and duration of therapy and is often cited by patients as a reason for dropout of clinical trials. Side-effects and toxicities include nausea, vomiting, malaise, anorexia, lethargy, sedation, ataxia, leukopenia, hypercholesterolemia, and hepatotoxicity. Some patients develop glucocorticoid deficiency during treatment. Metabolism of dexamethasone and fludrocortisone may be increased necessitating increased doses of these steroids given for adrenal replacement. Terzolo et al monitored plasma mitotane levels to guide treatment in patients with ACC (45). They showed that toxicity may be limited by adjusting the dose of mitotane according to plasma mitotane levels. They suggested a mitotane regimen of 3g daily for three months, monitoring of plasma mitotane levels on a monthly basis thereafter to permit dose adjustments, usually reductions to 1-2g daily, and then measures of mitotane concentrations at three month intervals to titrate the dose as needed.

Other agents that can be used to treat the hypercortisolism associated with ACC include aminoglutethimide, ketoconazole, and metyrapone (46).

The utility of external beam radiotherapy in the management of ACC is uncertain (47). Magee et al treated 10 patients with various stages of the disease (47). Nine patients received megavoltage radiotherapy to the primary site of their tumors and one received palliative radiotherapy to metastatic cervical nodes. Tumor doses ranged from 20 Gy to 30 Gy. Three patients received adjuvant mitotane. The 10-year survival rate in the nine patients who had radiotherapy to their primary site was 33%. These survivors were children who had had complete tumor resections followed by postoperative radiotherapy. In all other patients, radiotherapy was judged to be ineffective in controlling bulky disease.

Molecular events leading to development of ACC are poorly understood. A high incidence of ACC is seen in patients with Li-Fraumeni syndrome (25, 48, 49). This disorder, characterized by a familial susceptibility to a variety of cancers, is due to germline mutations in p53 (48). ACC has also been reported in patients with the Beckwith-Wiedemann syndrome (25). Genetic alterations in sporadic adrenocortical tumors are numerous and have been well catalogued by Latronico and Chrousos (25).

ADRENAL HYPERPLASIA

Bilateral adrenal hyperplasia (BAH) is an infrequent cause of Cushing's syndrome. BAH is characterized by an increase in the size and weight of the adrenal glands due to a nonneoplastic increase in the number of adrenocortical cells (15,50). The hyperplastic process may appear diffuse, nodular, or mixed. The nodular varieties may be classified according to nodule size as micronodular (< 1 cm), macronodular (> 1 cm), or massive macronodular (> 4 cm) **(Figure 7)**. In most cases, however, nodules of varying size are present in affected patients. Nodules may be discrete, yet unencapsulated, or ill-defined and apparently contiguous with adjacent hyperplastic adrenocortical tissue. Internodular tissue may be hyperplastic, normal, or even atrophic. The adrenal glands appear lobular and the normal architecture is often grossly distorted.

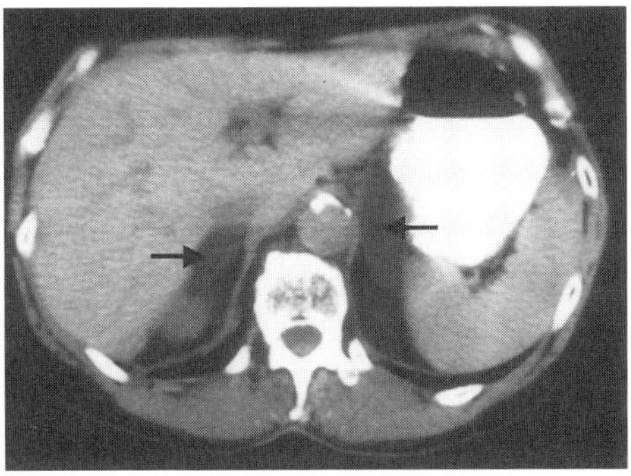

Figure 7. Bilateral massive macronodular hyperplasia in a patient with ACTH-independent Cushing's syndrome.

ACTH-independent macronodular hyperplasia is more common in men than in women (50). The age of presentation is higher than that compared with other forms of nodular adrenal hyperplasia (50). Adrenocortical tissue generally retains responsiveness to exogenous ACTH but resistance to dexamethasone is usually seen. Histologic examination of resected tissue generally reveals nodular hyperplasia with internodular adrenocortical atrophy (51). In vitro studies have demonstrated adrenocortical cells from patients with bilateral macronodular hyperplasia have high proliferative rates in the absence of ACTH and have high basal rates of ACTH independent cortisol secretion (52).

ACTH Hypersecretion

A number of different disorders may result in BAH. Perhaps the most well-studied of these is BAH in response to ACTH-secreting tumors (53-55). Patients with Cushing's disease usually have either diffuse or micronodular hyperplasia. Macronodular hyperplasia may be seen in 10-40% of patients and is especially common in those with long-standing pituitary tumors (53). The intervening nodular tissue is usually hyperplastic (51,53). These observations suggest that a continuum of morphologic alterations may result from a single disorder and that classification of the AH's based on size of the nodules may be of limited clinical usefulness.

In some patients with chronic ACTH hypersecretion, the hyperplastic adrenal tissues become semi-autonomous and are able to secrete cortisol independent of ACTH (53-55). As cortisol levels rise, ACTH levels fall to less than 15 pg/ml and may even become undetectable (54). Administration of high doses of dexamethasone may fail to lower the cortisol excretion rates in some of these patients (53-55). Due to these observations, patients may be erroneously diagnosed with ACTH-independent primary adrenal Cushing's.

Since most of these patients have ACTH-secreting tumors that express CRH receptors, administration of CRH may be followed by a rise in ACTH levels and thus permit differentiation of those who have Cushing's disease and secondary hyperplasia from those with primary BAH (56). Some patients, however, have not responded to CRH (54,55). Bilateral adrenalectomy often results in a more rapid rise in ACTH levels than usually seen in patients with primary adrenal Cushing's following surgery (54). Resection of the offending pituitary adenoma usually leads to regression in the nodular enlargement of the adrenal glands (57).

Familial Adrenal Hyperplasia

Familial variants of micronodular and macronodular adrenal hyperplasia have been described (58,59,60). Minami et al reported ACTH-independent macronodular hyperplasia in four first degree relatives; only two had Cushing's syndrome (60). The cause was unknown. Approximately two-fifths of patients with MEN-1 have macronodular changes in the adrenal glands but Cushing's syndrome is not often described (61). An ACTH receptor activating mutation has been described (62). McCune-Albright syndrome is characterized by fibrous dysplasia of bone, café-au-lait pigmentation, and endocrine dysfunction including precocious puberty, acromegaly, hyperthyroidism, and Cushing's syndrome (63). Rare patients have presented with Cushing's syndrome in infancy due to ACTH-independent BAH (63,64). In these patients, activating mutations of the stimulatory guanine nucleotide-binding protein (G_s) initiate and maintain steroidogenesis and result in a hyperplastic pattern of cellular proliferation (64). These mutations are present in nodular tissues but not in adjacent non-nodular normal appearing adrenal tissues from the same gland. Similar findings have been reported in skin, autonomous functioning thyroid nodules, and growth hormone-secreting pituitary lesions in patients with McCune-Albright syndrome (63-65).

Aberrant Expression of Membrane Receptors

Aberrant expression of various membrane receptors on adrenocortical cells may play a central role in the pathogenesis in some patients with BAH and Cushing's syndrome (66-68). Perhaps the best studied of these disorders is BAH attributed to the abnormal expression of GIP receptors (69-72). Affected patients express a membrane bound GIP receptor (71). Dispersed cells from the adrenal glands of these patients secrete cortisol in response to GIP in a dose dependent fashion (73). Lebrethon et al showed that GIP increases mRNA of the side chain cleavage, 17-hydroxylase, and 3-betahydroxysteroid dehydrogenase enzymes in cell cultures of adrenocortical cells from affected patients but not in cell cultures derived from normal adrenal glands (73). Physiologic rises in GIP following the ingestion of food stimulate cortisol synthesis and secretion by the adrenal glands (69-70). Chronic repeated stimulation is believed to result in BAH. Cortisol levels usually fall during an overnight fast (69). Dexamethasone does not prevent the food-induced rise in cortisol secretion (69). Octreotide, a long-acting somatostatin analogue, inhibits GIP secretion and, thus, inhibits cortisol synthesis and secretion (69,70).

Some cases of BAH may be due to expression of V1 arginine vasopressin receptors (74). Diadoh et al described a patient with ACTH-independent macronodular hyperplasia and hypersensitivity to vasopressin (74). Cortisol secretion increased in response to upright posture, furosemide administration, and following vasopressin administration. Dispersed adrenocortical cells secreted cortisol in response to vasopressin administration. Pretreatment with a selective V1 arginine vasopressin receptor antagonist reduced the mean 24h UFC by approximately 50%.

Expression of β-adrenergic receptors has been reported in a patient with macronodular adrenal hyperplasia and Cushing's syndrome (75). Plasma cortisol levels typically rise following maneuvers that increase endogenous catecholamine secretion including insulin-induced hypoglycemia, postural changes, and isoproterenol infusion. Rises in cortisol during postural change can be blunted by pretreatment with β-adrenergic antagonists. Similarly, treatment with β-adrenergic receptor antagonists decreases the 24h UFC in affected patients (66,75).

Abnormal expression of LH receptors on adrenocortical cells has been reported in a patient with Cushing's syndrome and BAH (76). Lacroix et al described a patient with Cushing's syndrome that was transient during pregnancy, presumably due to hCG, and sustained during menopause, when LH levels were persistently elevated. Plasma cortisol levels rose following injection of GnRH and hCG. The 24h UFC initially rose following the administration of leuprolide but then levels gradually declined and were normalized after one week of treatment.

Aberrant expression of serotonin receptors may also account for some cases of Cushing's syndrome and bilateral adrenal hyperplasia (67).

Primary Pigmented Nodular Adrenocortical Disease

Primary pigmented nodular adrenocortical disease (PPNAD) is an uncommon ACTH-independent form of Cushing's syndrome (77). In this disorder, the adrenal glands are usually of normal size but are studded with black, brown, or red nodules ranging in size from 4 mm to 3 cm (15). The internodular adrenal cortex is usually atrophic. Thin section high-resolution CT scans of the adrenal glands are often necessary to identify this disorder (78). The adrenal glands appear normal in 50% of patients. They may appear to look like a "string of beads" in younger patients whereas macronodular changes are often seen in older patients.

Approximately one-half of the cases of PPNAD are seen in the setting of the Carney complex (Table 8) while the remaining cases appear to be sporadic (5,79,81). The median age of presentation of the Carney complex is 14 years with a range of 5-33 years and a mean of 16.7 years (5). One-fifth of patients with Carney complex have PPNAD and Cushing's syndrome (5).

Table 8. The Carney Complex (adapted from references79&80).

Finding	Frequency (%)
Cardiac myxomas	61
Skin myxomas	37
Breast myxomas	20
Spotty pigmentation/Nevi	67
PPNAD	33
Sertoli-cell tumors	30
Pituitary tumors (Acromegaly/Prolactinoma)	11
Thyroid Tumors	10
Melanotic schwannoma	11

Carney complex is usually inherited as an autosomal dominant disorder. Sporadic cases contribute to one half of all patients (80). The gene for the Carney complex has been mapped to chromosome 2p16 based on linkage analysis of 11 kindreds (80). A few kindreds, however, have not had any linkage to that region (81). Goldstein et al identified a genetic locus at chromosome 17q2 thought to cause Carney complex in some families (82). These mutations appear to have a dominant gain of function effect and the gene appears to be involved in regulation of genomic stability, structure of telomeres in replicating chromosomes, and function of the mitotic apparatus (80). Circulating adrenal-stimulating immunoglobulins have been detected in some patients with the Carney complex with Cushing's syndrome and in some affected first degree relatives who did not have Cushing's syndrome (83). The pathogenetic role of these autoantibodies is not clear.

Overt Cushing's syndrome is present in 84% of patients with PPNAD (84). Many patients are asthenic. Osteoporosis is present in 39% of patients and many of them have short stature and growth failure as well (84). Young and Carney reviewed the literature and found that the mean age of diagnosis of Cushing's syndrome in PPNAD is 19.9 ±10.7 years with a range of 3.8-64 years (77). The median duration of symptoms prior to presentation was 4.3

±4 years with the range of 0.3-18 years. Nearly all (92%) of patients have low or undetectable ACTH levels. Interestingly, 24h UFC and 17-hydroxycorticosteroid levels rise paradoxically following the administration of high-dose dexamethasone (Figure 8) (85). An increase in 24h UFC by ≥50% on day six of the high-dose dexamethasone suppression test identified nine of thirteen patients with PPNAD and excluded patients with other forms of BAH. The cause of this paradoxical rise in cortisol secretion and excretion in response to dexamethasone is unknown. Periodic Cushing's syndrome has been described in some patients with PPNAD (77).

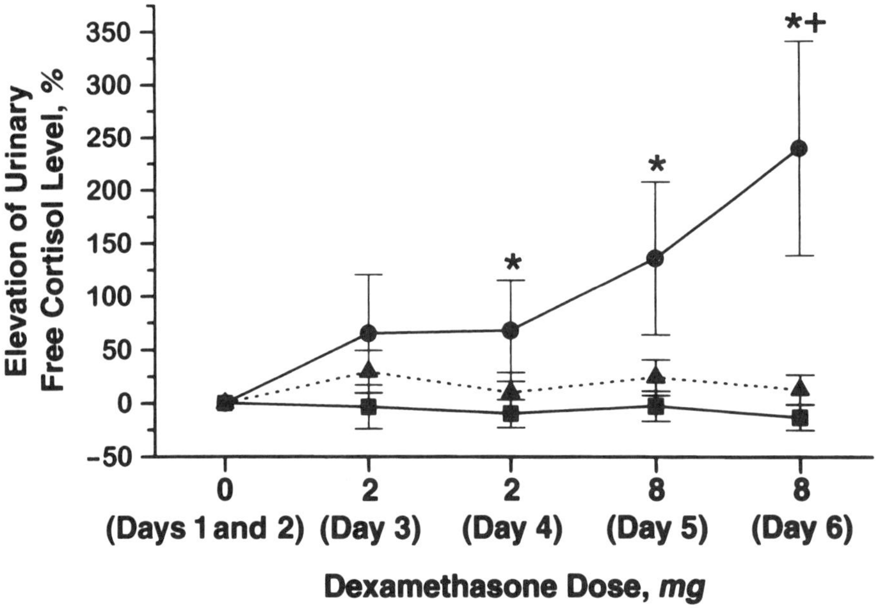

Figure 8. Mean percentage change in 24h UFC during the Liddle test. Circles represent patients with PPNAD, squares indicate patients with macronodular adrenocortical disease, and triangles designate patients with cortisol-producing adrenocortical adenomas. ***Reproduced with permission from reference 85.***

Treatment Considerations

Bilateral adrenalectomy remains the most effective form of treatment for patients with BAH. Patients with BAH due to aberrant expression of membrane-bound receptors may be amenable to long-term pharmacological therapies (69,70,74,75,76).

REFERENCES

1. Invitti C, Giraldi FP, de Martin M, et al. Diagnosis and management of Cushing's syndrome: Results of an Italian multicentre study. J Clin Endocrinol Metab 84:440-448, 1999.
2. Scott HW, Abumrad NN, Orth DN. Tumors of the adrenal cortex and Cushing's syndrome. Ann Surg 201:586-594, 1985.
3. Reincke M, Nieke J, Krestin GP, Saeger W, Allolio B, Winkelmann W. Preclinical Cushing's syndrome in adrenal "incidentalomas": Comparison with Adrenal Cushing's syndrome. J Clin Endocrinol Metab 75:826-832, 1992.
4. Daitch JA, Goldfarb DA, Novick AC. Cleveland Clinic experience with adrenal Cushing's syndrome. J Urol 158:2051-2055, 1997.
5. Carney JA. Carney complex:The complex of myxomas, spotty pigmentation, endocrine overactivity, and schwannomas. Semin Derm 14:90-98, 1995.
6. Morio H, Terano T, Yamamoto K, et al. Serum levels of dehydroepiandrosterone sulfate in patients with asymptomatic cortisol producing adenoma: Comparison with adrenal Cushing's syndrome and non-functional adrenal tumor. Endocr J 43:387-396, 1996.
7. Del Gaudio A, Del Gaudio GA. Virilizing adrenocortical tumors in adult women. Cancer 72:1997-2003, 1993.
8. Kelly WF, O'Hare MJ, Loizou S, Davies D, Laing I. Hypermineralocorticism without excessive aldosterone secretion: An adrenal carcinoma producing deoxycorticosterone. Clin endocrinol 17:353-361, 1982.
9. Irony I, Biglieri EG, Perloff D, Rubinoff H. Pathophysiology of deoxycorticosterone-secreting adrenal tumors. J Clin Endocrinol Metab 65:836-840, 1987.
10. Farge D, Chatellier G, Pagny JY, Jeunemaitre X, Plouin PF, Corvol P. Isolated clinical syndrome of primary aldosteronism in four patients with adrenocortical carcinoma. Am J Med 83:635-640, 1987.
11. Arteaga E, Biglieri EG, Kater CE, Lopez JM, Schambelan M. Aldosterone-producing adrenocortical carcinoma. Ann Intern Med 101:316-321, 1984.
12. Aiba M, Kawakami M, Ito Y, Fujimoto Y, Suda T, Demura H. Bilateral adrenocortical adenomas causing Cushing's syndrome. Arch Pathol Lab Med 116:146-150, 1992.
13. Kato S, Masunaga R, Kawabe T, et al. Cushing's syndrome induced by hypersecretion of cortisol from only one of bilateral adrenocortical adenomas. Metabolism 41:260-263, 1992.
14. Bertagna C, Orth DN. Clinical and laboratory findings and results of therapy in 58 patients with adrenocortical tumors admitted to a single medical center (1951 to 1978). Am J Med 71:855-875, 1981.
15. Tumors of the Adrenal Gland and Extra-adrenal Paraganglia. Ernest E. Lack ed. Armed Forces Institute of Pathology. Washington D.C. 1997.
16. Evans HL, Vassilopoulou-Sellin R. Adrenal cortical neoplasms. Am J Clin Pathol 105:76-86, 1996.
17. McLeod MK, Thompson NW, Gross MD, Bondeson AG, Bondeson L. Sub-clinical Cushing's syndrome in patients with adrenal gland incidentalomas. Am Surgeon 56:398-403, 1990.
18. Rossi R, Tauchmanova L, Luciano A, et al. Subclinical Cushing's syndrome in patients with adrenal incidentaloma: Clinical and biochemical features. J Clin Endocrinol Metab 85:1440-1448, 2000.
19. Perry RR, Nieman LK, Cutler GB, et al. Primary adrenal causes of Cushing's syndrome. Ann Surg 210:59-68, 1989.
20. Kawashima A, Sandler CM, Fishman EK, et al. Spectrum of CT findings in nonmalignant disease of the adrenal gland. Radiographics 18:393-412, 1998.

21. Boraschi P, Braccini G, Gigoni R, et al. Diagnosis of adrenal adenoma:Value of central spot of high-intensity hyperintense rim sign and homogenous isointensity to liver on gadolinium-enhanced fat-suppressed spin-echo MR images. J Magn Reson Imaging 9:304-310, 1999.
22. Beierwaltes WH, Sturman MF, Ry U, Ice RD. Imaging functional nodules of the adrenal glands with 131-I-19-iodocholesterol. J Nucl Med 15:246-251, 1974.
23. Abs R, Nobels F, Verhelst J, et al. Hyperfunctioning unilateral adrenal macronodule in three patients with Cushing's disease:Hormonal and imaging characterization. Acta Endocrinol 129:284-290, 1993.
24. Beuschlein F, Reincke M, Karl M, et al. Clonal composition of human adrenocortical neoplasms. Cancer Res 54:4927-4932, 1994.
25. Latronico AC, Chrousos GP. Adrenocortical tumors. J Clin Endocrinol Metab 82:1317-1324, 1997.
26. Chabre O, Liakos P, Vivier J, et al. Cushing's syndrome due to a gastric inhibitory polypeptide-dependentadrenal adenoma: Insights into hormonal control of adrenocortical tumorigenesis. J Clin Endocrinol Metab 83:3134-3143, 1998.
27. Willenberg HS, Stratakis CA, Mmarx C, Ehrhart-Bornstein M, Chrousos GP, Bornstein SR. Aberrant interleukin-1 receptors in a cortisol-secreting adrenal adenoma causing Cushing's syndrome. N Engl J Med 339:27-31, 1998.
28. De Herder WW, Hofland LJ, Usdin TB, et al. Ffood-dependent Cushing's syndrome resulting from abundant espression of gastric inhibitory polyeptide receptors in adrenal adenoma cells. J Clin Endocrinol Metab 81:3168-3172, 1996.
29. Horbia N, Suda T, Aiba M, et al. Lysine vasopressin stimulation of cortisol secretion in patients with adrenocorticotropin-independent macronodular hyperplasia. J Clin Endocrinol Metab 80:2336-2341, 1995.
30. Arnaldi G, Gasc JM, de Keyzer Y, et al. Variable expression of the V1 vasopressin receptor modulates the phenotypic response of steroid-secreting adrenocortical tumors. J Clin Endocrinol Metab 83:2029-2035, 1998.
31. Luton J-P, Cerdas S, Billaud L, et al. Clinical features of adrenocortical carcinoma, prognostic factors, and the effect of mitotane therapy. N Engl J Med 322:1195-1201, 1990.
32. Kasperlik-Zahuska AA, Migdalska BM, Zgliczynski S, Makowska AM. Adrenocortical Carcinoma. Cancer 75:2587-2591, 1995.
33. Kasperlik-Zahuska AA. Clinical results of the use of mitotane for adrenocortical carcinoma. Braz J Med Biol Res 33:1191-1196, 2000.
34. Lanigan D, Choa RG, Evans J. A feminizing adrenocortical carcinoma presenting with gynaecomastia. Post Grad Med J 69:483-485, 1993.
35. Gabriolove JL, Sharma DC, Woitz HH, Dorfman RI. Feminizing adrenocortical tumors in the male: a review of 52 cases, including a case report. Medicine 44:37-76, 1965.
36. Bellantone R, Ferrante A, Boscherini M, et al. Role of reoperation in recurrence of adrenocortical carcinoma: Results from 188 cases collected in the Italian National Registry for adrenocortical carcinoma. Surgery 122:1212-1218, 1997.
37. Lawrence HL, Gaudin PB, Brennan MF. Pathologic features of prognostic significance for adrenocortical carcinoma after curative resection. Arch Surg 134:181-185, 1999.
38. Jensen JC, Pass HI, Sindelar WF, Norton JA. Recurrent or metastatic disease in select patients with adrenocortical carcinoma. Arch Surg 126:457-461, 1991.
39. Schulick RD, Brennan MF. Long-term survival after complete resection and repeat resection in patients with adrenocortical carcinoma. Ann Surg Oncol 6:713-714, 1999.
40. Wooten MD, King DK. Adrenocortical Carcinoma. Epidemiology and treatment with mitotane and a review of the literature. Cancer 72:3145-3155, 1993.
41. Vassilopoulou-Sellin R, Guinee VF, Klein MJ, et al. Impact of adjuvant mitotane on the clinical course of patients with adrenocortical cancer. Cancer 71:3119-3123, 1993.

42. Bonacci R, Gigliotti A, Baudin E, et al. Cytotoxic therapy with etoposide and cisplatin in advanced adrenocortical carcinoma. Br J Cancer 78:546-549, 1998.
43. Williamson SK, Lew D, Miller GJ, Balcerzak SP, Baker LH, Crawford ED. Phase II evaluation of cisplatin and etoposide followed by mitotane at disease progression in patients with locally advanced or metastatic adrenocortical carcinoma. Cancer 88:1159-1165, 2000.
44. Berruti A, Terzolo M, Pia A, Angeli A, Dogliotti L. Mitotane associated with etoposide, doxorubicin, and cisplatin in the treatment of advanced adrenocortical carcinoma. Cancer 83:2194-2200, 1998.
45. Terzolo M, Pia A, Berruti A, et al. Low-dose monitored mitotane treatment achieves the therapeutic range with manageable side effects in patients with adrenocortical cancer. J Clin Endocrinol Metab 85:2234-2238, 2000.
46. Miller JW, Crapo L. The medical treatment of Cushing's syndrome. Endocr Rev 14:443-458, 1993.
47. Magee BJ, Gattamaneni R, Pearson D. Adrenal cortical carcinoma: survival after radiotherapy. Clin Radiol 38:587-588, 1987.
48. Li FP, Fraumeni JF, Mulvihill JJ, et al. A cancer family syndrome in twenty-four kindreds. Cancer Res 48:5358-5362, 1988.
49. Wagner J, Portwine C, Rabin K, Leclerc JM, Narod SA, Malkin D. High frequency of germline p53 mutations in childhood adrenocortical cancer. J Natl Cancer Inst 86:1707-1710, 1994.
50. Aiba M, Hirayama A, Iri H, et al. Adrenocorticotrophic hormone-independent bilateral adrenocortical macronodular hyperplasia as a distinct subtype of Cushing's syndrome. Am J Clin Pathol 96:334-340, 1991.
51. Swain JM, Grant CS, Schlinert RT, et al. Corticotropin-independent macronodular adrenal hyperplasia. Arch Surg 133:541-546, 1998.
52. Cheirlin RA, Westphal M, Cabrera CM, Fujii DK, Snyder J, Fitzgerald PA. Cushing's syndrome due to bilateral adrenal macronodular hyperplasia with undetectable ACTH: Cell culture of adenoma cells on extracellular matrix. Hormone Res 29:162-167, 1998.
53. Smals AGH, Pieters GFFM, van Haelst UJG, Kloppenborg PWC. Macronodular adrenocortical hyperplasia in long-standing Cushing's disease. J Clin Endocrinol Metab 58:25-31, 1984.
54. Hermus AR, Pieters GF, Smals AG, et al. Transition from pituitary-dependent to adrenal-dependent Cushing's syndrome. N Engl J Med 318:966-970, 1988.
55. Hocher B, Bahr V, Dorfmuller S, Oelkers W. Hypercortisolism with non-pigmented micronodular adrenal hyperplasia: Transition from pituitary-dependent to adrenal-dependent Cushing's syndrome. Acta Endocrinol 128:120-125, 1993.
56. Freda PU, Wardlaw SL, Bruce JN, Post KD, Goland RS. Differential diagnosis in Cushing's syndrome. Use of Corticotropin-releasing hormone. Medicine 74:74-82, 1995.
57. Fish HR, Sobel DO, Miegel CA. Macronodular adrenal hyperplasia with hypothalamic-pituitary-adrenal suppression by ultra-high-dose dexamethasone: Regression following hypophysectomy. Clin Neuropharmacol 9:303-308, 1986.
58. Findlay JC, Sheeler LR, Engeland WC, Aron DC. Familial adrenocorticotropin-independent Cushing's syndrome with bilateral macronodular adrenal hyperplasia. J Clin Endocrinol Metab 76:189-191, 1993.
59. Hodge BO, Froesch TA. Familial Cushing's Syndrome Arch Intern Med 148:1133-1136, 1988.
60. Minami S, Sugihara H, Sato J, et al. ACTH independent Cushing's syndrome occurring in siblings. Clin Endocrinol 44:483-488, 1996.
61. Skogseid B, Larsson C, Lindgren PG, et al. Clinical and genetic features of adrenocortical lesions in multiple endocrine neoplasia type 1. J Clin Endocrinol Metab 75:76-81, 1992.

62. Malchoff DM, Thorner MO, Orth D, et al. Functional consequences of a naturally occurring mutation of the type 2 melanocortin (ACTH) receptor. 10th International Congress of Endocrinology, Abstract P3-576, 1996.
63. Kirk JM, Brain CE, Carson DJ, Hyde JC, Grant DB. Cushing's syndrome caused by nodular adrenal hyperplasia in children with McCune-Albright syndrome. J Pediatr 134:789-792, 1999.
64. Boston BA, Mandel S, LaFranchi S, Bliziotes M. Activating mutation in the stimulatory guanine nucleotide-binding protein in an infant with Cushing's syndrome and nodular adrenal hyperplasia. J Clin Endocrinol Metab 79:890-893, 1994.
65. Weinstein LS, Shenker A, Gejman PV, Merino MJ, Friedman E, Spiegel AM. Activating mutations of the stimulatory G protein in the McCune-Albright syndrome. N Engl J Med 325:1688-1695, 1991.
66. Lacroix A, N'Diaye N, Mircescu H, et al. Abnormal expression and function of hormone receptors in adrenal Cushing's syndrome. Endocr Res 24:835-843, 1998.
67. Mircescu H, Jilwan J, N'Diaye N, et al. Are ectopic or abnormal membrane hormone receptors frequently present in adrenal Cushing's syndrome? J Clin Endocrinol Metab 85:3531-3536, 2000.
68. Lacroix A Mircescu H, Hamet P. Clinical evaluation of the presence of abnormal hormone receptors in adrenal Cushing's syndrome. The Endocrinologist 9:9-15, 1999.
69. Lacroix A, Bolte E, Tremblay J, et al. Gastric inhibitory polypeptide-dependent cortisol hypersecretion-A new cause of Cushing's syndrome. N Engl J Med 327:974-980, 1992.
70. Reznik Y, Allali-Zerah V, Chayvialle JA, et al. Food-dependent Cushing's syndrome mediated by aberrant adrenal sensitivity to gastric inhibitory polypeptide. N Engl J Med 327:981-986, 1992.
71. N'Diaye N, Hamet P, Tremblay J, Boutin J-M, Gaboury L, Lacroix A. Asynchronous development of bilateral nodular adrenal hyperplasia in gastric inhibitory polypeptide-dependent Cushing's syndrome. J Clin Endocrinol Metab 84:2616-2622, 1999.
72. N'Diaye N, Tremblay J, Hamet P, de Herder WW, Lacroix A. Adrenocortical overexpression of gastric inhibitory polypeptide receptor underlies food-dependent Cushing's syndrome. J Clin Endocrinol Metab 83:2781-2785, 1998.
73. Lebrethon MC, Reznik AY, Archambeaud F, et al. Food-dependent Cushing's syndrome: Characterization and functional role of gastric inhibitory polypeptide receptor in the adrenals of three patients. J Clin Endocrinol Metab 83:4514-4519, 1998.
74. Daidoh H, Morita H, Hanafusa J, et al. In vivo and in vitro effects of AVP and V1a receptor antagonist on Cushing's syndrome due to ACTH-independent bilateral macronodular adrenocortical hyperplasia. Clin Endocrinol 48:403-409, 1998.
75. Lacroix A, Tremblay J, Rousseau G, Bouvier M, Hamet P. Propranolol therapy for ectopic β-adrenergic receptors in adrenal Cushing's syndrome. N Engl J Med 337:1429-1434, 1997.
76. Lacroix A, Hamet P, Boutin J-M. Leuprolide acetate therapy in lutenizing hormone-dependent Cushing's syndrome. N Engl J Med 341:1577-1581, 1999.
77. Young WF, Carney JA. Cushing's syndrome: Primary pigmented nodular adrenal disease. Adv Endocr Metab 3:179-197, 1992.
78. Doppman JL, Travis WD, Nieman L, et al. Cushing's syndrome due to primary pigmented nodular adrenocortical disease:Findings at CT and MR imaging. Radiology 172:415-420, 1989.
79. Carney JA, Gordon H Carpenter PC, Shenoy V, Go VLW. The complex of myxomas, spotty pigmentation, and endocrine overactivity. Medicine 64:270-283, 1985.
80. Stratakis CA, Carney JA, Lin J-P, et al. Carney complex, a familial multiple neoplasia and lentiginosis syndrome: Analysis of 11 kindreds and linkage to the short arm of chromosome 2. J Clin Invest 97:699-705, 1996.

81. Milunsky J, Huang X-L, Baldwin CT, Farah MG, Milunsky A. Evidence for genetic heterogeneity of the Carney complex (familial atrial myxoma syndromes). Cancer Genet Cytogenet 106:173-176, 1998.
82. Goldstein MM, Casey M, Carney JA, Basson CT. Molecular genetic diagnosis of the familial myxoma syndrome (Carney complex). Am J Med Genet 86:62-65, 1999.
83. Young WF, Carney JA, Musa BU, Wulffraat NM, Lens JW, Drexhage HA. Familial Cushing's syndrome due to primary pigmented nodular adrenocortical disease. N Engl J Med 321:1659-1664, 1989.
84. Sarlis NJ, Chrousos GP, Doppman JL, Carney JA, Stratakis CA. Primary pigmented nodular adrenocortical disease: Reevaluation of a patient with Carney complex 27 years after unilateral adrenalectomy. J Clin Endocrinol Metab 82:1274-1278, 1997.
85. Stratakis CA, Sarlis N, Kirschner LS, et al. Paradoxical response to dexamethasone in the diagnosis of primary pigmented nodular adrenocortical disease. Ann Intern Med 131:585-591, 1999.

10

ABERRANT HORMONE RECEPTORS

André Lacroix, M.D., Johanne Tremblay, M.D. and Pavel Hamet, M.D.
Hôtel-Dieu du Centre hospitalier de l'Université de Montréal, Montréal, Québec, Canada

INTRODUCTION

The mechanisms by which cortisol is produced in adrenal Cushing's syndrome, where ACTH is suppressed, were previously unknown and believed to be "autonomous." Studies by several groups have now revealed that some of the cortisol-producing adrenal tumors or hyperplasias may actually be under the control of aberrant (or ectopic, illicit, or hyperactive) hormone membrane receptors (1, 2). This chapter will focus on *in vitro* and *in vivo* evidences of the abnormal expression or function of receptors for various hormones in primary adrenal Cushing's syndrome. The investigation strategy utilized to reveal the presence of such receptors will be presented. The identification of abnormal membrane hormone receptors in adrenal Cushing's syndrome has now opened the field of new therapeutic strategies to control hypercortisolism by interfering with ligand binding to these receptors and will also be reviewed.

Initial *in vitro* evidence of aberrant adrenal hormone receptors

ACTH is the main regulator of glucocorticoid production from the adrenal cortex. Its binding to G_s-coupled melanocortin type 2 receptor elicits short- (acute) and long-term (chronic) specific responses (Figure 1). Activation of the adenylyl cyclase /cAMP/ protein kinase A pathways leads to phosphorylation of proteins that regulate early and late steps of steroidogenesis (3-5). Several other hormones may also play minor roles in the regulation of adrenocortical function (6).

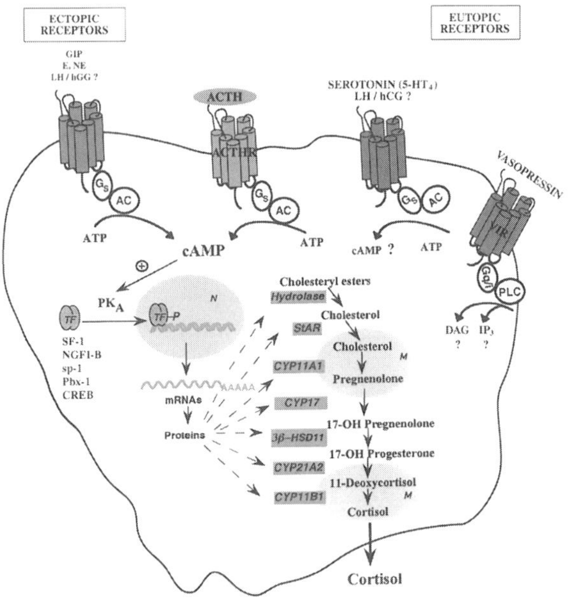

Figure 1.

The concept of ectopic adrenal membrane receptor expression was proposed initially by Robert Ney and his collaborators, while working at Vanderbilt University in 1971 (1). In studying the role of adenylyl cyclase in mediating the effects of ACTH in rat adrenal steroidogenesis, they noted that only ACTH was capable of stimulating adenylyl cyclase in normal cortex membrane preparations; however, in corticosterone-producing rat adrenocortical carcinoma 494, they demonstrated that adenylyl cyclase was stimulated by other hormones in addition to ACTH, such as epinephrine, norepinephrine, TSH, FSH, LH and prostaglandin E_1, but not by glucagon, insulin, vasopressin, PTH or calcitonin. Catecholamine effects on adenylyl cyclase were induced by β-, but not by α-adrenergic agonists. Propranolol was able to block the effects of catecholamines, but not of other hormones. The illicit hormones exerted no additive or synergistic actions, suggesting that the tumor possessed multiple specific receptors, which activated a common adenylyl cyclase.

Further *in vitro* studies indicated the presence of several membrane hormone receptors in some human adrenocortical benign and malignant tumors; this included receptors for TSH, LH, FSH, β-adrenergic agonists, prolactin, human placental lactogen, Ang-II, glucagon, and interleukin-I (1, 2, 7-9). Some androgen-secreting adrenal tumors were stimulated by LH and hCG (2, 10).

In vivo demonstration of the functionality of ectopic or abnormal membrane hormone receptors

The proposed concept of ectopic hormone receptors had been demonstrated *in vitro*, but had not been applied directly to the investigation of patients with adrenal Cushing's syndrome *in vivo* yet. We reasoned that if aberrant hormone receptors were present in the steroid secreting adrenal tumors, it should be possible to identify their presence *in vivo,* by monitoring steroid levels during tests that would modulate the endogenous levels of the ligands for the putative aberrant receptors. As endocrinology fellows at Vanderbilt University in the late 1970's, we investigated two patients with cortisol-secreting adrenal adenomas in whom we found no response of plasma cortisol following elevations of endogenous levels of TSH, prolactin, FSH and LH during stimulation tests with TRH and LHRH (unpublished observation). A decade later, the initial clinical manifestation of the presence of aberrant hormone receptors was suggested by the description of food-dependent Cushing's syndrome.

1. Food- and GIP-dependent Cushing's syndrome

Hamet et al. (11) were the first to identify "food-dependent" cortisol production in a 41 year-old male patient presenting with Cushing's syndrome secondary to a unilateral adrenal adenoma and periodic hormonogenesis. Plasma cortisol was consistently low in the morning or during fasting, but increased to abnormal levels following meals; food-induced elevations of plasma cortisol were not suppressed by high oral doses of dexamethasone. Adenylyl cyclase activity in the resected adrenal adenoma membrane preparation was stimulated 27% by ACTH and 62% by vasopressin, but not by FSH, glucagon or Ang-II; the effects of various gastrointestinal hormones were not examined in this case.

Two patients with bilateral ACTH-independent bilateral macronodular adrenal hyperplasia (AIMAH) and food-dependent cortisol production were studied in detail a few years later and allowed to clarify the pathophysiology of this syndrome (12, 13). A 48 year-old French-Canadian woman with Cushing's syndrome had low plasma cortisol levels fasting in the morning and higher levels during the day, whereas plasma ACTH was always suppressed. Plasma cortisol was stimulated by oral administration of glucose, lipid-rich or protein-rich meals, but not by intravenous glucose. In addition, the stimulatory effect of oral glucose on cortisol secretion was inhibited by pretreatment with somatostatin. A review of the various secretagogues of gastrointestinal hormones indicated that only GIP and glucagon-like peptide 1 were stimulated significantly by oral glucose and lipids and, to a lesser extent, by proteins. Plasma cortisol levels were correlated with plasma GIP concentrations during the various test meals. *In vivo,* GIP infusion, administered to reproduce physiological post-prandial concentrations, augmented cortisol production in the patient, but not in 4 normal controls. The aberrant presence of GIP receptors in adrenal tissues was supported by adrenal imaging after the injection of [^{123}I]-GIP *in vivo* (12). Following bilateral adrenalectomy, the incubation of dispersed adrenal cells confirmed GIP-mediated cortisol secretion in the patient's cells; no cortisol response to GIP was found in normal adult or fetal adrenal cells, or in other cortisol- or aldosterone-secreting adenomas (12). There was no stimulation of cortisol production in the patient's adrenal cells after incubation with fourteen other hormones. Our investigation strategy was shared with colleagues in France who were investigating a similar 49 year-old woman with Cushing's syndrome, AIMAH, and unusual diurnal fluctuations of plasma cortisol (13). The food-dependent nature of plasma cortisol secretion was also demonstrated in this patient and shown to be GIP-dependent. Chronic sc administration of octreotide up to 100 μg 3 times daily resulted in a temporary improvement of the clinical syndrome and a return of urinary free cortisol levels to the upper limit of normal. However, there was an eventual escape from octreotide after 5 months of therapy, requiring bilateral adrenalectomy (14). It was thus concluded that food-dependent cortisol secretion resulted from the abnormal responsiveness of adrenal cells to the physiologic secretion of GIP; it was proposed that ectopic GIP receptor expression on adrenal cells underlied this new etiology of Cushing's syndrome (12, 13).

Food- or GIP-dependent Cushing's syndrome has now been identified in 20 patients, including 13 with AIMAH (2, 12-18) and 7 with unilateral adenoma (11, 14, 19-22). At pathological examination, no distinctive features were reported, compared to non-GIP-dependent cortisol-secreting adenomas or bilateral macronodular hyperplasia, except in 1 possible case associated with Carney's syndrome features (16). In 2 cases of AIMAH, the patient initially presented with a unilateral lesion, and developed contralateral enlargement only later in time (15, Hermus et al. unpublished). Except for three patients, all other patients are females; adrenal Cushing's syndrome is more frequent in females, but it remains to be seen if an even higher female frequency will be found in GIP-dependent Cushing's syndrome and what molecular mechanism underlies this sex distribution. Average age at the time of diagnosis may be somewhat greater in patients with AIMAH than in patients with unilateral adrenal adenoma (23); the youngest patient with a unilateral adenoma was only 15 years-old. In GIP-dependent Cushing's syndrome, chronic GIP-induced hypercortisolism eventually leads to suppression of CRH and ACTH; this suppression, coupled with low GIP levels in the fasting state, is responsible for the decreased plasma cortisol levels, which can be accompanied by symptoms of relative cortisol insufficiency (13, 20). However, in certain patients, fasting plasma cortisol levels were not particularly low, indicating that GIP-dependent Cushing's syndrome should not be excluded without performing a test meal (2, 15, 17). In one patient with food-dependent AIMAH, but in whom fasting plasma cortisol was relatively elevated, Pralong et al. (17) reported that, in addition to GIP, leptin also aberrantly stimulated cortisol secretion in dispersed adrenal cells; thus, the potential presence of more than 1 abnormal receptor may modify the phenotypic appearance. The potential presence of ectopic GLP-1 receptors has been excluded to date by the lack of stimulation of cortisol production after GLP-1 administration, either *in vivo* or *in vitro* (15, 17, 21). In one patient with GIP-dependent AIMAH, plasma ACTH and cortisol responses to CRH were still preserved, presumably because the intermittent food-dependent stimulation of cortisol had not yet completely suppressed the HPA axis (18). In a female patient with hirsutism and a unilateral adenoma, both adrenal androgens and cortisol were found to be stimulated by food intake *in vivo* and GIP *in vitro* (22); hypercortisolism was modest and ACTH was not fully suppressed.

The ectopic expression of GIP receptor has now been demonstrated in affected adrenal tisues of GIP-dependent adrenal macronodular hyperplasias (14, 15, 18, 21, 24) and adenomas (14, 19-22, 24); it is not present in non-GIP-dependent Cushing's syndrome adrenal tissues (14, 24) or the human adrenocortical carcinoma cell line H295 (19). GIPR expression was detected, even in the early stages of adrenal hyperplasia (15). The small amount of GIPR mRNA sometimes found in normal fetal or adult adrenal tissues after amplification was not efficiently coupled to steroidogenesis (24), and may reflect a low number of GIPR in endothelial cells (25) rather than in adrenocortical cells. Sequence analysis of the full-length cDNA in GIP-dependent adrenal tissues revealed no mutation of GIP receptor (21, 24). Thus, the concept of functional ectopic receptors is valid in explaining the pathophysiology of GIP-dependent Cushing's syndrome (Figure 2).

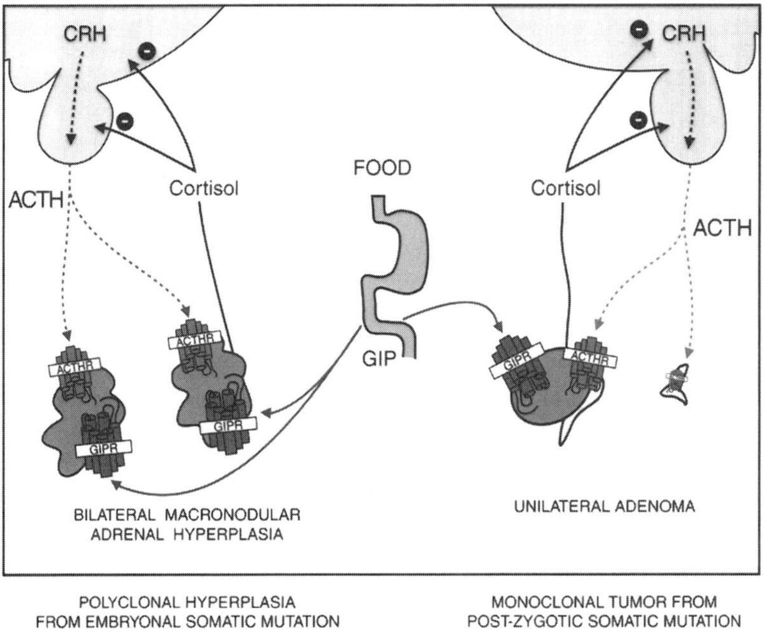

Figure 2.

The *in vitro* cortisol-stimulating effects of GIP are coupled to an increase of cAMP, but not of IP_3 production (14, 21). In studying GIP-dependent adrenal cells in primary culture, GIP receptor down regulation by its own ligand has been demonstrated (14). By stimulating steroidogenic enzyme activity, ACTH pretreatment of cells increased the GIP-induced cortisol response, but did not appear to modify GIPR expression directly (14).

Stimulation of thymidine incorporation into newly synthesized DNA by GIP was observed in primary cultures of adrenal cells from GIP-dependent Cushing's syndrome, but not in normal cells (21). Activation of p42-p44 mitogen-activated-protein kinases was observed after treatment of pathological cells with GIP (21). Such results suggest that GIP is possibly capable of regulating cell proliferation, in addition to steroidogenesis, in these tissues; however, cell growth stimulation by GIP has not yet been clearly demonstrated.

2. Vasopressin-responsive Cushing's syndrome

The binding of vasopressin to its V3-receptor (26) on corticotrophs of the anterior lobe stimulates the synthesis and maturation of POMC, leading to ACTH secretion (27). Similarly, a large proportion of pituitary corticotroph adenomas have been shown to augment their ACTH release after administration of lysine-vasopressin, resulting in increased plasma cortisol levels (28). In contrast, in adrenal Cushing's syndrome, where ACTH is suppressed, it is expected that plasma cortisol should not increase following administration of vasopressin (29). However, abnormal stimulation of cortisol secretion in response to exogenous arginine-vasopressin or lysine-vasopressin administration has been described in ACTH-independent canine (30) and human Cushing's syndrome, secondary to unilateral adrenal adenomas, carcinomas or AIMAH (2, 31-37). In a retrospective study of 26 patients with Cushing's syndrome secondary to unilateral cortisol-secreting tumors, Arnaldi et al. (34) observed an increase of plasma cortisol larger than 30 ng/ml after lysine-vasopressin testing in 27% of cases (5 adenomas and 2 carcinomas). In several patients, the abnormal adrenal response *in vivo* was confirmed directly by increased cortisol secretion in response to stimulation of dispersed adrenal cells with vasopressin *in vitro* (31-36).

Intracellular calcium flux was stimulated in dispersed tumor cells by vasopressin and was inhibited by a V1-vasopressin receptor antagonist (31, 35, 36). V2 vasopressin receptor agonists had no effect on steroidogenesis or calcium flux. Since V1-vasopressin receptor is present in the normal adrenal cortex and modulate modest effects of vasopressin on steroidogenesis, the exaggerated steroidogenic responses to vasopressin in these patients would be secondary to the abnormal function of a "eutopic" receptor-effector system, rather than to the presence of an ectopic receptor. V1-vasopressin receptor mRNA levels were found to be expressed either at higher (36) or similar (31, 34) levels, compared to normal control adrenal tissues. The binding affinity and dose-response of intracellular calcium flux for V1-vasopressin receptor noted in the adrenal AIMAH tissue were not different from those reported in other normal tissues (31). Thus, no evidence of ectopic receptor or gross overexpression of the eutopic V1-vasopressin receptor has been presented to date. The molecular mechanisms leading to the abnormal response of V1-vasopressin receptor or its effector system, which would increase the response to vasopressin, remain to be elucidated.

A woman with Cushing's syndrome and AIMAH presented the unusual association of orthostatic hypotension (31). The steroidogenic response was mediated by V1-vasopressin receptor, as shown by the effects of V1 receptor antagonists and the lack of effect of V2 receptor agonists. Alteration of the V1-receptor-effector system was not limited to the adrenal tissues of this patient, as there was also an abnormal, prolonged vascular vasoconstrictive response to vasopressin, compared to the arterioles of normal or hypertensive subjects. The persistence of decreased stimulation of plasma vasopressin and endothelin levels during postural hypotension, several months after correction of the hypercortisolism, also raised the possibility of an exaggerated V1-vasopressin receptor signal at the hypothalamic level in this patient.

The demonstration of an exaggerated cortisol response to pharmacological levels of exogenous vasopressin does not constitute direct evidence that fluctuations of endogenous vasopressin levels are the main regulator of steroidogenesis in these patients. This was illustrated in a male patient with AIMAH who was shown to have increased plasma cortisol in response to upright posture and administration of 10 IU arginine-vasopressin (38); however, the modulation of endogenous vasopressin levels by water dilution or hypertonic saline infusion did not modify plasma cortisol levels. In addition, *in vivo* administration of a V1-vasopressin receptor antagonist inhibited the response of cortisol to exogenous vasopressin, but not to upright posture. In fact, this patient was found to have ectopic β-adrenergic receptors (see next section) in his adrenal tissues; it is believed that pharmacological vasopressin levels stimulated catecholamine release, including from the adrenal medulla (39), explaining the cortisol release in this case. Further support came from the absence of V1-vasopressin receptor in his adrenal tissues (Lacroix et al. unpublished observation).

Daidoh et al. (35) studied a patient with very large bilateral AIMAH and severe Cushing's syndrome; i.v. injection of small amounts of arginine vasopressin (0.3 IU) increased plasma cortisol 3.7-fold without any detectable rise in ACTH. Similarly, insulin-induced hypoglycemia elevated plasma vasopressin and cortisol without any increase in plasma ACTH; catecholamine effects were not studied however. Upright posture augmented plasma vasopressin and cortisol. Oral administration of the V1-vasopressin receptor antagonist OPC-21268 for 8 days decreased urinary free cortisol levels, but potential spontaneous fluctuations of cortisol secretion were not evaluated for long periods. In another woman with Cushing's syndrome and AIMAH, plasma cortisol was stimulated by upright posture and 10 IU of arginine-vasopressin, but not by dDVAP (40); plasma cortisol was inhibited by water loading, and elevated during hypertonic saline infusion. These last 2 cases represent clear demonstrations of fluctuations in plasma cortisol levels in parallel with small physiological changes in endogenous vasopressin levels.

An abnormal increase of plasma cortisol in response to vasopressin administration was also noted in patients with preclinical bilateral macronodular adrenal hyperplasia (41). Recently, an exaggerated plasma cortisol response to lysine-vasopressin was seen in a 67 year-old woman with Cushing's syndrome and bilateral macronodular adrenal hyperplasia, whose brother had died after bilateral adrenalectomy for Cushing's syndrome and AIMAH (42); the precise nature of the abnormal hormone receptor implicated is unknown, but this constitutes the first demonstration of abnormal hormone responsiveness in familial AIMAH.

3. **Catecholamine-dependent Cushing's syndrome**

Catecholamines are known to modulate HPA activity. Activation of α_1-adrenoreceptors in the paraventricular nuclei leads to CRH release and increased plasma levels of ACTH and cortisol (43). Administration of β_1- or β_2-adrenergic agonists or antagonists has no effect on ACTH or cortisol secretion (44). Peripherally administered α_1-adrenoreceptor agonists fail to activate the HPA, as the blood-brain barrier prevents their access to the paraventricular nuclei. Direct adrenal stimulatory or inhibitory effects of catecholamines on glucocorticoid or mineralocorticoid secretion have been noted in several species, but are limited to aldosterone secretion in humans, where cortisol secretion is unaffected (6).

The abnormal presence of β-adrenergic receptors or the activation of adenylyl cyclase activity by catecholamines has been reported *in vitro* in several cases of human adrenal tumors associated with Cushing's syndrome

(7, 9, 45); no evidence of such receptors has been found in the normal adrenal cortex. However, the clinical expression of this abnormality was only appreciated recently in 2 patients. A 56 year-old French-Canadian man with AIMAH and Cushing's syndrome (38) was shown to have ACTH-independent overproduction of cortisol and aldosterone during elevations of endogenous catecholamines levels (upright posture, insulin-induced hypoglycemia, and the stress test). Augmented plasma cortisol during upright posture was decreased after pretreatment with the β-adrenergic antagonist, propranolol; in contrast, this did not occur after inhibition of the renin-angiotensisn system with captopril or losartan, or of AVP with a V1-AVPR antagonist. Isoproterenol infusion stimulated cortisol and aldosterone secretion in the patient, but not in normal subjects, in whom ACTH had been suppressed by dexamethasone. High affinity binding sites compatible with β_1- or β_2-adrenergic receptors were found in the adrenal tissues of the patient, but not in controls. They were efficiently coupled to steroidogenesis as shown by adenylyl cyclase stimulation with isoproterenol *in vitro* and catecholamine-induced steroidogenesis *in vivo* (38).

Another 50 year-old American woman with Cushing's syndrome and AIMAH (40) was found to have abnormal responses to catecholamines in addition to an exaggerated response to arginine-vasopressin. In this patient, plasma cortisol had risen after upright posture and exogenous vasopressin, but also after insulin-induced hypoglycemia, while ACTH remained suppressed. Infusion of isoproterenol for 30 min increased plasma cortisol which returned rapidly to baseline when the infusion was discontinued. Pretreatment of the patient with the angiotensin receptor type-1 antagonist losartan did not prevent the elevation of plasma cortisol during upright posture. It was concluded that cortisol secretion was mediated by the abnormal presence and function of β-adrenergic and V1-vasopressin receptor. Further molecular studies are needed to properly characterize the β-adrenergic receptor subtype expressed in hyperplastic adrenal tissues and to determine whether they are mutated or not.

4. LH/hCG-dependent Cushing's syndrome

The LH/hCG receptor (LH/hCGR) normally activates AC and PLC to stimulate gonadal steroidogenesis (46). The receptor is mainly expressed in gonadal tissues, but also in other tissues, including the uterus, fallopian tubes, placenta, brain, hypothalamus and prostate (47); recently, the presence of LH/hCGR was identified in the zona reticularis of the human adrenal (48) by immunohistochemistry and in situ hybridization. hCG stimulates DHEAS secretion in human fetal adrenal cells (49).

The aberrant expression and function of LH/hCG receptor was identified in a woman who presented with a history of transient Cushing's syndrome during pregnancies; hypercortisolism became persistent with bilateral macronodular adrenal hyperplasia only after the long-term physiological increase of LH levels 10 years after her menopause (50). Cortisol production was increased by the *in vivo* administration of GnRH, hCG, and recombinant human LH (hLH). Plasma free testosterone and estradiol were also augmented by hLH administration. Abnormal stimulation of cortisol, free testosterone and DHEAS production was also evoked in this patient by oral intake of cisapride and metoclopramide, two 5-HT_4R agonists (50). Administration of the long-acting GnRH analog leuprolide acetate initially increased LH and FSH secretion, which resulted in a rise in cortisol secretion; however, this was followed within 10 days by suppression of endogenous LH and FSH levels and normalization of cortisol production. Studies of normal adult controls did not indicate any coupling of LH/hCGR to adrenal synthesis of cortisol or DHEAS. The aberrant adrenal expression of the LH/hCG receptor (Figure 3) was confirmed by RT-PCR and sequencing in biopsy material (51). Abnormal stimulation of plasma cortisol after GnRH and LH administration was also found in one woman with bilateral macronodular adrenal hyperplasia and normal urinary cortisol levels, which did not suppress normally with dexamethasone (41). This suggests that the same abnormality can be present in preclinical bilateral macronodular adrenal hyperplasia.

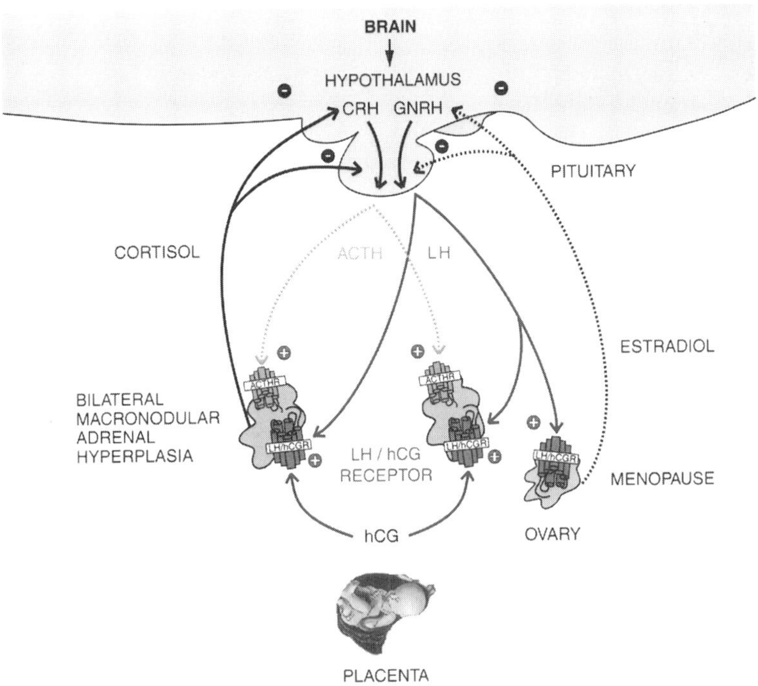

Figure 3.

In some cases of pregnancy and Cushing's syndrome secondary to adrenal adenomas (52) or large bilateral macronodular adrenal hyperplasia (53), the clinical syndrome regressed after abortion or delivery. When these patients were tested after delivery, biochemical evidence of residual abnormal cortisol secretion was still present, despite substantial improvement, and was fully corrected only after surgical removal of the adenoma. Transient ACTH-independent Cushing's syndrome during pregnancy with complete resolution after spontaneous abortion or delivery was described in 2 patients with mild bilateral adrenal hyperplasia (54, 55) or unknown adrenal pathology (56). A paradoxical increase in cortisol excretion during dexamethasone administration in pregnancy completely returned to normal following delivery (55). One patient developed severe biochemical and clinical evidence of Cushing's syndrome at each of 3 pregnancies, and ACTH or cortisol levels were not stimulated by vasopressin administration (54); there was a transient period of hypocortisolism after each delivery, followed by complete clinical regression. In one case, short-term hCG administration elevated urinary 17-

hydroxycorticosteroid levels, while sequential estrogen and progestogen administration had no effect (56).

It is thus possible that some of these patients with transient Cushing's syndrome during pregnancy, or in whom hypercortisolism increased during pregnancy, also expressed ectopic LH receptors in their adrenal adenomas or in their adrenal cortex. Specific testing of the regulation of steroidogenesis with LH or estrogens in future cases of transient Cushing's syndrome during pregnancy will help in elucidating the pathophysiology. It must be pointed out that spontaneous remission of Cushing's syndrome after delivery has also been reported in a patient with ACTH-dependent Cushing's syndrome of probable pituitary origin (57); the mechanisms involved in this regression have not been elucidated.

5. LH-dependent adrenal androgen-secreting tumors

Although the regulation of cortisol secretion by LH in adrenal Cushing's syndrome was only demonstrated recently *in vivo*, there have been several reports indicating that the regulation of steroidogenesis in some rare, pure, androgen-secreting tumors was stimulated by hCG or GnRH (2, 10). As plasma LH was relatively suppressed in some of these patients, the role of endogenous LH in maintaining androgen production may be uncertain. In some cases, suppression of endogenous LH levels by administration of estrogens or by ACTH stimulation of GC (58) inhibited androgen production. In other cases, estrogens were unsuccessful in depressing androgen production (59). It has been suggested that gonadal cells localized in the adrenals could explain this phenomenon; however clear evidence of adrenal origin of the tumors were identified in certain cases (10).

6. Serotonin-responsive Cushing's syndrome

Serotonin (5-HT) is produced by intra-adrenal mast cells in humans and can regulate corticosteroid production via a paracrine mechanism (60, 61); these effects are mediated by the 5-HT_4R subtype which is expressed mainly in adrenal zona glomerulosa, but also present in zona fasciculata cells (62, 63). 5-HT_4 receptor agonists are potent stimulators of aldosterone secretion in humans; they are weak stimulators of cortisol secretion by human adrenocortical cells *in vitro*, but not of plasma cortisol in normal subjects (60).

In the patient with LH/hCG-dependent Cushing's syndrome (50), cisapride and metoclopramide, two 5-HT_4R agonists, produced large elevations in plasma cortisol after their oral administration. Plasma corticotropin levels remained undetectable during cisapride and metoclopramide testing. Stimulation of plasma cortisol in this patient after treatment with cisapride and metoclopramide was proportional to their respective affinity for 5-HT_4 receptor (60); no such response to cisapride was found in 5 other patients with bilateral adrenal hyperplasia, 11 with unilateral adenoma and 1 with carcinoma and Cushing's syndrome (40). Another patient with Cushing's syndrome and AIMAH was found to increase plasma cortisol in response to cisapride as well as to lysine-vasopressin, despite suppression of ACTH (64). Recent observations in patients with bilateral macronodular adrenal hyperplasia and pre-clinical hypercortisolism also documented marked stimulation of cortisol secretion upon administration of cisapride, vasopressin and LH/hCG (41).

The exaggerated cortisol responses to cisapride in these patients could be secondary to the increased zona fasciculata expression or abnormal function of an "eutopic" 5-HT_4 receptor-effector system, rather than to the presence of an ectopic receptor. The presence of a 5-HT_4 receptor has been detected by RT-PCR in the adrenal tissues in some of these patients and was similar to that found in normal adrenal cortex; however, full receptor sequencing and adrenal zone distribution have not been performed (51, 65).

7. Steroid-responsive Cushing's syndrome

Caticha et al. (66) described a woman who developed transient and reversible clinical and biochemical signs of ACTH-independent Cushing's syndrome during 3 pregnancies and during intake of oral contraceptives. Her adrenal histology was described as being compatible with primary nodular dysplasia, but there were no comments on pigmentation of her adrenal nodules; there was also no family history of adrenal disease and no other features of Carney's complex. Paradoxical increases in cortisol production were noted during oral dexamethasone suppression tests. Dose-responsive stimulation of cortisol secretion occurred after bilateral adrenalectomy when the cells were exposed to estradiol; the *in vitro* addition of dexamethasone was not reported, nor were the effects of anti-estrogens.

Paradoxical increases in plasma cortisol and urinary free cortisol were observed during the last 2 days of classical Liddle's 4-day low- and high-dose oral dexamethasone tests in patients with PPNAD with or without Carney's complex (67). We found no evidence of ectopic membrane hormone receptors in 2 patients with PPNAD, who showed an increase in cortisol secretion during prolonged dexamethasone administration; GC receptors appeared to be highly expressed by immunohistochemistry in PPNAD micronodules, compared to the adjacent internodular atrophic adrenal or to the normal control adrenal cortex (68). Similar paradoxical elevations of cortisol production during dexamethasone have been reported in several cases of Cushing's syndrome during pregnancy (53, 55).

8. Other abnormal hormone responses in adrenal Cushing's syndrome

Hashimoto et al. (69) described a 51-year-old male with large bilateral AIMAH in whom plasma cortisol increased during insulin-induced hypoglycemia, while ACTH remained at undetectable levels; *in vitro*, cortisol secretion was stimulated in dispersed adrenal cells by ACTH, but not by insulin, catecholamines, vasopressin or Ang-II. A very similar patient with AIMAH studied by the same group (70) also displayed elevated plasma cortisol during insulin-induced hypoglycemia and combined lysine-vasopressin-CRH tests, while plasma ACTH remained undetectable; *in vitro* studies were not performed in this case. It remained unclear whether insulin itself, a factor increased during hypoglycemia, or sub-detectable rises in plasma ACTH were responsible for the regulation of cortisol secretion in these cases.

Leptin synthesis is stimulated by glucocorticoids (71), and their plasma levels are elevated in patients with Cushing's syndrome. Leptin receptors are expressed in normal adrenals, as well as in adrenocortical adenomas and carcinomas (72); leptin normally inhibits cortisol secretion in the adrenal (72, 73). Pralong et al. (17) recently reported a 36-year-old woman with AIMAH and Cushing's syndrome in whom a mixed meal heightened plasma cortisol levels, and this effect was decreased by octreotide pretreatment. GIP was not infused, but GIP stimulated cortisol secretion in vitro. Leptin (single dose of 100 nM) aberrantly increased cortisol secretion *in vitro*. GIP or leptin receptors were not measured directly. Thus, this case raises the possibility of paradoxical stimulation of steroidogenesis by leptin in some cases of AIMAH, but more detailed studies are required in other similar cases to confirm this possibility.

INVESTIGATION STRATEGY TO DETECT ABERRANT RECEPTORS *IN VIVO*

Initial clinical screening protocol

A protocol has been developed to screen patients with adrenal Cushing's syndrome for the presence of ectopic/abnormal adrenal hormone receptors (74); the strategy is based on monitoring plasma levels of steroids during various tests which transiently modulate the levels of ligands for the potential aberrant receptors. Serial measurements of plasma ACTH, cortisol and other steroids or hormones are performed during the course of various tests, at 30-60 min intervals during 2-3 hours (Table 2). All tests are performed after an overnight fast and in a supine posture for at least 1 hour. Initial screening includes on the first day, a posture test performed as a 2-h supine position, followed by a 2-h ambulatory period to evaluate potential modulation of aberrant receptors by angiotensin-II, vasopressin, catecholamines, or atrial natriuretic peptides. This is followed by a standard mixed meal to evaluate the potential presence of adrenal receptors for gastrointestinal hormones. The administration of 250 μg ACTH 1-24 i.v., serves as a reference test for the presence of a functional ACTH receptor. On another day, the administration of 100 μg GnRH i.v. examines the potential modulation by FSH, LH, or GnRH; this is followed by 200 μg TRH i.v. to search for aberrant receptors for TSH, prolactin or TRH. Potential aberrant adrenal responses to 1 mg glucagon i.v., 10 IU arginine-vasopressin i.m. and 10 mg metoclopramide orally (a serotonin 5-HT_4 receptor agonist) are tested sequentially on the third day. A change of less than 25% of plasma cortisol is arbitrarily defined as no response, a 25 to 49% change, as a partial response, and a change of 50% or greater, as a positive response. If a partial or positive cortisol response is found, the test is repeated to verify its consistency and to determine if other steroids, such as aldosterone, DHAS, testosterone and estradiol are also modified. At the same time, fluctuations of ligand hormones of interest (i.e. catecholamines, vasopressin, renin/angiotensin-II and ANP during a posture test) are measured. If a prolonged response to a test masks the evaluation of the following test, it is repeated separately.

Further characterization of abnormal hormone receptors

After initial screening, other tests are performed to confirm the responses or to elucidate which hormone is implicated (Figure 4). For example, if cortisol stimulation by upright posture is found, the inverse effect, i.e. suppression by assuming a supine posture after ambulation is verified. The respective contributions of vasopressin, catecholamines, angiotensin-II, or atrial natriuretic peptides need to be distinguished. An exaggerated cortisol response to pharmacological levels of exogenous vasopressin is followed by evaluation of whether physiological fluctuations of endogenous vasopressin would modify plasma cortisol levels. Endogenous plasma vasopressin levels can be modulated by a 20 cc/kg water load, followed by infusion of NaCl 3% at 0.1 cc/kg/min for 120 min. The expected result would be an initial suppression of vasopressin and cortisol during water loading, followed by an increase of vasopressin and cortisol levels. To determine whether the vasopressin receptor involved in this response is a V1, V2 or V3 receptor type, 2.5 μg desmopressin, a preferential V2 receptor agonist, is administered s.c. (31,40); the absence of a response to desmopressin would suggest a V1 or V1b/V3 receptor-mediated response. Pretreatment with a specific V1 receptor antagonist (SR 49049) has been used to demonstrate *in vivo* the involvement of the V1 receptor in this response (38). In case of no response to exogenous vasopressin, the role of angiotensin-II is assessed by repeating the posture test after administration of an AT-1 receptor antagonist or by direct infusion of angiotensin-II. If a catecholamine response is suspected, endogenous catecholamine stimulation is produced by insulin-induced hypoglycemia, and, if positive, by isoproterenol infusion (38). An attempt to block the response and to treat the patient with a ß-blocker would be conducted if the stimulation of cortisol production is reproduced.

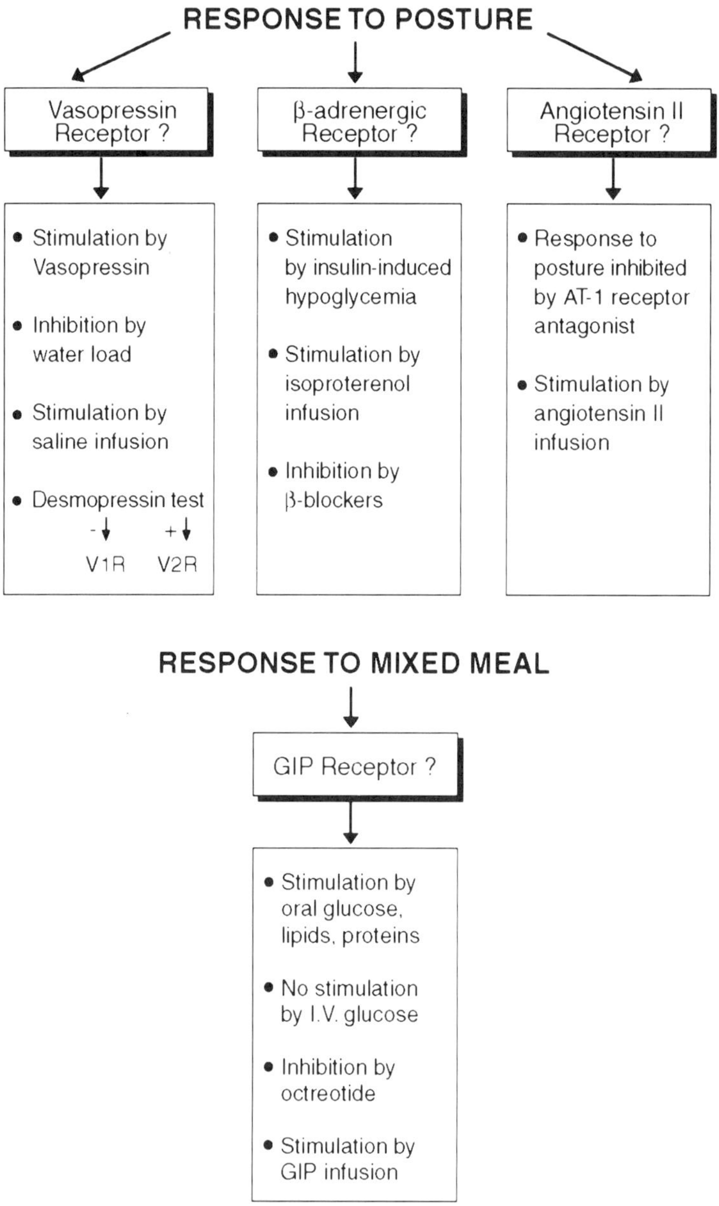
RESPONSE TO POSTURE
Vasopressin Receptor ?
β-adrenergic Receptor ?
Angiotensin II Receptor ?
• Stimulation by Vasopressin
• Inhibition by water load
• Stimulation by saline infusion
• Desmopressin test
- V1R
+ V2R
• Stimulation by insulin-induced hypoglycemia
• Stimulation by isoproterenol infusion
• Inhibition by β-blockers
• Response to posture inhibited by AT-1 receptor antagonist
• Stimulation by angiotensin II infusion
RESPONSE TO MIXED MEAL
GIP Receptor ?
• Stimulation by oral glucose, lipids, proteins
• No stimulation by I.V. glucose
• Inhibition by octreotide
• Stimulation by GIP infusion

Figure 4a.

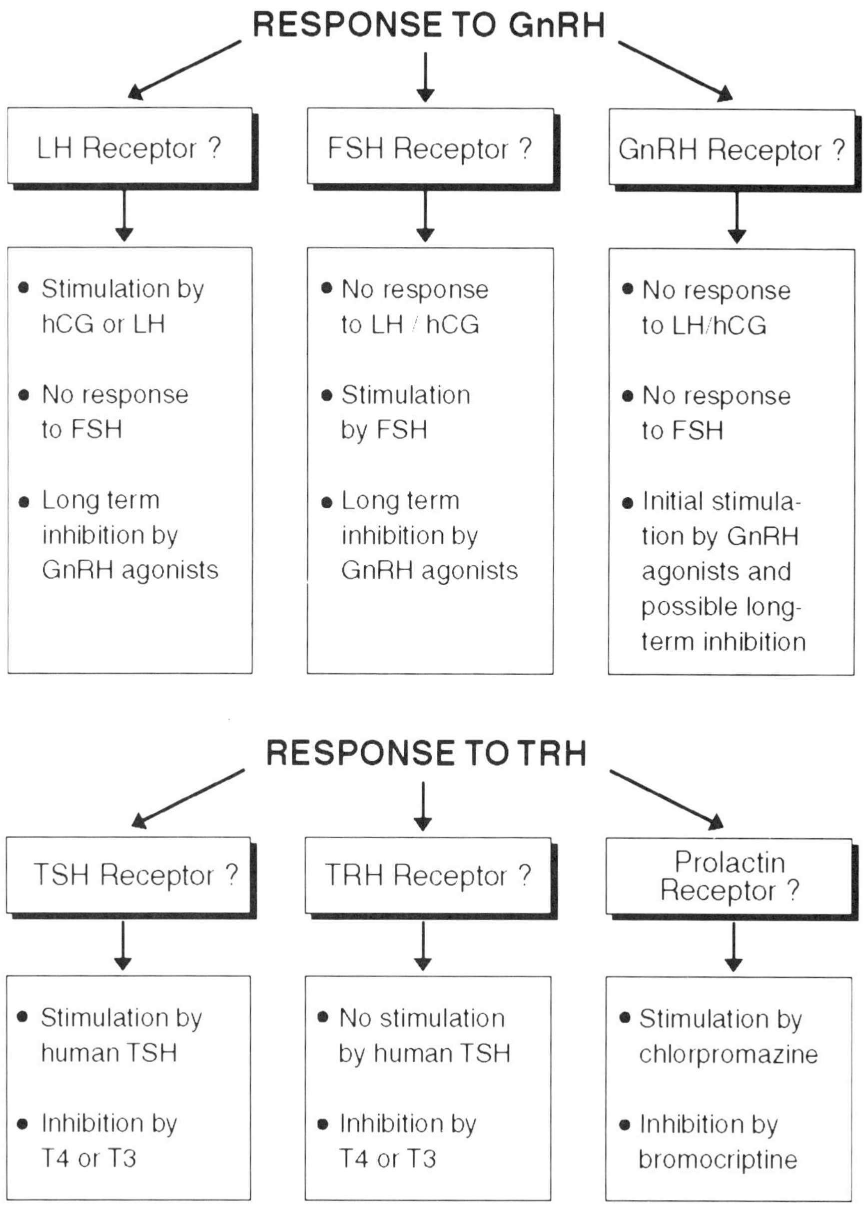
RESPONSE TO GnRH
LH Receptor ?
FSH Receptor ?
GnRH Receptor ?
• Stimulation by hCG or LH
• No response to FSH
• Long term inhibition by GnRH agonists
• No response to LH / hCG
• Stimulation by FSH
• Long term inhibition by GnRH agonists
• No response to LH/hCG
• No response to FSH
• Initial stimulation by GnRH agonists and possible long-term inhibition
RESPONSE TO TRH
TSH Receptor ?
TRH Receptor ?
Prolactin Receptor ?
• Stimulation by human TSH
• Inhibition by T4 or T3
• No stimulation by human TSH
• Inhibition by T4 or T3
• Stimulation by chlorpromazine
• Inhibition by bromocriptine

Figure 4b.

In the case of a response to a mixed meal, confirmation and identification of a specific gastrointestinal hormone involvement are based on evaluation of the effects of carbohydrates, proteins, or lipids on cortisol secretion. Patients ingest, at 3-h intervals, 75-g oral glucose, an isocaloric protein-rich meal or a lipid-rich meal (12, 15). Plasma cortisol, ACTH, GIP and insulin levels are measured at regular intervals during these tests. The absence of plasma cortisol response to the i.v. administration of 25 g glucose, or when 100 μg octreotide is administered s.c. 60 min prior to repeating the oral 75 g glucose challenge, confirms the role of a gastrointestinal hormone (12, 15). As only GIP and GLP-1 respond well to oral glucose, lipids and partially to proteins, human GIP is infused at a rate of 0.6 μg/kg/min during the administration of 150 cc/h of 10% glucose; this is compared to the response to GLP-1 infused at a rate of 0.75 pmol/kg/min, also under 10% glucose (15). The pattern of response to the various secretagogues would be different in the presence of abnormal receptors for gastrointestinal hormones other than GIP. Various candidate hormones would then be infused to confirm the identity of the steroidogenesis modulator.

Stimulation of cortisol production after GnRH administration could result from the abnormal adrenocortical presence of receptors for LH/hCG, FSH, or GnRH itself. The cortisol response after the administration on different days, of hCG 10,000 U i.m., purified human FSH 150-300 U i.m., and recombinant LH 300 U i.v. can be compared (50). A response to GnRH coupled to an absence of response to FSH, LH, and hCG would suggest an ectopic GnRH receptor; various analogs and antagonists of this receptor are available for testing the hypothesis. In addition, the response to an acute dose of GnRH should persist despite the suppression of endogenous gonadotropins by the administration of supraphysiologic doses of gonadal steroids or the use of long-acting GnRH analogs. In the presence of an ectopic LH/hCG receptor, a response to exogenous hCG or LH but not to exogenous FSH should be evident; the response to acute GnRH administration should disappear when the LH response is abolished by exogenous gonadal steroids or after the chronic administration of long-acting GnRH analogs. Therapy with long-acting GnRH analogs should produce eventual suppression of the endogenous LH ligand and normalize cortisol production, as demonstrated recently by our group in one patient (50). In the presence of an ectopic FSH receptor, there should be no response to hCG or LH, but cortisol production should be increased after the administration of purified FSH. Here again, long-acting GnRH analogs should suppress the biologically active ligand and correct the hypercortisolism.

Stimulation of cortisol synthesis after TRH administration has not yet been reported. However, adenylyl cyclase stimulation by TSH has been demonstrated in adrenocortical adenomas *in vitro* (1). Thus, a response to TRH would suggest the possibility of an ectopic receptor either for TSH, TRH, or prolactin. The prolactin receptor does not belong to the family of G-coupled 7-transmembrane receptors which could mimic the ACTH receptor and activate adenylyl cyclase. However, adrenocortical stimulation by prolactin has been described *in vitro*, and the presence of this receptor in adrenal tumors has been confirmed (75). Elevation of endogenous prolactin levels after a chlorpromazine test and its inhibition by a bromocryptin test would easily clarify the role of endogenous prolactin. The potential presence of an ectopic TSH receptor would be assessed directly by the administration of purified human TSH and by inhibiting endogenous TSH production with exogenous thyroxin. The lack of an ectopic TRH receptor would be confirmed by disappearance of the cortisol response when the response of TSH to TRH has been suppressed by thyroxin administration.

The *in vitro* response of adenylyl cyclase to glucagon has been demonstrated previously in a cortisol-secreting adenoma, but a clinical case has not been reported yet. If a response to 1 mg of exogenous glucagon was found, it would be necessary to show that fluctuations of endogenous glucagon levels during insulin-induced hypoglycemia, fasting, or oral administration of glucose correlate well with fluctuations of cortisol levels.

The oral administration of 10-mg metoclopramide, a 5-HT_4 receptor agonist, is expected to induce a large increase in aldosterone, but not in cortisol levels in normal individuals (63). If a cortisol response to metoclopramide is found, a response to other 5- HT_4 receptor agonists such as zacopride should be found, but not to specific 5-HT-1,2,3 agonists. Although some specific 5-HT_4 receptor antagonists are currently under investigation, we have not been able to obtain any yet, but they should become very valuable to confirm the role of this abnormally expressed receptor.

Systematic clinical screening for ectopic/abnormal hormone receptors

This protocol has been used to perform the systematic clinical screening of 20 consecutive patients with adrenal Cushing's syndrome secondary to either bilateral macronodular adrenal hyperplasia (n=6), unilateral adenoma (n=13) or carcinoma (n=1) for the presence of aberrant hormone receptors (40). All 6 patients with AIMAH had a positive response to at least 1 test, in addition to ACTH 1-24: 2 patients, to the mixed meal (GIP-dependent); 1 patient, to GnRH (LH/hCG receptor) and cisapride (5-HT_4 receptor); and 3 patients, to the upright posture and vasopressin (1 β-adrenergic receptor, 1 V1-vasopressin receptor, 1 β-adrenergic receptor and V1- vasopressin receptor). In patients with unilateral adenoma, only 3 out of 13 patients had a response either to upright posture, mixed meals, or vasopressin, but were not further characterized. In a patient with adrenocortical carcinoma or in 2 patients with micronodular adrenal dysplasia (68), plasma cortisol was not modified by any of the tests.

In 6 patients with subclinical cortisol secretion and incidentally identified AIMAH, all 6 were found to express at least, one and up to three aberrant receptors including V_1-vasopressin, LH/hCG and 5-HT_4 (41).

Initial experience suggests that the adrenal expression of various ectopic or abnormal hormone receptors is frequently implicated in the pathophysiology of bilateral macronodular adrenal hyperplasia (40), but less frequently in unilateral adenoma (34). It must be noted that the initial protocol utilized to date did not screen for many other G-protein-coupled membrane receptors, such as those for PTH, calcitonin, acetylcholine, dopamine, opiates, prostaglandins, etc; it may thus become pertinent to investigate these other potential abnormal receptors in the future.

Molecular mechanisms of aberrant hormone receptors

The hormonal regulation of adrenal cortex development and function requires the tissue-specific expression of hormone receptors. This implies the existence of fine-tuned mechanisms of regulation that involve cis-acting regulatory elements (promoters) and trans-acting factors (transcription factors) for these receptors. The molecular mechanisms responsible for the ectopic or aberrant expression and function of membrane receptors in adrenal Cushing's syndrome have not been identified yet. In fact, the important question of regulation of the tissue-specific expression of genes is raised by this new pathophysiology of adrenal Cushing's syndrome. A mutation in the promoter of the hormone receptor could generate an appropriate binding site for an adrenocortical-specific transcription factor/co-activator complex, leading to ectopic expression. Alternatively, altered expression of hormone receptors could be achieved as a consequence of mutations in transcription factors, their co-activators or co-repressors. In the case of overexpression of a single receptor, any of these hypotheses is plausible. In contrast, when more than one aberrant adrenal receptor is present, it is improbable that acquired mutations of several unrelated promoters would occur simultaneously. A mutation in a factor regulating cell differentiation or in a transcription factor common for the different receptors may be more likely. Another interesting hypothesis emerges from a recent study by Kero et al. (76), who observed that transgenic mice expressing bLHβ-CTP (a chimeric protein of β sub-units fragments of bovine LH and hCG) in their pituitary develop adrenal Cushing's syndrome in addition to polycystic ovaries and ovarian tumors following chronically elevated serum LH levels. It was shown that this resulted from ectopic expression of LH/CG receptors in the adrenal cortex, not detectable or functional in control mice. Since this induction is abolished by gonadectomy, it was proposed that elevated estrogens and prolactin levels were responsible for inducing the illicit expression of the LH/CG receptor in the adrenal cortex. This observation would thus raise the possibility that the "ectopic expression" of a receptor may not require a mutation of cis- or trans-acting regulators, but may result from exaggerated stimulation of a gene, which is normally silent.

The presence of aberrant membrane hormone receptors in unilateral cortisol-secreting adenomas may arise from the monoclonal expansion of a primary adrenocortical cell which acquired a somatic mutation, leading to the abnormal expression and function of that receptor at the post-zygotic stage of adrenal cortex development; most studies confirm the monoclonal composition of human adrenocortical tumors (77, 78). In patients with aberrant membrane hormone receptors in bilateral macronodular adrenal hyperplasia, the mutational event must have occurred very early during embryogenesis so that every cortical cell of both adrenals would be affected by the defect, which is polyclonal. There have been rare reports of familial AIMAH (42, 79-81), and in only one case, an abnormal response to lysine-vasopressin was demonstrated in a sibling (42); thus, abnormalities of receptor expression in these syndromes are probably frequently the consequence of somatic mutations, but could also be germline mutations in some cases. In the case of a very early mutational event resulting in AIMAH, the abnormal expression could affect diverse tissues so that polymorph aberrant manifestations would be expected. This was the case in the patient with vasopressin-dependent Cushing's syndrome (31) who also displayed an abnormal vascular response to vasopressin and postural hypotension.

The majority of ectopic or abnormal hormone receptors in adrenocortical tumors or hyperplasias belong to the G-protein-coupled receptor superfamily (1, 2). Studies of the second messengers implicated in ectopic/abnormal hormone receptors in adrenal Cushing's syndrome suggest that they regulate steroidogenesis by mimicking the cellular events triggered normally by ACTH receptor activation (Figure 1). It is thus expected that only ectopic or eutopic hormone receptors capable of coupling efficiently to the intracellular signaling systems present in adrenocortical cells (i.e. those for ACTH, V1-AVP, etc.) will be able to regulate steroidogenesis aberrantly.

What is the role of abnormal hormone receptors in altered cell growth and tumorigenesis? One could postulate that the primary event is a mutation resulting in aberrant adrenal expression of the receptor, leading to increased proliferation and eventually to increased hormone production. Alternatively, it can be proposed that the primary event is an unknown proliferative one resulting in cell dedifferentiation with resultant expression of "embryonal" type genes, including one or several hormone receptors. In either hypothesis, it is clear that a relatively long time period is necessary before phenotypic expression of the abnormal hormone receptor becomes evident. This is particularly true for AIMAH, as several decades are necessary before the hyperplasia and hyperfunction become clinically manifest. This may be secondary to the transient occupation of the receptor by the ligand, as illustrated by the cases of GIP and LH-dependent Cushing's syndrome (Figures 2 and 3). In GIP-dependent Cushing's syndrome, the adrenal tissues are stimulated only briefly but repeatedly following each food ingestion; in LH/hCG dependent Cushing's syndrome, the hyperplasia and hyperfunction occurred only after intense and prolonged exposure to the ligands i.e. during pregnancy for hCG, or after menopause for LH. Reversal of the hyperplasia between pregnancies would favor the hypothesis that the ectopic receptor is a primary event rather than one which is secondary to another proliferative event; however, this awaits clear demonstration of adenoma or AIMAH regression following complete blockade of the ectopic receptor. There is also indication, based on cases of preclinical cortisol production in bilateral macronodular disease, that steroidogenesis can be relatively inefficient, despite significant proliferation. This suggests poor steroidogenic enzyme activities in the adrenal lesions or that the low expression of abnormal receptors is better coupled to proliferative signals than to hormone synthesis.

The concept of abnormal G-protein-coupled receptors and/or post-receptor events leading to increased cAMP and proliferation is now well established especially in somatotroph and thyroid cells (82-86). We hypothesize that the aberrant expression of any G-protein-coupled receptor could induce the stimulation of adrenal cells by trophic factors lacking regulatory negative feedback by cortisol. This stimulus may lead to increased function and confer a proliferative advantage. The event provides a gain of function to adrenocortical cells; thus, this category of mechanism, i.e. the abnormal tissue-specific expression of membrane hormone receptors should be added to the list of abnormalities of hormone receptors implicated in certain human diseases. GIP has been shown to stimulate the cAMP production and DNA synthesis in GIP-dependent, cortisol-secreting adenoma cells in a manner similar to ACTH (21). Hormone-stimulated LH/hCG receptor can act as an adrenocortical tumor promoter when ectopically expressed in the adrenal cortex of gonadectomized mice transgenic for the inhibin α-subunit promoter/simian virus 40 T-antigen fusion gene (87); it remains to be seen whether the expression of ectopic adrenocortical receptors, in the absence of other oncogenic events, is sufficient for adrenal overgrowth. Future animal models such as transgenic mice expressing ectopic membrane hormone receptors in the adrenal cortex will be informative in this regard. This is already supported by the demonstration of bilateral adrenal hyperplasia and Cushing's syndrome in the mice transgenic for bLHβ-CTP with ectopic adrenal expression of LH/CGR (76).

What is the cell of origin in which the receptor is expressed abnormally? Based on the profile of steroids produced, it appears that it can occur in well-differentiated cells of the fasciculata-reticularis (pure cortisol- or mixed cortisol-/androgen-secreting adenoma), and in cells from the reticularis (pure androgen-secreting adenoma); the 3 classes of adrenal steroids are sometimes secreted in macronodular hyperplasia, suggesting that all zonae are affected. It remains to be seen whether some cases of unilateral adenomas or bilateral hyperplasia in primary hyperaldosteronism can also be secondary to ectopic hormone receptors (Figure 5).

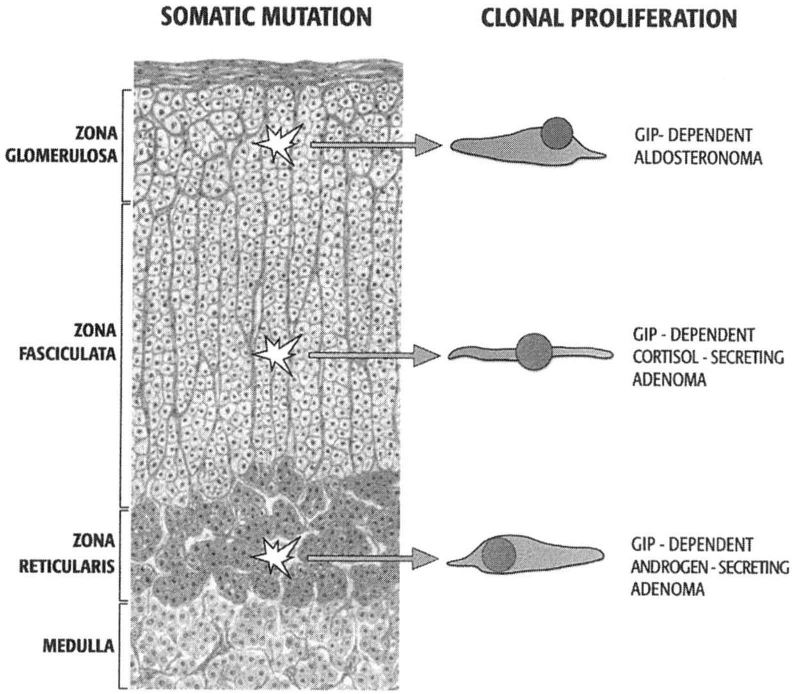

Figure 5.

No constitutive activating mutations of the ACTH receptor have been found yet in adrenocortical neoplasms or hyperplasias (88). Recent studies suggested that ACTH receptor could act as a tumor suppressor gene in adrenal tumorigenesis (89) in a way similar to p53, which is involved in many tumor types, including adrenocortical tumors (90). Loss of heterozygocity of the ACTH receptor gene was shown to be associated with high malignancy or the absence of secretion in a subset of human adrenocortical tumors. Furthermore, lower expression of ACTH receptor was found in adrenocortical carcinomas compared to adrenocortical adenomas from patients with Cushing's syndrome (91, 92). ACTH is known to be a differentiating factor with low potential for promotion of cell proliferation, as demonstrated by *in vitro* experiments. It has thus been speculated that a defect in the ACTH receptor signal cascade could result in de-differentiation and increased cell proliferation (93). Obviously, much work remains to be done to better understand the mechanisms underlying tumorigenesis of the adrenal cortex.

An opportunity for new pharmacological therapeutic strategies

The identification of aberrant adrenal hormone receptors in cortisol-secreting hyperplasias or tumors provides new opportunities to use specific pharmacological therapies as alternatives to adrenalectomy. This was initially suggested by the short-term improvement of hypercortisolism when triiodothyronine was administered briefly to a patient prior to resection of an adrenal adenoma in which adenylyl cyclase was stimulated by TSH (1); however, these data were very preliminary and could not clearly distinguish if the changes in cortisol levels reflected spontaneous fluctuations of cortisol secretion, or were truly the result of endogenous TSH suppression.

Pharmacological blockade of post-prandial GIP release with octreotide was attempted in a few patients with GIP-dependent Cushing's syndrome as an alternative to surgery (13, 18, 20). During the first months of s.c. octreotide administration before each meal, clinical and biological improvements were documented, but long-term treatment proved to be ineffective. It is presumed that the escape of octreotide efficacy was secondary to down- regulation of somatostatin receptors in GIP-secreting intestinal cells. Thus, adrenalectomy remains the long-term treatment of choice for this syndrome until specific GIP receptor antagonists become available. Short-term use of the oral V1-vasopressin receptor antagonist OPC-21268 for 8 days decreased urinary free cortisol levels in a patient with vasopressin-responsive AIMAH and Cushing's syndrome (35).

In the patient with catecholamine-dependent Cushing's syndrome and bilateral AIMAH (38), initial treatment with propranolol up to 320 mg daily was able to considerably reduce cortisol secretion; however, urinary cortisol levels remained approximately twice the upper limit of normal, and it was decided to remove one of the two very large adrenals surgically. It then became possible, upon restoration of propranolol administration, to completely normalize cortisol production. Interestingly, the control of hypercortisolism was followed by a decreased requirement in the dosage of the β-blocker from 320 mg to 20 mg of propranolol daily, as higher doses were causing adrenal insufficiency. Glucocorticoids are known to stimulate β_2-adrenergic receptor transcription (94) via glucocorticoid-responsive-elements located in promoters of the target genes (95). The normalization of cortisol levels may have decreased β- adrenergic receptor density, which would explain the lower requirement for the antagonist. Propranolol therapy did not reduce the size of the remaining adrenal even after 3 years of follow-up; however, the minimal dose of propranolol necessary to maintain normal cortisol production was administered, without blocking the receptors completely. This constituted the first example of long-term pharmacological blockade of an ectopic adrenal membrane hormone receptor.

In the patient with LH/hCG-dependent AIMAH and Cushing's syndrome, the suppression of endogenous LH levels with chronic, long-acting leuprolide acetate controlled the hypercortisolism and avoided bilateral adrenalectomy (50). Leuprolide acetate, a long-acting GnRH agonist, initially stimulated gonadotropins release which increased cortisol production for one week; this was followed by suppression of endogenous LH levels and normalization of cortisol production. Despite complete suppression of endogenous LH levels, the patient did not present cortisol insufficiency. It is possible that basal cortisol production was maintained by serotonin stimulation, since there was also evidence of abnormal 5-HT_4 receptor function in the same adrenals. The absence of regression of bilateral adrenal hyperplasia, despite chronic suppression of endogenous LH, indicates that its size was maintained by abnormal function of 5-HT_4 receptor, or that aberrant receptors regulate steroidogenesis but not cell proliferation. It will be interesting to study the effects of a specific 5-HT_4 recptor antagonist in these patients when it becomes available. A GnRH analog has previously been used successfully in long-term suppression of testosterone secreting ovarian tumor (96).

Further studies will probably identify a larger diversity of hormone receptor abnormalities and should eventually allow the use of new pharmacological tools to inhibit either the production of endogenous ligands or block the receptors with appropriate specific antagonists (Table 3). Since it is also possible to detect the presence of ectopic/abnormal hormone receptors at the stage of pre-clinical steroid hormone production (41), it will be of great interest to investigate whether the progression of adrenal tumors or hyperplasias can be prevented by these new pharmacological approaches.

SUMMARY AND CONCLUSIONS

Taken together, the results of *in vitro* and *in vivo* studies indicate that a wide diversity of abnormal adrenocortical membrane hormone receptors can be present in adrenal Cushing's syndrome. These may include ectopic hormone receptors, such as those for GIP, β-adrenergic agonists, LH/hCG, or other receptors capable of coupling to G-proteins, adenylyl cyclase, and steroidogenesis. There is evidence that the IL-1R receptor, which do not belong to the 7-transmembrane receptor family and do not use the same signaling pathway as the ACTH receptor, may also become coupled to steroidogenesis. A similar outcome may result from increased or altered activity of eutopic receptors, such as those for vasopressin (V1-receptor) or serotonin. The presence of aberrant receptors places adrenal cells under stimulation by a trophic factor, which is not under the main regulatory negative feedback exerted by glucocorticoids. This constitutes an unregulated new trophic stimulus, which leads to increased function and possibly to hyperplasia and proliferative advantage. The molecular mechanisms responsible for the ectopic expression of hormone receptors or to increased activation of the signaling cascade and steroidogenesis are still largely unknown. Characterization of the pathophysiology of adrenal hyperplasias or tumors can eventually lead to diverse pharmacological therapies as alternatives to adrenalectomy. Further studies will probably identify a larger diversity of hormone receptor abnormalities in adrenal and other endocrine and non-endocrine tissues. Elucidation of the molecular mechanisms leading to abnormal hormone receptor expression will probably contribute to our understanding of the regulation of tissue-specific expression of genes.

ACKNOWLEDGMENTS

We wish to acknowledge the contributions of Drs N. N'Diaye, I. Bourdeau, H. Mircescu, J.-M. Boutin, E. Bolté, E.L. Schiffrin, M. Bouvier, the patients and their referring physicians, CHUM Endocrine Division members, nurses, and laboratory technicians for their contributions to this work. The assistance of Mrs. Victoria Baranga in preparing the manuscript and of Mr. Bruno St-Pierre for medical illustrations is also appreciated.

REFERENCES

1. Hingshaw HT, Ney RL. Abnormal Control in the Neoplastic Adrenal Cortex. In: McKerns KW, editor. Hormones and Cancer. New York: Academic Press, 1974: pp 309-327.
2. Lacroix A, N'Diaye N, Tremblay J, Hamet P. Ectopic and Abnormal Hormone Receptors in Adrenal Cushing's Syndrome. Endocr Rev 2001; 22(1):75-110.
3. Simpson ER, Waterman MR. Regulation of the synthesis of steroidogenic enzymes in adrenal cortical cells by ACTH. Annu Rev Physiol 1988; 50:427-440.
4. Miller WL. Molecular biology of steroid hormone synthesis. Endocr Rev 1988; 9(3):295-318.

5. Stocco DM, Clark BJ. Regulation of the acute production of steroids in steroidogenic cells. Endocr Rev 1996; 17(3):221-244.
6. Ehrhart-Bornstein M, Hinson JP, Bornstein SR, Scherbaum WA, Vinson GP. Intraadrenal interactions in the regulation of adrenocortical steroidogenesis. Endocr Rev 1998; 19(2):101-143.
7. Hirata Y, Uchihashi M, Sueoka S, Matsukura S, Fujita T. Presence of ectopic beta-adrenergic receptors on human adrenocortical cortisol-producing adenomas. J Clin Endocrinol Metab 1981; 53(5):953-957.
8. Willenberg HS, Stratakis CA, Marx C, Ehrhart-Bornstein M, Chrousos GP, Bornstein SR. Aberrant interleukin-1 receptors in a cortisol-secreting adrenal adenoma causing Cushing's syndrome. N Engl J Med 1998; 339(1):27-31.
9. Katz MS, Kelly TM, Dax EM, Pineyro MA, Partilla JS, Gregerman RI. Ectopic beta-adrenergic receptors coupled to adenylate cyclase in human adrenocortical carcinomas. J Clin Endocrinol Metab 1985; 60(5):900-909.
10. Leinonen P, Ranta T, Siegberg R, Pelkonen R, Heikkila P, Kahri A. Testosterone-secreting virilizing adrenal adenoma with human chorionic gonadotrophin receptors and 21-hydroxylase deficiency. Clin Endocrinol (Oxf) 1991; 34(1):31-35.
11. Hamet P, Larochelle P, Franks DJ, Cartier P, Bolte E. Cushing syndrome with food-dependent periodic hormonogenesis. Clin Invest Med 1987; 10(6):530-533.
12. Lacroix A, Bolte E, Tremblay J et al. Gastric inhibitory polypeptide-dependent cortisol hypersecretion--a new cause of Cushing's syndrome. N Engl J Med 1992; 327(14):974-980.
13. Reznik Y, Allali-Zerah V, Chayvialle JA et al. Food-dependent Cushing's syndrome mediated by aberrant adrenal sensitivity to gastric inhibitory polypeptide. N Engl J Med 1992; 327(14):981-986.
14. Lebrethon MC, Avallet O, Reznik Y et al. Food-dependent Cushing's syndrome: characterization and functional role of gastric inhibitory polypeptide receptor in the adrenals of three patients. J Clin Endocrinol Metab 1998; 83(12):4514-4519.
15. N'Diaye N, Hamet P, Tremblay J, Boutin JM, Gaboury L, Lacroix A. Asynchronous development of bilateral nodular adrenal hyperplasia in gastric inhibitory polypeptide-dependent Cushing's syndrome. J Clin Endocrinol Metab 1999; 84(8):2616-2622.
16. Archambeaud-Mouveroux F, Comas JM, Tessier MP. Food-dependent Cushing's syndrome associated with the Carney complex. In : Program & Abstracts. The 10th International Congress of Endocrinology, San Francisco, CA, 1996. Abstract P3-158, p 899.
17. Pralong FP, Gomez F, Guillou L, Mosimann F, Franscella S, Gaillard RC. Food-dependent Cushing's syndrome: possible involvement of leptin in cortisol hypersecretion. J Clin Endocrinol Metab 1999; 84(10):3817-3822.
18. Croughs RJ, Zelissen PM, Van Vroonhoven ThJ et al. GIP-dependent adrenal Cushing's syndrome with incomplete suppression of ACTH. Clin Endocrinol (Oxf) 2000; 52:235-240.
19. Luton JP, Bertherat J, Kuhn JM, Bertagna X. [Aberrant expression of the GIP (Gastric Inhibitory Polypeptide) receptor in an adrenal cortical adenoma responsible for a case of food-dependent Cushing's syndrome]. Bull Acad Natl Med 1998; 182(9):1839-1849.
20. de Herder WW, Hofland LJ, Usdin TB et al. Food-dependent Cushing's syndrome resulting from abundant expression of gastric inhibitory polypeptide receptors in adrenal adenoma cells. J Clin Endocrinol Metab 1996; 81(9):3168-3172.
21. Chabre O, Liakos P, Vivier J et al. Cushing's syndrome due to a gastric inhibitory polypeptide-dependent adrenal adenoma: insights into hormonal control of adrenocortical tumorigenesis. J Clin Endocrinol Metab 1998; 83(9):3134-3143.

22. Tsagarakis S, Tsigos C, Vassiliou V et al. Food-Dependent Androgen and Cortisol Secretion by a Gastric Inhibitory Polypeptide-Receptor Expressive Adrenocortical Adenoma Leading to Hirsutism and Subclinical Cushing's Syndrome: In Vivo and in Vitro Studies. J Clin Endocrinol Metab 2001; 86(2):583-589.
23. Stratakis CA, Kirschner LS. Clinical and genetic analysis of primary bilateral adrenal diseases (micro- and macronodular disease) leading to Cushing syndrome. Horm Metab Res 1998; 30(6-7):456-463.
24. N'Diaye N, Tremblay J, Hamet P, de Herder WW, Lacroix A. Adrenocortical overexpression of gastric inhibitory polypeptide receptor underlies food-dependent Cushing's syndrome. J Clin Endocrinol Metab 1998; 83(8):2781-2785.
25. Usdin TB, Mezey E, Button DC, Brownstein MJ, Bonner TI. Gastric inhibitory polypeptide receptor, a member of the secretin-vasoactive intestinal peptide receptor family, is widely distributed in peripheral organs and the brain. Endocrinology 1993; 133(6):2861-2870.
26. de Keyzer Y, Auzan C, Lenne F et al. Cloning and characterization of the human V3 pituitary vasopressin receptor. FEBS Lett 1994; 356(2-3):215-220.
27. Bertagna X. Proopiomelanocortin-derived peptides. Endocrinol Metab Clin North Am 1994; 23(3):467-485.
28. Raux MC, Binoux M, Luton JP, Gourmelen M, Girard F. Studies of ACTH secretion control in 116 cases of Cushing's syndrome. J Clin Endocrinol Metab 1975; 40(2):186-197.
29. Arnaldi G, de Keyzer Y, Gasc JM, Clauser E, Bertagna X. Vasopressin receptors modulate the pharmacological phenotypes of Cushing's syndrome. Endocr Res 1998; 24(3-4):807-816.
30. van Wijk PA, Rijnberk A, Croughs RJ, Wolfswinkel J, Selman PJ, Mol JA. Responsiveness to corticotropin-releasing hormone and vasopressin in canine Cushing's syndrome. Eur J Endocrinol 1994; 130(4):410-416.
31. Lacroix A, Tremblay J, Touyz RM et al. Abnormal adrenal and vascular responses to vasopressin mediated by a V1- vasopressin receptor in a patient with adrenocorticotropin-independent macronodular adrenal hyperplasia, Cushing's syndrome, and orthostatic hypotension. J Clin Endocrinol Metab 1997; 82(8):2414-2422.
32. Horiba N, Suda T, Aiba M et al. Lysine vasopressin stimulation of cortisol secretion in patients with adrenocorticotropin-independent macronodular adrenal hyperplasia. J Clin Endocrinol Metab 1995; 80(8):2336-2341.
33. Iida K, Kaji H, Matsumoto H et al. Adrenocorticotrophin-independent macronodular adrenal hyperplasia in a patient with lysine vasopressin responsiveness but insensitivity to gastric inhibitory polypeptide. Clin Endocrinol (Oxf) 1997; 47(6):739-745.
34. Arnaldi G, Gasc JM, de Keyzer Y et al. Variable expression of the V1 vasopressin receptor modulates the phenotypic response of steroid-secreting adrenocortical tumors. J Clin Endocrinol Metab 1998; 83(6):2029-2035.
35. Daidoh H, Morita H, Hanafusa J et al. In vivo and in vitro effects of AVP and V1a receptor antagonist on Cushing's syndrome due to ACTH-independent bilateral macronodular adrenocortical hyperplasia. Clin Endocrinol (Oxf) 1998; 49(3):403-409.
36. Perraudin V, Delarue C, de Keyzer Y et al. Vasopressin-responsive adrenocortical tumor in a mild Cushing's syndrome: in vivo and in vitro studies. J Clin Endocrinol Metab 1995; 80(9):2661-2667.
37. Yamakita N, Murai T, Ito Y et al. Adrenocorticotropin-independent macronodular adrenocortical hyperplasia associated with multiple colon adenomas/carcinomas which showed a point mutation in the APC gene. Intern Med 1997; 36(8):536-542.

38. Lacroix A, Tremblay J, Rousseau G, Bouvier M, Hamet P. Propranolol therapy for ectopic beta-adrenergic receptors in adrenal Cushing's syndrome. N Engl J Med 1997; 337(20):1429-1434.
39. Gallo-Payet N, Guillon G. Regulation of adrenocortical function by vasopressin. Horm Metab Res 1998; 30(6-7):360-367.
40. Mircescu H, Jilwan J, N'Diaye N et al. Are ectopic or abnormal membrane hormone receptors frequently present in adrenal Cushing's syndrome? J Clin Endocrinol Metab 2000; 85(10):3531-3536.
41. Bourdeau I, D'Amour P, Hamet P, Boutin JM, Lacroix A. Abnormal membrane hormone receptors in subclinical bilateral macronodular adrenal hyperplasia. In: Program & Abstracts. The 82nd Annual Meeting of The Endocrine Society, June 21-24, 2000. Toronto, Ont. Abstract 1934, p 467.
42. Grunenberger F, Noal E, Bachellier P, Weber JC, Jaeck D, Schlienger JL. Hypercorticisme familial non ACTH dépendant par hyperplasie macronodulaire bilatérale des surrénales. Ann Endocrinol (Paris) 1999; Abstract A04, pp 60-290.
43. Grossman A. Corticotropin-Releasing Hormone: Basic Physiology and Clinical Applications. In: DeGroot LJ, Besser GM, Burger HG et al., editors. Endocrinology. Philadelphia: W.B. Saunders Company, 1999; pp 341-354.
44. Al Damluji S, Perry L, Tomlin S et al. Alpha-adrenergic stimulation of corticotropin secretion by a specific central mechanism in man. Neuroendocrinology 1987; 45(1):68-76.
45. Matsukura S, Kakita T, Sueoka S et al. Multiple hormone receptors in the adenylate cyclase of human adrenocortical tumors. Cancer Res 1980; 40(10):3768-3771.
46. Segaloff DL, Ascoli M. The lutropin/choriogonadotropin receptor ... 4 years later. Endocr Rev 1993; 14(3):324-347.
47. Rao CV. The Beginning of a New Era in Reproductive Biology and Medicine: Expression of Low Levels of Functional Luteinizing Hormone/Human Chorionic Gonadotropin Receptors in Nongondal Tissues. J Physiol Pharmacol 1996; 47:41-53.
48. Pabon JE, Li X, Lei ZM, Sanfilippo JS, Yussman MA, Rao CV. Novel presence of luteinizing hormone/chorionic gonadotropin receptors in human adrenal glands. J Clin Endocrinol Metab 1996; 81(6):2397-2400.
49. Seron-Ferre M, Lawrence CC, Jaffe RB. Role of hCG in regulation of the fetal zone of the human fetal adrenal gland. J Clin Endocrinol Metab 1978; 46(5):834-837.
50. Lacroix A, Hamet P, Boutin JM. Leuprolide acetate therapy in luteinizing hormone-dependent Cushing's syndrome. N Engl J Med 1999; 341(21):1577-1581.
51. N'Diaye N, Cartier D, Bourdeau I et al. Characterization of aberrant LH/hCG and serotonin 5-HT4 receptors in adrenal Cushing's syndrome. The 83rd Annual Meeting of The Endocrine Society, June 20-23, 2001. Denver, Colorado. Abstract
52. Reschini E, Giustina G, Crosignani PG, D'Alberton A. Spontaneous remission of Cushing syndrome after termination of pregnancy. Obstet Gynecol 1978; 51(5):598-602.
53. Calodney L, Eaton RP, Black W, Cohn F. Exacerbation of Cushing's syndrome during pregnancy: report of a case. J Clin Endocrinol Metab 1973; 36(1):81-86.
54. Wallace C, Toth EL, Lewanczuk RZ, Siminoski K. Pregnancy-induced Cushing's syndrome in multiple pregnancies. J Clin Endocrinol Metab 1996; 81(1):15-21.
55. Close CF, Mann MC, Watts JF, Taylor KG. ACTH-independent Cushing's syndrome in pregnancy with spontaneous resolution after delivery: control of the hypercortisolism with metyrapone. Clin Endocrinol (Oxf) 1993; 39(3):375-379.
56. Wieland RG, Shaffer MBJ, Glove RP. Cushing's syndrome complicating pregnancy. A case report. Obstet Gynecol 1971; 38(6):841-843.
57. Aron DC, Schnall AM, Sheeler LR. Spontaneous resolution of Cushing's syndrome after pregnancy. Am J Obstet Gynecol 1990; 162(2):472-474.

58. Givens JR, Andersen RN, Wiser WL, Coleman SA, Fish SA. A gonadotropin-responsive adrenocortical adenoma. J Clin Endocrinol Metab 1974; 38(1):126-133.
59. de Lange WE, Pratt JJ, Doorenbos H. A gonadotrophin responsive testosterone producing adrenocortical adenoma and high gonadotrophin levels in an elderly woman. Clin Endocrinol (Oxf) 1980; 12(1):21-28.
60. Lefebvre H, Contesse V, Delarue C, Vaudry H, Kuhn JM. Serotonergic regulation of adrenocortical function. Horm Metab Res 1998; 30(6-7):398-403.
61. Lefebvre H, Gonzalez KN, Contesse V, Delarue C, Vaudry H, Kuhn JM. Effect of prolonged administration of the serotonin4 (5-HT4) receptor agonist cisapride on aldosterone secretion in healthy volunteers. Endocr Res 1998; 24(3-4):749-752.
62. Lefebvre H, Contesse V, Delarue C et al. Serotonin-induced stimulation of cortisol secretion from human adrenocortical tissue is mediated through activation of a serotonin4 receptor subtype. Neuroscience 1992; 47(4):999-1007.
63. Lefebvre H, Contesse V, Delarue C et al. Effect of the serotonin-4 receptor agonist zacopride on aldosterone secretion from the human adrenal cortex: in vivo and in vitro studies. J Clin Endocrinol Metab 1993; 77(6):1662-1666.
64. Bonnin C, Monsaingeon M, Bex V et al. Hypercorticism by bilateral adrenal hyperplasia with several paradoxical responses. Program & Abstracts. The 82nd Annual Meeting of The Endocrine Society. June 21-24, 2000. Toronto, Ont., Abstract 1936, p 467.
65. Bonnin C, Monsaingeon M, Bex V et al. Hypercorticism by bilateral adrenal hyperplasia with several paradoxical responses. Program & Abstracts. The 82nd Annual Meeting of The Endocrine Society. June 21-24, 2000. Toronto, Ont. Abstract 1936, p 467.
66. Caticha O, Odell WD, Wilson DE et al. Estradiol stimulates cortisol production by adrenal cells in estrogen-dependent primary adrenocortical nodular dysplasia. J Clin Endocrinol Metab 1993; 77(2):494-497.
67. Stratakis CA, Sarlis N, Kirschner LS et al. Paradoxical response to dexamethasone in the diagnosis of primary pigmented nodular adrenocortical disease. Ann Intern Med 1999; 131(8):585-591.
68. Bourdeau I, Caron P, Schürch W, N'Diaye N, Antakly T, Lacroix A. Paradoxical response to dexamethasone correlates with high expression of glucocorticoid receptors in primary pigmented nodular adrenocortical disease. In: Program & Abstracts. The 82nd Annual Meeting of the Endocrine Society. June 21-24, 2000. Toronto, Ont. Abstract 1935, p 467.
69. Hashimoto K, Kawada Y, Murakami K et al. Cortisol responsiveness to insulin-induced hypoglycemia in Cushing's syndrome with huge nodular adrenocortical hyperplasia. Endocrinol Jpn 1986; 33(4):479-487.
70. Makino S, Hashimoto K, Sugiyama M et al. Cushing's syndrome due to huge nodular adrenocortical hyperplasia with fluctuation of urinary 17-OHCS excretion. Endocrinol Jpn 1989; 36(5):655-663.
71. Slieker LJ, Sloop KW, Surface PL et al. Regulation of expression of ob mRNA and protein by glucocorticoids and cAMP. J Biol Chem 1996; 271(10):5301-5304.
72. Glasow A, Bornstein SR, Chrousos GP, Brown JW, Scherbaum WA. Detection of Ob-receptor in human adrenal neoplasms and effect of leptin on adrenal cell proliferation. Horm Metab Res 1999; 31(4):247-251.
73. Pralong FP, Roduit R, Waeber G et al. Leptin inhibits directly glucocorticoid secretion by normal human and rat adrenal gland. Endocrinology 1998; 139(10):4264-4268.
74. Lacroix A, Mircescu H, Hamet P. Clinical Evaluation of the Presence of Abnormal Hormone Receptors in Adrenal Cushing's Syndrome. The Endocrinologist 1999; 9:9-15.
75. Glasow A, Haidan A, Gillespie J, Kelly PA, Chrousos GP, Bornstein SR. Differential expression of prolactin receptor (PRLR) in normal and tumorous adrenal tissues:

separation of cellular endocrine compartments by laser capture microdissection (LCM). Endocr Res 1998; 24(3-4):857-862.
76. Kero J, Poutanen M, Zhang FP et al. Elevated luteinizing hormone induces expression of its receptor and promotes steroidogenesis in the adrenal cortex. J Clin Invest 2000; 105(5):633-641.
77. Gicquel C, Bertagna X, Le Bouc Y. Recent advances in the pathogenesis of adrenocortical tumours. Eur J Endocrinol 1995; 133(2):133-144.
78. Latronico AC, Chrousos GP. Extensive personal experience: adrenocortical tumors. J Clin Endocrinol Metab 1997; 82(5):1317-1324.
79. Findlay JC, Sheeler LR, Engeland WC, Aron DC. Familial adrenocorticotropin-independent Cushing's syndrome with bilateral macronodular adrenal hyperplasia. J Clin Endocrinol Metab 1993; 76(1):189-191.
80. Minami S, Sugihara H, Sato J et al. ACTH independent Cushing's syndrome occurring in siblings. Clin Endocrinol (Oxf) 1996; 44(4):483-488.
81. Cooper RJ, Udelsman R, Resta C, Tollin SR. Familial Cushing's syndrome secondary to bilateral macronodular hyperplasia. Program and Abstracts. The 80th Annual Meeting of the Endocrine Society, 1998. New Orleans, LA, Bethesda, MD. Abstract P2-397, pp 333.
82. Dhanasekaran N, Heasley LE, Johnson GL. G protein-coupled receptor systems involved in cell growth and oncogenesis. Endocr Rev 1995; 16(3):259-270.
83. Billestrup N, Swanson LW, Vale W. Growth hormone-releasing factor stimulates proliferation of somatotrophs in vitro. Proc Natl Acad Sci U S A 1986; 83(18):6854-6857.
84. Dumont JE, Jauniaux JC, Roger PP. The cyclic AMP-mediated stimulation of cell proliferation. Trends Biochem Sci 1989; 14(2):67-71.
85. Farfel Z, Bourne HR, Iiri T. The expanding spectrum of G protein diseases. N Engl J Med 1999; 340(13):1012-1020.
86. Spiegel AM. Mutations in G proteins and G protein-coupled receptors in endocrine disease. J Clin Endocrinol Metab 1996; 81(7):2434-2442.
87. Rilianawati, Paukku T, Kero J et al. Direct luteinizing hormone action triggers adrenocortical tumorigenesis in castrated mice transgenic for the murine inhibin alpha-subunit promoter/simian virus 40 T-antigen fusion gene. Mol Endocrinol 1998; 12(6):801-809.
88. Light K, Jenkins PJ, Weber A et al. Are activating mutations of the adrenocorticotropin receptor involved in adrenal cortical neoplasia? Life Sci 1995; 56(18):1523-1527.
89. Reincke M, Mora P, Beuschlein F, Arlt W, Chrousos GP, Allolio B. Deletion of the adrenocorticotropin receptor gene in human adrenocortical tumors: implications for tumorigenesis. J Clin Endocrinol Metab 1997; 82(9):3054-3058.
90. Reincke M, Karl M, Travis WH et al. p53 mutations in human adrenocortical neoplasms: immunohistochemical and molecular studies. J Clin Endocrinol Metab 1994; 78(3):790-794.
91. Reincke M, Beuschlein F, Latronico AC, Arlt W, Chrousos GP, Allolio B. Expression of adrenocorticotrophic hormone receptor mRNA in human adrenocortical neoplasms: correlation with P450scc expression. Clin Endocrinol (Oxf) 1997; 46(5):619-626.
92. Arnaldi G, Mancini V, Costantini C et al. ACTH receptor mRNA in human adrenocortical tumors: overexpression in aldosteronomas. Endocr Res 1998; 24(3-4):845-849.
93. Reincke M, Mora P, Beuschlein F, Arlt W, Chrousos GP, Allolio B. Deletion of the adrenocorticotropin receptor gene in human adrenocortical tumors: implications for tumorigenesis. J Clin Endocrinol Metab 1997; 82(9):3054-3058.
94. Mak JC, Nishikawa M, Barnes PJ. Glucocorticosteroids increase beta 2-adrenergic receptor transcription in human lung. Am J Physiol 1995; 268(1 Pt 1):L41-L46.

95. Malbon CC, Hadcock JR. Evidence that glucocorticoid response elements in the 5'-noncoding region of the hamster beta 2-adrenergic receptor gene are obligate for glucocorticoid regulation of receptor mRNA levels. Biochem Biophys Res Commun 1988; 154(2):676-681.
96. Barnes RB, Ehrmann DA. Long-term suppression of testosterone after treatment with a gonadotropin-releasing hormone agonist in a woman with a presumed testosterone secreting ovarian tumor. J Clin Endocrinol Metab 1997; 82(6):1746-1748.

11
TRANSSPHENOIDAL SURGERY FOR CUSHING'S DISEASE

Robert J. Weil, M.D. and George S. Allen, M.D., PhD.
Department of Neurosurgery, Vanderbilt University School of Medicine, Nashville, TN, USA

INTRODUCTION

Transsphenoidal resection of an adrenocorticotrophic hormone (ACTH)-secreting pituitary adenoma has the potential to cure most patients with Cushing's disease (CD). Cure rates following surgery range from 20-90%, depending upon the institution, the accuracy of the pro-operative diagnosis, and the experience of the surgeon. Although diagnostic methods for defining a pituitary adenoma have improved in the past several decades, there remain challenges in the surgical management of these tumors. This chapter will focus on the indications for surgery, preoperative and intra-operative evaluation and surgical management of these tumors, outcome of surgery in the microsurgical era, potential complications, management of persistent or recurrent tumor, and long-term outcome with surgery.

INDICATIONS FOR SURGERY

Common indications for surgery include control of hypercortisolism, relief of mass effect on the sellar and parasellar structures, and, more rarely, to make a diagnosis (6, 29, 41). Persistent, unrelieved excess cortisol secretion is a debilitating and potentially fatal condition; surgical extirpation of a pituitary adenoma is the best method to induce remission of CD and obviate the deleterious effects of hypercortisolism (2, 28, 30). Unlike other pituitary adenomas, ACTH-secreting macroadenomas (greater than 1cm) are rare, and constitute 5-20% of the tumors in most series; this is likely because the effects of hypersecretion become symptomatic while these tumors are small. However, like other macroadenomas, ACTH-secreting tumors may grow and impinge upon the optic nerves or chiasm, cavernous sinus, hypothalamus, third ventricle, brain stem or periventricular structures causing visual field abnormalities, endocrinopathies, hydrocephalus or cranial neuropathies, brainstem compression with long tract signs, or invade bone locally into the paranasal sinuses, clivus or other portions of the skull base (25, 43).

PRESURGICAL EVALUATION

The first and most important step in evaluating a patient with CD for surgery is the identification and localization of the source of excessive cortisol production (3, 4, 14, 18, 28, 30, 32, 33, 40, 41). This has been described in detail in an earlier chapter.

Once endocrinological evaluation has defined the pituitary as the source of the hypercortisolism, neuroimaging studies are performed. Although computerized tomographic scanning (CT) continues to possess utility, particularly to reveal bony abnormalities of importance in guiding either the transsphenoidal approach or to reveal evidence of bony invasion at the skull base, magnetic resonance imaging (MR) has become the study of choice (7,14, 20, 48). MR imaging, without and with contrast enhancement and special sequences, reveals sellar abnormalities in 40-80% of patients with CD; localization of a microadenoma consistent with the intraoperative findings occurs in 30-70% of cases, depending upon the series. Although MR imaging can be helpful to identify the location of a tumor, "positive" MRs are not invariably confirmed at surgery (14, 20, 48). This has led some groups to employ intra-operative ultrasound examination of the sella and pituitary gland; this method will be discussed below (38, 48).

In the face of uninformative or equivocal neuroimaging, inferior petrosal sinus sampling (IPSS) can be performed (26, 34). IPSS has been shown to confirm the pituitary basis for the hypercortisolemic state; and, in some instances, to help lateralize a tumor to one side of the pituitary gland. IPSS is performed by catheterizing the femoral veins and guiding them into the inferior petrosal sinuses. Here, the venous drainage from the cavernous sinuses may be sampled to compare the concentration of ACTH on either side of the pituitary gland and the peripheral blood. A central-to-peripheral gradient of ≥ 2 is approximately 90% sensitive and 100% specific for CD; with the addition of CRH stimulation, a ratio ≥ 3 is nearly 100% sensitive. Lateralization ratios ≥ 1.4 between sinuses predict the correct side of tumor in positive transsphenoidal explorations in 55-70% of cases; by contrast, some studies suggest that IPSS lateralization may be more helpful in cases of negative adenoma exploration, where selective hypophysectomy can be performed on the side favored by IPSS. IPSS sampling is an important adjunct in the evaluation of CD where neuroimaging has failed to reveal a tumor and we would recommend that all patients in whom MR has failed to uncover a tumor should undergo IPSS pre-operatively (26, 34, 37, 48).

Finally, as noted above, neuroimaging—both CT and MR—is vital to evaluate the bony and soft tissue structures in the region of the sellar and parasellar structures and paranasal sinuses. MR is particularly helpful to reveal local invasion, such as into the cavernous sinuses, which may alter the approach or the aggressiveness of surgery. CT delineates the paranasal sinus anatomy, which is critical in the trans-nasal dissection: for example, septations of the sphenoid sinus are common and may be used to help guide dissection to the sellar floor and avoid inadvertent exposure of parasellar structures such as the carotid artery, which could prove dangerous, or incomplete sellar exposure, which may lead to incomplete exploration and failure to find an adenoma (7, 16, 37).

CHOICE OF SURGICAL APPROACH

Transsphenoidal microsurgical resection is the treatment of choice in over 95% of all pituitary adenomas, particularly those causing CD (6, 16, 29, 37, 38). The majority of pituitary tumors causing CD are microadenomas, many of them 5 mm or less in size. In cases with a microadenoma or macroadenoma without significant parasellar or skull base invasion (most tumors <4cm), or in neuroimaging-negative cases, transsphenoidal exploration is preferred. In the instance of MR-negative or MR positive microadenomas, selective adenomectomy is the goal of surgery. Most macroadenomas causing CD can similarly be targeted for selective resection.

Macroadenomas that extend above the diaphragma sella are rare; most, however, can be totally or almost completely resected via a transsphenoidal approach. It may be necessary to wait a few weeks or months for any remaining tumor to descend into the sella, which can then be re-explored and resected. Supradiagphragmatic tumor that does not appear likely to descend—such as a tumor that possess a dumbell shape above and below the diaphragma sella—or tumor with a large or invasive extrasellar component that is unlikely to be cured by transsphenoidal surgery and may require transcranial surgery (25, 43). This may be done using a subfrontal or pterional approach; furthermore, craniotomy may be combined with transsphenoidal surgery. However, most such invasive tumors are not likely to be cured by surgical means. The goals of surgery are to achieve maximal decompression—of the pituitary, hypothalamus, optic apparatus, venticular system and brain stem, as necessary—and effect cytoreduction so that the tumor may be more responsive to adjuvant therapies, both medical and those involving radiation, whether fractionated radiotherapy or stereotactic radiosurgery.

TRANSSPHENOIDAL SURGERY

Since the transsphenoidal approach has been detailed in numerous monographs and publications (25, 16, 37, 38, 43), we will not describe the anatomic and technical variations of the surgery in detail here. Rather, we will describe our general approach and philosophy with respect to transsphenoidal surgery of CD, the main operative and perioperative issues of importance in management of these patients, recommendations for the management of persistent or recurrent disease, and those features which are of special relevance for the management of CD in children and adolescents.

Perioperative anesthetic and airway management possess unique features in transsphenoidal surgery. Although airway changes are commoner in patients with acromegaly, the nature of the surgery requires the surgeon and the anesthetist to share access to the airway and to be vigilent about the development of airway difficulties during and immediately after surgery. Similarly, common perioperative issues such as fluid and electrolye balance similarly require special attention on the part of the surgeon and anesthetist during surgery.

Patients receive endotracheal intubation and general anesthesia. Most surgeons tend to operate from the patient's right side, so the endotracheal tube must be brought out the left corner of the mouth and secured carefully yet avoid the operative field over the lips and nose. As the patient's head is generally elevated only slightly from horizontal, venous air embolism is not common and monitoring the end tidal CO_2 is usually sufficient; greater vertical elevation may necessitate precordial doppler monitoring and pulmonary or central venous access. A pharyngeal pack to prevent blood reaching the stomach is optional. All invasive monitoring lines are placed, if possible, on the patient's left side. A nasogastric tube is generally employed.

The specific features of anesthetic technique in the perioperative period are discussed in several excellent neuroanesthetic monographs and will not be detailed here (1, 12). However, a few specific concerns related to anesthetic manangement of the patient with CD undergoing transsphenoidal surgery include the following points.

First, as obesity and gastrointestinal reflux are common, H2-blockers and antacids are useful adjuncts; rapid sequence induction may reduce the risk of aspiration. Second, hypertension and left ventricular hypertrophy are common in adult patients with CD and make fluid and pain management more stringent. Arterial lines ease evaluation and treatment of electrolyte disturbances and blood pressure alterations. Third, although not as common as in the patient with acromegaly, sleep apnea may be present in some patients with CD and requires special attention to airway issues, although fiberoptic intubation is not usually required. Fourth, venous access may be difficult and may necessitate central line placement (1, 12).

For transsphenoidal surgery, patients are placed supine on the operating table. We prefer three-point pin fixation of the skull, as this permits easy alignment of the head to allow a direct line of vision to the sella. Flouroscopy is preferred to localize the sella. The nasal mucosa and sublabial regions are infiltrated with 1% lidocaine and cotton pledgets soaked in 4% cocaine solution are placed intranasally for vasoconstriction. Although some surgeons perform both the transnasal approach and tumor resection, we prefer to utilize the expertise of rhinological surgeons to gain access to the sellar floor. Few contraindications to transspheonoidal surgery exist: active, untreated bacterial sinusitis and medical instability precluding elective general anesthesia are the only common contraindications.

A transnasal-transseptal or sublabial transseptal approach may be selected to approach the posterior sphenoid sinus and sellar floor. We prefer a sublabial approach, with the incision at least 1 cm above the teeth to minimize sensory abnormalities. Fluoroscopy confirms the approach. Recent advances in frameless stereotaxy also permit use of CT-guided, three-dimensional tracking of the approach and intra-sellar manipulations. Once the rhinologist has identified the sellar floor, the neurosurgeon begins removal of the bony sella. While some surgeons remove bone with a biting instrument or chisel, we favor use of a high speed drill to reduce the risk of injury to the parasellar and periorbital bones and minimize risk of injury to the optic nerve. Drilling is often mandated in children, in whom the sphenoid is incompletely pneumatized. We open the sellar floor widely to facilitate complete examination of the pituitary and sella. Wax is applied to the bony edges to facilitate hemostasis.

The dura is coagulated and opened carefuly. Venous channels may require tamponade or coagulation; meticulous hemostasis is essential to thorough exploration of the gland and sella. After durotomy, we inspect the gland surface carefully. If no tumor or tumor bulge is identified, we begin exploration. Usually, the tumor is grey or white-yellow in color and slightly softer than the gland and may be enclosed in a pseudocapsule. Attention is directed first to the side identified by MR; or, if neuroimaging was negative, to the side suggested by IPSS lateralization. Vertical incisions about 1-1.5 mm apart are made, in parallel with the blood supply, which runs in a cranial-to-caudal direction. Incisions are carried back to the anterior-posterior pituitary interface, and lateral to medial; then, as needed, to the contralateral gland. The interface between the anterior and posterior pituitary is also examined. After, this, the junction of the gland and the infundibulum is explored. Next, the wall of the cavernous sinus ipsilateral to the IPSS gradient is examined; the contralateral sided is then viewed. If no tumor is found, selective hemihypophysectomy (approxamtely 50-60% of the gland) is performed, including portions of the central aspect of the gland where many tumors are found, and the lateral aspect of the anterior gland ipsilateral to the IPSS gradient.

If an adenoma is identified upon exploration, we use blunt microinstruments and small ring curettes to dissect the psuedocapsule around the tumor to accomplish complete resection. If it is friable--which many tumors are--this may not be possible. Then, we use ring curettes to remove the tumor and, if necessary, a small rim of normal-appearing tissue around it to be more certain of achieving complete resection. A gelatin-sponge soaked in absolute ethanol may be placed in the cavity to promote hemostais and eradicate residual tumor cells; while advocated by some surgeons, this is not our practice. If a cerebrospinal fluid (CSF) leak has occurred this step must be omitted. Although we do send frozen section if the tumor size permits, we rely most heavily on the appearance of the tumor to assess whether resection has been accomplished.

Some investigators have employed a small, 5-Hz hand-held ultrasound probe to evaluate the sella and pituitary gland (38, 48). In a recent study of CD patients undergoing transsphenoidal surgery in whom MR imaging did not demonstrate a tumor, who were compared to a matched control group in whom intraoperative ultrasound (IOUS) was not performed, IOUS helped raise the rate of positive explorations from 72% to 91%. Remission occurred in 95% of the IOUS group compared with 87%; furthermore, only 5 patients (7.5 %) required hemihypophysectomy, compared to 15 (22%) in the control group (48).

When the resection has been completed, and there has been no CSF leak, a portion of the vomer bone previously removed is placed over the sellar floor and secured beneath the bony edges of the sellar floor. If a few drops of CSF have been encounted, fat may be placed in the sellar cavity and the back of the sphenoid sinus. If there is a large communication with the subarachnoid space, fat packing may be supplemented by fascia lata, fibrin glue, prolonged nasal packing, and CSF diversion for 3-5 days using a lumbar drain. Nasal packing is maintained ordinarily for 2 -3 days after uncomplicated surgery and prophylactic antibiotics and coritcosteroids are administered. Steroids are slowly decreased until, at the time of discharge, the patient may be sent home on physiological doses, such as hydrocortisone 20 mg every morning and 10 my every evening at dinner time. Close monitoring of serum electrolytes as well as serum osmolality and urine specific gravity, is carried out in the postoperative period. Most patients are discharged on the second or third post-operative day, after their nasal packing is removed. It is important to remember that some patients may have electrolyte imbalances that may present up to 7-14 days after surgery (35).

RESULTS OF SURGERY

Serum and urine cortisol concentrations usually fall to low or unmeasurable levels within the first 1-2 days after surgery and steroid replacement is required for 3-12 months until normal hypothalmic-pituitary-adrenal axis function returns. At present, there are no consensus criteria that predict long-term remission, but current strategies are outlined below (2, 3, 6, 8, 16, 45, 46). Laboratory evidence of low or unmeasurable serum or urinary cortisol levels, however, do predict immediate outcome and can help determine whether surgery should be re-considered (see below). In general, plasma cortisol levels less than 5-7 μg/dL in the first 2-5 days after surgery predicts remission. Similarly, a deficient response to provocative testing with CRH stimulation in the immediate post-operative period suggests a favorable response to surgery. In contrast, early recovery of HPA activity after initial low values frequently signals recurrent or residual disease. Several recommendations regarding recurrent or persistent disease will be outlined in the following paragraphs.

Recent studies recommend early repeat surgery for persistent CD (16, 37). In one large series, 17 patients underwent repeat transsphenoidal surgery 7-46 days after the initial operation; 71% had immediate remission of hypercortisolism. In some cases an adenoma was found upon re-exploration; in cases where no adenoma was found, selective hypophysectomy was carried out. The risk of partial hypopitutarism, however, was increased compared with the initial surgery, especially after hemihypophysectomy (16).

Furthermore, transsphenoidal surgery should be considered strongly in cases where CD recurs some interval after initial remission. Friedman et al reported that remission of hypercortisolism occurred in 73% of patients undergoing repeat transsphenoidal surgery. Remission was greatest (95%)

when an adenoma was identified at surgery; it was 41% in the face of a negative exploration and hemihypophysectomy (16).

Although its current role is not completely delineated, radiotherapy, particularly stereotactic radiosurgery may be a reasonable alternative to repeat surgery, especially when neuroimaging fails to demonstrate a tumor after the initial surgery (9, 42). This avoids some of the morbidity involved with a second surgery, although its effect may be delayed compared to re-exploration and the risk of partial or complete pituitary dysfunction may exceed that of repeat exploration (9, 17, 42, 47).

In the setting of persistent or recurrent disease after HH, several possibilities exist. First, if IPSS was not previously performed, is important to consider performing it at this time to rule out an ectopic source of ACTH secretion or psuedo-Cushing's syndrome. Second, neuroimaging may now reveal a tumor in the remnant anterior pituitary, the intermediate or posterior lobes, along the stalk or in an extrapituitary location. Third, one may consider, with its attendant risks, total hypophysectomy. Fourth, if a tumor was identified at initial surgery and on pathological examination, there may be residual tumor, multiple tumors, or invasion of the parasellar structures not previously suspected. These situations may be addressed by re-operation, medical suppression or radiotherapy, as appropriate. Finally, if there is corticotroph hyperplasia rather than a discrete adenoma, total hypophysectomy may be required to effect a cure; this is a rare cause of CD.

About 5-15% of patients who initially go into remission will recur, often at many years' interval. Re-evaluation with biochemical and neuroimaging studies may reveal re-growth of tumor. Re-operation, as noted above, is the initial therapy of choice. If there is evidence of invasive tumor, focused radiotherapy--whose effects on hypersecretion may take 1-5 years to reach maximal effect--is used, either alone, or combined with cytoreductive surgery. Smaller tumors are more responsive to radiation than larger tumors, and tumors which abut the optic apparatus may be better treated with fractionated radiation than with radiosurgery to minimize the exposure of the visual apparatus to radiation. Until the radiation takes effect, suppressive medical therapy is required. Bilateral adrenalectomy is reserved for failure of all other measures; it is associated with a 1-% mortality rate and a 10% incidence of Nelson's syndrome. In Nelson's syndrome, there is pituitary adenoma growth, excessive ACTH secretion with hyperpigmentation, and development of visual field deficits, oculomotor palsies, and persistent headaches, among other symptoms (reviewed in 25, 43).

Long-term results in several series from the microsurgical era suggest that 70-90% of patients who undergo transsphenoidal surgery will experience a remission. It is highest when a discrete adenoma is identified and removed *in toto*. Causes of failure include invasive tumor, ectopic sources of ACTH secretion or pseud-Cushing's syndrome (for example, in alcoholism or depression), failure to identify or completely resect the tumor, extra-pituitary location of an adenoma, or corticotroph hyperplasia. Table 1 reviews the results from several large series (2, 6, 8,10, 19, 22, 31, 36, 37, 39, 41, 44, 45, 46).

Table 1: Outcome after trans-sphenoidal surgery in adults: Contemporary Microsurgical Series

Series, year	# Pts	Remission (%)	Recurrence (%)[1]	Mean F/U (months)
Hardy, 1982	75	84.0	0.0	21
Chandler et al, 1987	34	70.5	N/S	N/S
Nakane et al, 1987	100	86.0	9.3	38
Guilhaume et al, 1988	61	68.9	14.3	24
Mampalam et al, 1988	216	79.2	5.3	46
Arnott et al, 1990	28	85.7	12.5	22
Burke et al, 1990	54	81.5	4.5	56
Post & Habas, 1990	37	78.4	3.4	N/S
Tindall et al, 1990	53	86.8	2.1	57
Robert & Hardy, 1991	78	76.9	8.3	77
Tahir & Sheeler, 1992	45	75.6	20.6	69
Trainer et al, 1993	48	81.2	7.7	N/S
Ram et al, 1994	222	92.3	N/S	N/S
Bochicchio et al, 1995	668	76.3	12.7	46
Gsponer et al, 1999	15	93.3	6.8	N/S

[1]Rate of recurrence in those patients who entered sustained remission after their initial surgery.
Pts, patients; F/U, follow-up; N/S, not stated.

Unlike adults, Cushing's syndrome in children is nearly always (90-95% of cases) caused by a pituitary adenoma. Presentation of CD is marked, to a degree not seen with adults, by more profound central obesity, growth retardation, slowing or loss of secondary sexual characteristic development, hypertension, and psychological or behavioral alterations (21, 28, 29). Nonetheless, transsphenoidal surgery is the treatment of choice for pediatric CD. Recent studies suggest long-term remission rates of 70-95% with a significant chance to promote resumption of normal growth, restoration of menses and sexual development, and reduction or diminution of other signs and symptoms of CD (see table 2) (references, 13, 15, 24, 27, 28, 40). These effects are more prononced in pre-pubertal children. Recurrent tumor is best treated with repeat surgery, either adenomectomy or selective hypophysectomy based on IPSS. Total hypophysectomy--in both children and women of child-bearing age--is to be avoided.

Table 2: Outcome after trans-sphenoidal surgery in children and adolescents: Contemporary Microsurgical Series

Series	# Pts	Age (y)	Remission %	Recurrence %	Time to Recur (y)	Mean F/U (y)	Cure rate (%)
Dyet et al, 1994	33	<16	70.0	18.0	N/S	N/S (.8-14)	78.0
Knapper & Ludecke, 1995	55	<19	96.0	16.4	1.4 (0-5.0)	4.5 (1.2-15.5)	91.0
Leinung et al, 1995	22	<19	73.3	32.0	7.0 (.6-12.5)	6.7 (1.3-15.0)	47.3
Savage & Besser, 1996	11	<18	72.7	0.0	N/S	N/S (0.5-10.7)	72.7
Devoe et al, 1997	42	<20	83.3	27.0	4.2 (0.75-6.2)	7.2 (1.5-13.6)	73.0

Pts, patients; y, year(s); F/U, follow-up.

Similarly, although proponents of radiotherapy suggest that 75-80% of patients primarily treated may experience remission, the delay in resolving hypercortisolism, the frequency of delayed hypopituitarism (25-50%) and the risk of late recurrence suggest that conventional radiation therapy is not the first-line treatment of choice (9, 10, 17, 21, 23, 28, 42, 47). However, radiation, particularly stereotactic radiosurgery may be considered for patients who fail to go into remission or who recur after initial success but in whom a tumor cannot be defined on repeat neuroimaging or in whom the risks of surgery are high. Adrenalectomy should be reserved for complete treatment failures; Nelson's syndrome may also be more common in children than in adults. Finally, given that recurrences may occur many years after successful adenomectomy, children, like adults, should be monitored periodically, and indefinitely, with laboratory tests for serum or urine cortisol hypersecretion.

COMPLICATIONS OF SURGERY

Transsphenoidal surgery for pituitary adenomas in general, and for CD in particular, is a relatively safe operation. The principal determinants of both surgical success and perioperative complications are the presence and size of tumor, the regional anatomy, and the experience of the surgeon. Results of several large studies designed to monitor complications of transsphenoidal surgery for pituitary tumors suggest that the commonest complications are partial (0-15%) or complete (0-4%) anterior pituitary insufficiency; cerebrospinal fluid leak (0-15%); diabetes insipidus, temporary (10-20%) or permanent (0-5%); meningitis (1-2%); injury to the

hypothalmus, carotid or basilar arteries, optic apparatus, or cranial nerves occur in fewer than 2% of patients (5, 11). Perioperative mortality is less than 1%. In general, the risk of endocrinopathy, CSF leak, diabetes insipidus and other morbidities is slightly greater for larger tumors, recurrent surgeries, or those cases where selective hypophysectomy is performed in the face of negative gland exploration. Finally, the chronic debilities—such as diabetes mellitus, hypertension, and myocardial deficits—induced by persistent hypercortisolism place patients with CD at a higher risk for serious complications such as meningitis, myocardial infarction, or other problems than patients with other types of pituitary tumors who undergo transsphenoidal surgery (5, 11).

CONCLUSIONS

Transsphenoidal microsurgery is the treatment of choice for CD. It permits selective resection of an adenoma, with minimal morbidity, and the best chance for longterm remission or cure. The most difficult aspect of surgery is identification of a discrete adenoma; useful adjuncts include newer MR imaging techniques, IPSS, and, in some instances, intraoperative ultrasound. Remission rates are 75-90% in most large series and may be even higher in children, where the likelihood of misdiagnosis of a pituitary source for Cushing's syndrome is lower than in adults. Repeat transsphenoidal surgery is the initial treatment option for recurrent disease, and may be used singly or in combination with radiation therapy. Radiotherapy may be utilized when there is unresectable, invasive tumor, recurrent tumor, or surgical failure. Total hypophysectomy may be a useful therapy in some adults, particularly the elderly where the effects of hypopituitarism may not be as pronounced.

REFERENCES

1. Bochicchio D, Losa M, Buchfelder M, and the European Cushing's disease survey group. J Clin Endocrinol Metab 1995; 80: 3114-3120.
2. Magiakou MA, Mastorakos G, Chrousos GP. Final stature in patients with endogenous Cushing's syndrome. J Clin Endocrinol Metab 1994; 79: 1082-1085.
3. Sonino N, Zielezny M, Fava GA, Fallo F, Boscaro M. Risk factors and long-term outcome in pituitary-dependent Cushing's disease. J Clin Endocrion Metab 1996; 81: 2647-2652.
4. Arnott RD, Pestell RG, McKelvie PA, Henderson JK, McNEill PM, Alford FP. A critical evaluation of trasnssphenoidal surgery in the treatment of Cushing's disease: prediction of outcome. Acta Endocrinol (Copenh) 1990; 123: 423-430.
5. Magiakou MA, Mastorakos G, Oldfield EH, Gomez MT, Doppman JL, Cutler, Jr, GB, Nieman LK, Chrousos GP. Cushing's syndrome in children and adolescents: presentation, diagnosis, and therapy. N Engl J Med 1994; 331: 629-636.
6. Mampalam TJ, Tyrell JB, Wilson CB. Transsphenoidal microsurgery for Cushing's disease: a report of 216 cases. Ann Intern Med 1988; 109: 487-493.
7. Krischt AF, Tindall GT, eds. Pituitary disorders: comprehensive management. Baltimore: Lippincott Williams & Wilkins, 1999.
8. Tindall GT, Barrow DL. Disorders of the pituitary. St. Loius: C.V. Mosby, 1986.
9. Averginos PC, Chrousos GP, Nieman LK, Oldfield EH, Loriaux DL, Cutler GB Jr. the corticotropin-releasing hormone test in the postoperative evaluation of patients with Cushing's syndrome. J Clin Endocrnol Metab 1987; 65: 906-912.
10. Averginos PC, Yanovski JA, Oldfield EH, Nieman LK, Cutler, Jr, GB. The metyrapone and dexamethasone suppression tests for the differential diagnosis of the adrenocorticotropin-dependent Cushing syndrome: a comparison. Ann Int Med 1994; 121: 318-327.
11. Dwyer AJ, Frank JA, Doppman JL, Oldfield EH, Hickey AM, Cutler GB, Loriaux DL, Schiable TF. Pituitary adenomas in patients with Cushing disease: initial experience with Gd-DTPA-enhanced MR imaging. Radiology 1987; 163: 421-426.
12. Gomez MT, Malozowski S, Winterer J, Vamvakopoulos NC, Chrousos GP. Urinary free cortisol values in normal children and adolescents. J Pediatr 1991; 118: 256-258.
13. Nieman LK, Cutler GB Jr, Oldfield EH, Loriaux DL, Chrousos GP. The ovine corticotropin-releasing hormone (CRH) stimulation test is superior to the human CRH stimulation test for the diagnosis of Cushing's disease. J Clin Endocrinol Metab 1989; 69: 165-169.
14. Nieman LK, Oldfield EH, Wesley R, Chrousos GP, Loriaux DL, Cutler GB Jr. A simplified morning ovine corticotropin-releasing hormone stimulation test for the differential diagnosis of adrenocorticotropin-dependent Cushing's syndrome. J Clin Endocrinol Metab 1993; 77: 1308-1312.
15. Savage MO and Besser GM. Cushing's disease in childhood. Trends Endocrinol Metab 1996; 7: 213-216.

16. Buchfelder M, Nistor R, Fahlbusch R, et al. The accuracy of CT and MR evaluation of the sella turcica for detection of adrenocorticotrophic hormone secreting adenomas in Cushing's disease. Am J Neuroradiol 1993; 14: 1183-1190.
17. Hall WA, Luciano MG, Doppman JL, et al. Pitutary magnetic resonance imaging in abnormal human volunteers: occult adenomas in the general population. Ann Intern Med 1994; 120: 817-820.
18. Watson JC, Shawker TH, Nieman LK, DeVroom HL, Doppman JL, Oldfield EH. Localization of pituitary adenomas by using intraoperative ultrasound in patients with Cushing's disease and no demonstrable pituitary tumor on magnetic resonance imaging. J Neurosurg 1998; 89: 927-932.
19. Ram Z, Shawker TH, Bradford MH, et al. Intraoperative ultrasound-directed resection of pituitary tumors. J Neurosurg 1995; 83: 225-230.
20. Landolt AM, Schubinger O, Maurer R, et al. The value of inferior petrosal sinus sampling in diagnosis and treatment of Cushing's disease. Clin Endocrinol (Oxf) 1994; 40: 485-492.
21. Oldfield EH, Doppman JL, Nieman LK, Chrousos GP, Miller DL, Katz DA, Cutler, Jr, GB, Loriaux DL. Petrosal sinus sampling with and without corticotropin-releasing hormone for the differential diagnosis of Cushing's syndrome. N Engl J Med 1991; 325: 897-905.
22. Ram Z, Neiman LK, Cutler Jr GB, Chrousos GP, Doppmann JL, Oldfield EH. Early repeat surgery for persistent Cushing's disease. J Neurosurg 1994; 80: 37-45.
23. Friedman RB, Oldfield EH, Nieman LK, Chrousos GP, Doppman JL, Cutler , Jr, GB, Loriaux DL. Repeat transsphenoidal surgery for Cushing's disease. J Neurosurg 1989; 71: 520-527.
24. Albin MS, ed. Textbook of Anesthesia. New York: McGraw-Hill, 1997.
25. Cottrell JE, Smith DS, eds. Anesthesia and Neurosurgery, 3rd edition. St Louis: Mosby-Yearbook, 1994.
26. Olson BR, Gumowski J, Rubino D, Oldfield EH. Pathophysiology of hyponatremia after transsphenoidal surgery. J Neurosurg 1997; 87: 499-507.
27. Burke CW, Adams CBT, Esiri MM, Morris C, Bevans JS. Transsphenoidal surgery for Cushing's disease: does what is removed determine the endocrine outcome? Clin Endocrinol (Oxf) 1990; 33: 525-537.
28. Trainer PJ, Lawrie HS, Verheist J, et al. Transsphenoidal resection in Cushing's disease: undetectable serum cortisol as the definition of successful treatment. Clin Endocrinol (Oxf) 1993; 39: 73-78.
29. Tyrell JB, Wilson CB. Cushing's disease: therapy of pitutary adenomas. Endocrinol and Metab Clinics of N Am. 1994; 23: 925-938.
30. Cappa M, Stoner E, DiMartino, et al. Recurrence of Cushing's disease in childhood after radiotherapy-induced remission. Am J Dis Child 1987; 141: 736-740.
31. Thoren M, Rahn T, Hallengren B, et al. Treatment of Cushing's disease in childhood and adolescence by stereotactic pituitary irradiation. Acta Pediatr Scand 1986; 75: 388-395.
32. Grigsby PW, Thomas PR, Simpson JR, Fineberg BB. Long-term results of radiotherapy in the treatment of pituitary adenomas in children and adolescents. Am J Clin Oncol 1988; 607-611.
33. Wara WM, Richards GE, Grumbach MM, et al. Hypopituitarism after irradiation in chlidren. Int J Radiat Oncol Biol Phys 1977; 2: 549-552.
34. Chandler WF, Schteingart DE, Lloyd RV, et al. Surgical treatment of Cushing's disease. J Neurosurg 1987; 66: 204-212.
35. Gsponer J, deTribolet N, Deruaz J-P, et al. Diagnosis, treatment, and outcome of pituitary tumors and other abnormal intrasellar masses: retrospective analysis of 353 patients. Medicine 1999; 78: 236-269.
36. Hardy J. Cushing's disease: 50 years later. Can J Neurol Sci 1982; 9: 375-380.

37. Nakane T, Kuwanayam A, Watanabe M, et al. Long term results of trasnssphenoidal adenomenctomy in patients with Cushing's disease. Neurosurgery 1987; 21: 218-222.
38. Post KD, Habas JE. Comparison of long-term results between prolactin-secreting adenomas and ACTH-secreting adenomas. Can J Neurol Sci 19990; 17: 74-77.
39. Robert F, Hardy J. Cushing's disease: a correlation of radiological, surgical, and pathological findings with therapeutic results. Pathol Res Pract 1991; 187: 617-621.
40. Tindall GT, Herring CJ, Clark RV, Adams DA, Watts NB. Cushing's disease: results of transsphenoidal microsurgery with emphasis on surgical failures. J Neurosurg 1990; 72: 363-369.
41. Hammer LD, Kraemer HC, Wilson DM, Ritter PL, Dornbusch SM. Standardized percentile curves of body-mass index for children and adolescents. Am J Dis Child 1991; 145: 259-263.
42. Devoe DJ, Mille WL, Conte FA, Kaplan SL, Grumbach MM, Rosenthal SM, Wilson CB, Gitelman SE. Long-term outcome in children and adolescents after transsphenoidal surgery for Cushing's disease. J Clin Endocrinol Metab 1997; 82: 3196-3202.
43. Dyer EH, Civit T, Visot A, Delalande O, Derome P. Transsphenoidal surgery for pituitary adenomas in children. Neurosurgery 1994; 34: 207-212.
44. Knappe UJ and Ludecke DK. Transnasal microsurgery in children and adolescents with Cushing's disease. Neurosurgery 1996; 39: 484-493.
45. Leinung MC, Kane LA, Scheithauer BW, Carpenter PC, Laws, Jr, ER, Zimmerman D. Long term follow-up of transsphenoidal surgery for treatment of Cushing's disease in childhood. J Clin Endocrinol Metab 1995; 80: 2475-2479.
46. Jennings AS, Liddle GW, Orth DN. Results of treating childhood Cushing's disease with pituitary irradiation. N Engl J Med 1977; 297: 957-962.
47. Black PMcL, Zervas NT, Candia GL. Incidence and management of complications of transsphenoidal operation for pituitary adenomas. Neurosurgery 1987; 20: 920-924.
48. Ciric I, Ragan A, Baumgartner C, Pierce D. Complications of transsphenoidal surgery: results of a national survey, review of the literature, and personal experience. Neursurgery 1997; 40: 225-237.

12
PERIOPERATIVE EVALUATION AND MANAGEMENT

T. Joseph McKenna, M.D., FRCPI
St. Vincent's University Hospital, University College, Dublin.

INTRODUCTION

While the investigative procedures leading to a diagnosis of Cushing's syndrome and to identification of the precise disorder giving rise to the syndrome and the nature and technical details of the therapeutic interventions are well described, the perioperative management and intermediate follow up to surgical intervention are less well established. In this chapter that portion of the pre-operative assessment which bears particularly on the post-operative management will be considered. In the immediate post-operative period the need for glucocorticoid replacement will be considered. Fundamental to this will be an assessment as to whether or not a cure has been achieved. It is useful at this time to compare pre-operative and post-operative pituitary function other than that involving the hypothalamic-pituitary-adrenal axis (HPAA). Where the HPAA is found to be suppressed, which almost always accompanies successful operative intervention, assessment of subsequent recovery will be needed. When there is failure of the operative procedure to affect a cure consideration will be given to additional treatment. Furthermore, although treatment may be effective for months or years in follow-up, recurrence is seen in some patients. Emphasis will be placed on management difficulties not infrequently encountered in patients in whom Cushing's syndrome has been effectively treated who develop "the steroid withdrawal syndrome". While patients are reassured that they have normal cortisol levels and can no longer be said to have Cushing's syndrome, they may complain bitterly of symptoms which may seem to be worse than those present prior to "cure". The management of this syndrome, which may persist for months or years is difficult and patient support groups may be very useful.

PRE-OPERATIVE EVALUATION

Impact of Glucocorticoid Excess

The diagnosis of pituitary ACTH excess, adrenal tumor or ectopic ACTH syndrome and their respective treatments have been discussed elsewhere in this volume. However, it is useful at this point to emphasise evaluation of the extent of the disease and particularly evidence of pituitary dysfunction other than that involving the HPAA and any pre-operative treatment which may be effective in reducing the effects of cortisol (Table 1).

A patient with Cushing's disease usually has a pituitary microadenoma which does not give rise to pressure effects on the normal pituitary gland. However, glucocorticoid excess modulates pituitary function. This is most clearly expressed by the development of hypogonadotropic hypogonadism (1,2). While certainly suppression of the hypothalamic-pituitary-gonadal axis may in part be related to effects of excess adrenal androgens and their conversion to estrogen, there is a powerful effect of pure glucocorticoids in suppressing gonadotropin secretion (3). This is seen most clearly demonstrated by the suppression of the hypothalamic--pituitary-gonadal axis which accompanies high dose pure glucocorticoid treatment as used in the treatment of severe inflammatory arthritis, asthma etc. Indeed, androgen levels are very variable in each of the different forms of Cushing's syndrome being most severely elevated in adrenal carcinoma and perhaps least elevated when associated with an adrenal adenoma (4). Expression of androgen excess in patients with ectopic ACTH may be severe to minimal and probably reflects variable metabolism of the pro-opiomelanocortin (POMC) molecule in this syndrome. Glucocorticoid excess and gonadal suppression have profound effects on bone turnover leading to the development of osteoporosis (5). Establishing bone mineral density pre-operatively provides a useful baseline against which to monitor response to cure of Cushing's syndrome and to additional treatment e.g. calcium, vitamin D supplementation and or bisphosphonate as deemed necessary (6). More precise details on bone turnover will be obtained from measurement of specific indices of bone resorption (deoxypyridinoline, telopeptide) and indices of bone formation (e.g. bone specific alkaline phosphatase and osteocalcin) and 25-hydroxyvitamin D levels (7). The suppressive effects of glucocorticoids on growth hormone and IGF-1 production result in growth retardation in children (8-10) and probably also contribute to insulin resistance-related disturbances of lipids, fat distribution and psychological well being.

Hypertension is frequently encountered in patients with Cushing's syndrome but it is of variable intensity. The mechanisms whereby hypertension occur in Cushing's syndrome include the mineralocorticoid effect of cortisol, increased circulating levels of 11-deoxycorticosterone which has predominately mineralocorticoid activity, glucocorticoid-induced angiotensinogen secretion by the liver and increased peripheral vascular resistance. Volume depletion with sodium-losing and potassium-sparing diuretics are useful in the treatment of this condition. However, the hypertension associated with Cushing's syndrome may prove surprisingly resistant to calcium channel blockers and successful treatment of Cushing's syndrome corrects the hypertension in the majority of patients effected in this way (11).

A portion of patients with Cushing's syndrome demonstrates diabetes mellitus (12). The extent of glucose intolerance if any should be checked by measurement of fasting and 2 hour post prandial blood glucose and haemoglobin Alc levels prior to operative intervention. If diabetes is present specific treatment will include diet with or without oral hypoglycaemic agents. Insulin sensitizers e.g. metformin and the thiazolidinediones may be particularly helpful possibly used with a sulphonylurea. In some patients treatment with insulin will be indicated. The establishment of normal blood glucose levels in the preoperative angiotensin-converting enzyme inhibitors. Successful treatment of period is an important component in optimising preparation for surgical intervention with the main aim of reducing post-operative infection.

Macroadenoma

In patients with the more unusual finding of Cushing's disease associated with a pituitary macroadenoma the possibility of hypopituitarism should be explored. Distinguishing between physical disruption and functional suppression of gonadotropin and growth hormone secretion can be accomplished by measuring the gonadotropin and the growth hormone responses to GnRH and growth hormone releasing hormone respectively, but this is not always necessary. However, baseline measurements of LH, FSH,

Table 1. Glucocorticoid-Induced Complications Of Cushing's Syndrome: Mechanism and baseline assessments prior to treatment.

Cushing's syndrome Associated problems	Mechanism	Pre-operative assessment
Diabetes mellitus	Insulin resistance; Increased glycogenolysis	Fasting and 2 hr post prandial blood glucose; Haemoglobin A1c
Osteoporosis	Increased bone resorption	Bone mineral density; Bone turnover biomarkers
Hypertension	Mineralocorticoid effect; glucocorticoid-induced releaseof angiotensinogen; increased vascular resistance	Blood pressure monitoring
Dyslipidemia	Insulin resistance	Lipid profile
Short stature (children)	Growth hormone deficiency; IGF-1 deficiency	Growth-hormone, IGF-I
Hypogonadism	Suppression of LH secretion	Estradiol (women) Testosterone (men)
Psychological disturbances	Direct glucocorticoid effect; Growth hormone deficiency	Clinical"Quality of Life" questionaire
Suscepbbility to infection	Immune Suppression; Diabetes	Clinical monitoring
Hypothyroidism	TSH deficiency (TSH response to TRH)	Free T4 TSH

testosterone, growth hormone and insulin-like growth factor 1 (IGF1) levels should be obtained in the pre-operative period. Circulating thyroxine (T4) and TSH levels should always be measured. While glucocorticoid excess does suppress TSH levels this is not usually associated with suppressed circulating free T4 values (13). There is an advantage to measuring free T4 rather than total T4 levels as glucocorticoid excess suppresses thyroxine binding globulin values. Therefore, it is possible for a patient to exhibit suppressed total T4 levels while the biologically important Free T4 fraction remains within the normal range. The finding of a suppressed Free T4 value in the absence of an elevated TSH level in a patient with a pituitary macroadenoma causing Cushing's syndrome is strong evidence of physical disruption of the pituitary thyrotroph cells. Confirmation can be obtained by demonstrating a suppressed serum TSH response to stimulation with thyrotrophin releasing hormone. Preoperative thyroid hormone replacement is indicated if hypothyroidism is present. Disorders of prolactin secretion are not a feature of Cushing's syndrome but in view of suppression of the hypothalamic-pituitary-gonadal axis and the importance of the recovery of this axis in the post-operative phase, it is useful to have all potentially relevant baseline information established. Thus, pre-operative measurement of prolactin is advised.

The symptom of polyuria is not unusual in patients with Cushing's syndrome and this may also be accompanied by nocturnal thirst. The genesis of these symptoms include hyperglycaemia, hypokalaemia associated with mineralcorticoid excess and hypercalciuria sometimes associated with renal stone disease which is also a feature of Cushing's syndrome. A large pituitary tumor very rarely gives rise to diabetes insipidus. Evaluation for this disorder can be carried out most conveniently when blood glucose and potassium levels have been returned to normal. Measurements of urine and serum osmolality following a period of abstinence from fluids to a point at which thirst is appreciated will usually provide the answer. The finding of urine osmolality which is less than serum osmolality when serum osmolality is in excessive of 295mOsm/kg, is strongly indicative of diabetes insipidus; measurement of vasopressin may also be helpful (14). The possibility of renal tubular dysfunction e.g. associated with hypokalaemia, may be examined by assessing the ability of the renal tubule to respond to exogenous antidiuretic hormone or its analogue 1-Desamino, 1 D-arginine vasopressin (DDAVP or syndesmopressin). In a patient who is highly symptomatic because of thirst and polyuria, overnight fluid deprivation should be avoided as patients who have full blown diabetes insipidus will not be able to tolerate this procedure. The finding of a urine osmolality which is greater than serum osmolality indicates at least some anti-diuretic hormone activity. Urine osmolality, which is 50% in excess of serum osmolality virtually excludes significant diabetes insipidus.

Extent of Disorders Causing Cushing's Syndrome

The extent of the disease causing Cushing's syndrome should also be established. The important distinctions between the effects of *pituitary microadenomata* and *macroadenomata,* as indicated by tumor diameters of equal to or less than lcm and those greater than lcm in diameter, have been discussed. Very rarely a *pituitary carcinoma* may have evidence of intracranial metastases at the time of diagnosis although this usually becomes evident during follow up to initial surgery. The possibility that an *adrenal tumor is* malignant is suggested by rapidly progressive severe symptoms and the finding of a tumor in excess of 5cm in diameter. The possibility of intra-abdominal e.g. hepatic, or pulmonary metastases should be specifically addressed in this case. Likewise, when dealing with ectopic ACTH secretion knowledge of the extent of the disease will dictate the treatment plan. If a localised tumor is identified and can be resected e.g. a bronchial carcinoid, no other treatment may be necessary. However, in other instances where either the primary tumor is not amenable to surgery or there is evidence of wide spread metastases, bilateral adrenalectomy may be the treatment of choice. This should only be undertaken when assessment indicates that the patient's well being and long-term survival is more threatened by glucocorticoid excess than by other effects of the tumor.

When a question arises as to the patient's ability to tolerate the procedures planned in a patient presenting with advanced ravages of glucocorticoid excess, the possibility of using medical treatment should be explored (14). Agents such as o^{1},p^{1},DDD, metyrapone, aminogluthetimide and ketoconazole may be useful alone or in combination in reducing glucocorticoid excess. A period during which glucocorticoid levels are lowered pre-operatively is likely to improve post-operative wound healing and reduce the likelihood of infection. Such treatment is usually only indicated for very frail patients with severe disease resulting in an advanced catabolic state.

POST-OPERATIVE EVALUATION

Cushing's Syndrome

In the immediate post-operative period glucocorticoid cover will usually be required. Since the normal corticotrophs will have been suppressed by feedback of excess glucocorticoid pre-operatively. Where pituitary adenomectomy has been successful post-operative cortisol levels will be low and remain so for several months (16, 17). It is therefore necessary to provide supraphysiological levels of glucocorticoids to cover the stress of the immediate postoperative period. Normal or elevated endogenous cortisol levels at this time indicate failure of the procedure to achieve a cure.

Traditionally high doses of glucocorticoids e.g. hydrocortisone 100mg 6 hourly, for the 3-4 immediate postoperative days have been used. The level was then reduced gradually to baseline glucocorticoid replacement e.g. hydrocortisone 20-30mg per day given in divided doses by mouth, as needed. However, it is now appreciated that the glucocorticoid secretion rate even during stress in normal individuals, is significantly lower than previously accepted and the need for routine glucocorticoid replacement following pituitary adenomectomy has been questioned (18). The physiological cortisol secretion is currently estimated to be approximately 6 mgs per square metre body surface area (19). In addition, the use of high dose hydrocortisone which has significant mineralcorticiod activity has potential for side-effects e.g. hypertension, salt and water retention, hypokalaemia, that can be avoided by using a steroid with less mineralacorticoid effect e.g. prednisolone, dexamethasone. Since the renin-angiotensin-aldosterone axis is not suppressed in patients with Cushing's disease, specific mineralacorticoid replacement is not required in the postoperative phase. Fluid replacement prior to the patient resuming oral intake should consist of approximately 1 litre of normal saline and 1 litre of 5% dextrose alternating over 24 hours. Daily measurement of urea and electrolytes will guide specific supplementation e.g. with potassium. Even using nearly pure glucocorticoids there may be a tendency to develop hypokalaemia and each litre of fluid replacement may be supplemented with 20-40 mmol of potassium chloride.

Glucocorticoid replacement may be in the traditional form of hydrocortisone 50-100 mg or prednisolone 10-20 mg or dexamethasone 1-2 mg each given intravenously every 8 hours. In the absence of a postoperative febrile state this should be reduced by 50% every 48 hours. Once the patient has resumed normal eating glucocorticoid replacement should be by mouth. Since the bioavailability of steroids is greater given intravenously than orally, during the period when glucocorticoid dosage is being reduced by 50% every 48 hours, the same dosage should be given orally for the subsequent 48 hours as had been given intravenously immediately beforehand. When glucocorticoid treatment has been reduced to replacement doses assessment of the HPAA should be undertaken. Equivalent replacement doses are approximately hydrocortisone 10-20 mg per day, prednisolone 3-5 mg per day and dexamethasone 0.3-0.5 mg. Hydrocortisone and prednisolone should be given in divided doses with breakfast and lunch. Dexamethasone because of its longer duration of action may be given once daily. Replacement treatment with dexamethasone is limited because of the routine availability of only small 0.5 mg tablets. However, a liquid preparation containing 0. 1 mg per ml can be prepared and is very useful.

Assessing the status of the hypothalamic-pituitary-adrenal axis (Table 2)

A great deal of information is available shortly after surgical treatment of Cushing's syndrome by measurement of circulating endogenous cortisol levels in blood. However, in the immediate postoperative period interpretation can be confused by interference by exogenous glucocorticoid. When hydrocortisone is used it should be remembered that since it's half-life is 120 minutes and hydrocortisone is cortisol, there will be significant exogenous cortisol still circulating 6 hours after the previous treatment with hydrocortisone. Furthermore, supraphysiological doses of glucocorticoid will bring about suppression of the HPAA particularly if given at night. Therefore, the ideal time to assess the axis is in the morning prior to treatment with glucocorticoid on that day and 12 hours since administration of hydrocortisone or prednisolone and 24 hours

Table 2. Post-operative Evaluation Of Cushing's Disease Patients

Test	Result	Interpretation
Plasma cortisol	< 100 nmol/l	Cure; HPAA insufficiency.
	100 - 200 nmol/l	Probable cure; Probable HPAA insufficiency.
	200 - 400 nmol/1	Equivocal-stimulation testing of HPAA required;
	> 400	Failed surgery; Glucocorticoid replacement redundant
*Dexamethasone test plasma cortisol response	> 200	Persistent Cushing's disease
Insulin-induced hypoglycaemia plasma cortisol	< 500	Variable degrees of HPAA suppression
Metyrapone test 30mg/kg at mid-night	Sub-normal response - Plasma cortisol < 200 nmol/l and Plasma 11-deoxycortisol < 200 nmol/1	Variable degrees of HPAA suppression

*Dexamethasone - I mg at midnight or O.5mg every 6 hours for 48 hours.

To convert plasma cortisol nmol/1 to μg/dl divide by 27.6

To convert plasma 11-deoxycortisol nmol/l to μg/ dl divide by 28.9.

since administration of dexamethasone. Under these circumstances a sub-physiological early morning plasma cortisol i.e. less than 100 nmol, is consistent with both a successful outcome to the therapeutic procedure and suppressed activity of the HPAA. The lower the value present at this time the greater will be the likelihood of a long term success (16, 17). In the immediate post~operative period there is little additional information to be gained from stimulation tests. If such testing is to take place only procedures which test the entire HPAA will be of value e.g. insulin induced hypoglycaemia or metyrapone administration. Adrenal stimulation tests using ACTH e.g. 1-24 αACTH (Synachten) will almost inevitably give misleading information in the weeks following pituitary surgery for Cushing's disease. In this situation the adrenal glands have been primed to respond by years of over-stimulation by endogenous ACTH. In the absence of endogenous ACTH for a short period partial involution of the adrenals may have occurred. However, normal or exuberant responses to administration of the usual pharmacological doses of Synacthen, 250,μg IV, may occur (20, 21). The use of low dose ACTH stimulation testing using 1 μg intravenously has been demonstrated to provide little advantage and uncertainty exists about appropriate cut-off points for the plasma cortisol responses (22, 23). Some improved sensitivity depends on using a high cut-off point e.g. plasma cortisol of 600 nmol/1, 30 minutes following 1 μg Synacthen, but this is associated with a false positive response of approximately 20% (24). An early morning plasma cortisol level between 200 and 400 nmol/L is an equivocal range both regarding whether or not cure has been achieved and whether or not the HPAA is suppressed. Certainly the presence of this amount of circulating endogenous cortisol indicates significant activity of the axis and on initial review suggests a disappointing outcome. The use of insulin-induced hypoglycaemia or metyrapone in this situation will provide further information. A normal or excessive response is most consistent with failure of the surgical procedure and indicates that further glucocorticoid replacement therapy is *not* indicated (25). Similar inferences may be drawn from the finding of an early morning pre-treatment plasma cortisol level in excess of 400 nmol/L (26). In patients whose serum cortisol levels have demonstrated resistance to overnight 1 mg dexamethasone administration or the formal low dose dexamethasone test i.e. 0.5 mg by month 6 hourly for 48 hours prior to surgical intervention, it is useful to repeat these tests where plasma cortisol levels are detectable post-operatively. Persistence of failure of cortisol levels to suppress after administration of dexamethasone to less that 200 nmol/L and probably to less than 100 nmol/L, confirms that cure has not been achieved.

General hypothalamic-pituitary assessment

The clinician must be aware of the possible development of diabetes insipidus in the immediate post-operative phase following pituitary fossa exploration (27). This is certainly more likely to be associated with surgical resection of a macroadenoma than of a microadenoma but it is also seen in the latter case. Routine measurement of urine output, serum and urine osmolality and urea and electrolytes at least daily in the first 48 hours is indicated. Plasma vasopressin levels while theoretically useful in fully exploring the control of water balance are seldom if ever available to guide daily decisions for treatment. Generally a high urine output is indicative of a high intravenous fluid input. Reducing the intravenous fluid input to 2 litres per day if continued intravenous fluid is required, will be helpful. Most patients can resume oral intake within the first 24 hours of pituitary adenomectomy. The finding of serum osmolality levels within the high normal to elevated range (reference range 285-295 mOsm/kg) associated with a urine osmolality of a lower value is indicative of the presence of diabetes insipidus which if not present prior to surgery can be assumed to be of central rather than nephrogenic origin. In this situation diabetes insipidus is almost always transient indicative of intraoperative pressure on the cells of the posterior lobe of the pituitary rather than long term damage to the pituitary stalk or to the hypothalamus. Recovery of posterior pituitary function may take days or in some instances a few months. Treatment is with DDAVP an analogue of anti-diuretic hormone which has the benefit of longer duration of action and is devoid of the vasopressor and peristaltic activity of the native hormone. DDAVP may be given intramuscularly or orally. Its duration of action is from 8-12 hours and is usually given twice daily with one of those doses being given prior to bedtime to facilitate uninterrupted sleep. Increasing thirst is a useful indicator for the next dose of DDAVP. A trap of which the clinician should be wary is to administer DDAVP merely because the patient passes large volumes of urine. Since DDAVP increases fluid reabsorption by the collecting ducts of the kidney, this may result in excessive fluid retention. When the effect of DDAVP wears off there is the opportunity to off load fluid in the form of urine. Interpreting this as an indication of the immediate need for more DDAVP should be modulated by the patient's sensation of thirst and measurement of serum sodium, urea and osmolality. The finding of hyponatremia or low levels of urea indicate excessive water retention and the finding of a low serum osmolality would indicate the absence of a physiological stimulus to anti-diuretic hormone secretion. DDAVP should not be ordered routinely more frequently than 12 hourly and probably on an "as required" basis so that the patient will only receive the medication if thirsty. This prescription will avoid the occurrence of water intoxication but will ensure symptoms of diabetes insipidus will be promptly relieved. It is usual that the symptoms gradually disappear over days to weeks until DDAVP is no longer required. Should symptoms persist for over a period of 3 months formal retesting should occur and the requirement for long-term treatment examined.

Hyponatremia may also occur as a consequence of inadequate glucocorticoid replacement. The full expression of the mineralocorticoid effect depends on a physiological glucocorticoid environment. Therefore, although the renin-angiotensin-aldosterone axis would usually function normally in patients who have undergone either pituitary adenomectomy or unilateral adrenalectomy, mineralocorticoid deficiency leading to the development of hyponatremia and volume depletion may occur in patients with inadequate glucocorticoid replacement. It is therefore important to ensure adequate glucocorticoid replacement until full recovery of the HPAA has been achieved.

Other pituitary function testing should be delayed for approximately 4-8 weeks following pituitary exploration. This should include measurement of Free T4, estradiol levels in women and testosterone levels in men. Should these prove to be normal no further evaluation is necessary. In the case where these hormone levels are sub-normal further evaluation including measurements of the TSH and the LH and FSH response to stimulation with the hypothalamic releasing hormones TRH and gonadotropin releasing hormone will be useful. Appropriate substitution therapy should be introduced as required. Treatment should be withdrawn after 3-6 months and re-evaluated 1 month later. Persistence of any hormonal deficiency is likely to be permanent and the patient should be advised accordingly. In the unusual circumstances that gonadotropin activity has been compromised and if fertility is desired, the patient should be advised that gonadotropin treatment will probably be necessary.

Recovery of the hypothalamic-pituitary-adrenal axis

Patients who have undergone successful pituitary adenomectomy frequently demonstrate profound suppression of the residual HPAA. This probably results both from functional suppression of the hypothalamus and the corticotrophs and to damage to the normal pituitary corticotrophs during resection of the ACTH-secreting pituitary adenoma. In order to encourage recovery of the axis no more than physiological glucocorticoid replacement doses should be used. Re-establishment of the ordinary relationship between ACTH and cortisol secretion will require up to one year to elapse and occasionally longer. However, even after physiological endogenous cortisol levels have been achieved, patients may continue to complain of symptoms suggestive of inadequate replacement. During this time patients may report highly unpleasant symptoms consistent with inadequate glucocorticoid replacement. This is termed "the withdrawal syndrome" and occurs on rapid reduction of glucocorticoid levels from excessive to normal values whether the excess values have occurred spontaneously or as a result of iatrogenic intervention. This syndrome is dealt with further in the following section of this chapter. Use of supra-physiological glucocorticoid dosage in an attempt to relieve these symptoms is best avoided as it will prolong exposure to glucocorticoid excess and delay final recovery from the symptoms of the withdrawal syndrome.

Plasma cortisol (Table 2)

Early morning plasma cortisol levels prior to glucocorticoid treatment on that day, in excess of 400 nmol/L indicate recovery of the HPAA (26). At this point it is reasonable to withdraw glucocorticoid treatment and further evaluation is not indicated. However, patients should be advised that under conditions of stress e.g. severe infection or undergoing elective or emergency surgery, their clinician should consider the need for glucocorticoid cover. Plasma cortisol levels in the range 200-400 nmol/L indicate partial recovery of the HPAA. Sub-physiological or low physiological replacement dose of glucocorticoid will be required only until patients achieve an early morning plasma cortisol level in excess of 400 nmol/L or a normal response to tests examining the entire HPAA has been recorded. Hydrocortisone 10 mg or prednisolone 2.5 mg are useful for this purpose. However, dexamethasone should be avoided because of its longer duration of action and the possibility of suppressing HPAA recovery even when the dosage is administered once daily in the morning only. Patients with plasma cortisol levels less than 400 nmol/L may benefit from a more extensive evaluation of the axis using stimulation tests.

Synacthen (α 1-24ACTH)

The use of Synacthen for detection of secondary adrenal insufficiency is not recommended (21-23). The rational for the use of this test by those who promote it as providing useful information on activity of the entire axis, is that sub-physiological activity of the axis results in hyporesponsiveness of the adrenal gland to stimulation. While this does occur in the long term when there is profound suppression of HPAA, misleadingly normal responses to adrenal stimulation by ACTH are seen in approximately 10-30% of patients with sub-optimal activity, of the axis
when evaluated by tests which examine the entire axis (21, 28) This occurs either because testing with ACTH occurs before involution of the adrenal gland has been achieved in patients with profound secondary adrenal insufficiency, or because there is partial axis activity which although inadequate to respond normally in stressful situations does maintain the adrenal primed so that responsiveness to exogenous ACTH is retained. Several reports exist of serious clinical consequences arising where the clinician was misled by an erroneously normal response to ACTH in patients

with secondary adrenal insufficiency (20, 29) Since there are readily available tests of the entire HPAA which provide the information required, the use of Synacthen stimulation tests is discouraged.

Insulin-induced hypoglycaemia (Table 2)

This test has been promoted widely as the examination of choice for the entire HPAA. It has the advantage of also stimulating growth hormone and prolactin. Thus, if the clinician wishes to assess pituitary function in addition to the HPAA, insulin-induced hypoglycaemia probably is the test of choice. The test consists of administering soluble insulin, 0.1-0.15 U/Kg body weight intravenously and monitoring blood glucose, cortisol and ACTH levels after 30, 45, 60 and 90 minutes. The test should be carefully monitored as profound hypoglycaemia may occur. When the blood glucose level falls to less than 2.2 mmol/L (40 mg/dl) an adequate stimulus has be achieved and patients may eat at this point. If patients become drowsy or responsiveness is impaired, administration of intravenous glucose is indicated as a matter of urgency as grand mal seizures may result. A normal result indicating normal axis activity is associated with plasma cortisol levels rising to approximately 500 nmol/L depending on the assay used. Although the test is seldom associated with a major complication, it is unpleasant for patients and demanding on medical resources. Unfortunately, deaths have been reported as a result of hypoglycaemia and the resuscitative procedures (30). This test is sometimes promoted as the only validated test of the entire HPAA. However, similar information can be obtained from the metyrapone test which makes fewer demands on the patients and medical resources.

Metyrapone test. (Table 2)

Metyrapone is a drug which inhibits activity of the enzyme which the final step in cortisol biosynthesis, the conversion of 11-deoxycortisol. Administration of metyrapone is associated with a fall in plasma cortisol levels which will normally stimulate the hypothalamus to produce corticotrophin releasing factor which in turns stimulates ACTH release from the pituitary in an attempt to raise the cortisol levels. This results in increasing 11-deoxycortisol levels. Metyrapone is administered, 30 mg per kg body weight by mouth at midnight with a snack. A blood sample for measurement of 11-deoxycortisol and cortisol is obtained between 8.00-9.00 hours of the following morning. A normal response is indicated by a plasma 11-deoxycortisol value of greater than 200 nmol/L. The normal 11-deoxycortisol level seen in patients who have <u>not</u> received metyrapone is less than 30 nmol/L. Cortisol levels are measured to provide an index of the effectiveness of metyrapone in impairing cortisol production. Therefore, for a clear indication of sub-optimal axis response to metyrapone it is necessary to establish that metyrapone has been effective in reducing the coristol level to a point at which there is adequate endogenous stimulation for the hypothalamus and pituitary. This overnight version of the metyrapone test is the procedure of choice in the author's experience with over 800 metyrapone tests in which only infrequently minor side-effects have occurred e.g. dizziness, vivid dreams (31). Although worsening adrenal insufficiency has been reported this has not been the authors experience and hospital admission is not required. The temporary lowering of already low cortisol levels to somewhat lower cortisol levels is unlikely to be associated with the emergence of symptoms. Two factors have limited the widespread use of this test in clinical practice: 1, assays for measurement of 11-deoxycortisol are not widely available; and 2, in some parts of the world metyrapone availability is limited. Re-agents for ll-deoxycortisol immunoassay are immediately available and the assay poses no particular technical problems. Since the metyrapone has been used throughout the world it is likely that its availability will respond to demand. It is the author's opinion that since metyrapone is safe, convenient and reliably provides similar information as that of other tests of the entire HPAA, it is the test of choice.

Corticotrophin releasing factor (CRF)

This hypothalamic factor may be used to stimulate pituitary-adrenal activity. Following its administration blood samples are obtained for measurement of ACTH and cortisol. The test is again based on the rationale that when the axis is suppressed the ability to respond normally to stimulation with CRF will be absent. It is theoretically possible that the hypothalamus may be more suppressed than the pituitary or the adrenal and that normal ACTH and cortisol responses may be achieved to exogenous CRF when the hypothalamus is unable to release endogenous factor normally. The exact role of this test in assessing HPAA activity in the post pituitary adenomectomy situation or on the timing of glucocorticoid withdrawal has not been established but it is unlikely to have a major role in the future (32-34).

Withdrawal Syndrome (Table 3)

This term is used to indicate the clinical picture which arises in patients during reduction in the exposure to glucocorticoids. The steroid withdrawal syndrome referred to here requires normal HPAA activity, or treatment with physiological replacement doses of glucocorticoid (35). Where the HPAA has not recovered or where glucocorticoid replacement when required is sub-physiological, symptoms may be attributed to inadequate glucocorticoid exposure. It is frequently seen following treatment of Cushing's syndrome either of pituitary or adrenal origin. Typically, the symptoms arise during reduction of glucocorticoid dosages used in the post-operative period as the levels approach physiological replacement. Symptoms include weakness, lethargy, diffuse body aching, headaches, weight loss and occasionally fever. While the precise mechanism involved in this syndrome has not been elucidated symptoms appear to be related to re-adapatation of tissues to physiological glucocorticoid doses following prolonged exposure to supra-physiological doses. That the condition is not related to other than a glucocorticold effect is indicated by the emergence of a similar syndrome in patients receiving exogenous therapeutic pure glucocorticold when the dosage is rapidly tapered. These symptoms prove frustrating and disappointing for the patients affected. For this reason it is useful to warn patients prior to surgical intervention that such symptoms may occur. However, they can be confidently advised that the symptoms are transient and will clear eventually although several months or up to 2 years may be required. Severe symptoms can be temporarily relieved by increasing doses of glucocorticold replacement but this then should be gradually reduced to a physiological level at which time the symptoms, while persisting, may be lesser in degree. Maintenance of supra-physiological doses of glucocorticoids means persistence of glucocorticoid excess and delays the point at which HPAA recovery will occur and glucocorticoids may be withdrawn.

It is not unusual for a patient 1 year following "hormonal cure" to report continuing unpleasant symptoms of fatigue, weakness and aching only, to return 3 months later to report that these symptoms are no longer troublesome and the patient cannot remember when exactly they had last occurred. However, when the syndrome is symptomatic support for the patient is important. Patients' support groups are active in several countries. While it is useful for patients to learn of other patients experiences at the time of diagnosis and while undergoing investigation in preparation for treatment, the greatest value of these groups may be obtained by patients who experience "the withdrawal syndrome".

Table 3. Features of the Glucocorticoid Withdrawal Syndrome

Clinical -

Weakness
Diffuse aching
Weight loss
Fever (mild)
Depression

Adrenal function

Plasma cortisol levels > 400 nmol/L off glucocorticoid treatment., or
Metyrapone test – normal,, or
Cortisol response to hypoglycaemia – normal, or
Physiological glucocorticoid replacements

Prevention

Warn patients of possibility of "withdrawal syndrome" prior to treatment.
Gradual weaning off glucocorticoid replacement

Treatment

Support and reassurance
Engage patient support groups
Rarely glucocorticoid treatment if symptoms intolerable

Learning of other patients' experiences is helped particularly by reassurance that the syndrome is self-limiting as is encouragement to avoid using excessive glucocorticoid replacement. A study of the experiences of patients who had been treated for Cushing's syndrome identified facets of care which were particularly helpful including provision of related information, reinforcement of "hope", family involvement and interaction with support groups (36). Patients who had noted difficulties with memory, thought processes, emotional instability and depression found such support of great assistance. When responding to questionnaires patients indicated that the median time to full recovery was 12 months. However, 20% of patients had symptoms persisting for more than 18 months. The addresses for sample patient support groups in the United States and the United Kingdom are given at the end of this chapter.

Treatment failure and disease recurrence

Where it is clear soon after treatment that success has not been achieved further treatment options should be discussed with the patient. These include repeat pituitary exploration, the use of radiotherapy or bilateral adrenalectomy. Where pituitary exploration may have originally been performed by an operator with limited experience in Cushing's syndrome, a repeat imaging procedure and referral to an operator with particular expertise in pituitary tumors is certainly worthwhile. However, when an experienced operator performed the initial surgery there may be little point in a "second look" although this is advised by some operators (37). Patients should be advised that radiotherapy is given in divided doses daily over approximately a 4 week period using conventional radiation and that while benefit is usual it may take over 2 years to accrue (38). In contrast bilateral adrenalectomy is associated with a cure of Cushing's syndrome although substituting adrenal insufficiency. The possible occurrence of Nelson's syndrome, i.e. increasing ACTH secretion associated with an enlarging pituitary adenoma and deepening skin pigmentation, occurs in 10-20% of patients (39). Bilateral adrenalectomy may be performed laproscopically thus avoiding some of the immediate consequences of the major procedure involved in open adrenalectomy (40). Before discussing these points with the patient it is important that the clinician has a preferred treatment plan in mind.

While recurrence of Cushing's disease does not fall within the remit of this chapter dealing with peri-operative management directly, it is worthwhile alerting patients who have a successful immediate outcome to treatment that there remains an approximately 10-20% chance of recurrence.

Adrenal tumor

Resection of a benign adrenal tumor causing Cushing's syndrome is usually the least complex of the treatment modalities for the various forms of Cushing's syndrome. The usual concerns include wound infection and transient adrenal insufficiency. Full recovery of the HPAA usually occurs within 6-12 months. Previous comments dealing with assessments of the HPAA and the occurrence of the withdrawal syndrome also apply to patients following resection of adrenal tumors. In the case of adrenal carcinoma where surgical resection has not been possible either of the complete primary tumor or of metastases, treatment with agents which limit steroidogenesis should be introduced in the early post-operative period. If additional surgical intervention is planned this strategy should be discussed with the patient once it is appropriate to do so taking their psychological and physical well being into consideration.

Ectopic ACTH secretion

In patients who have undergone successful resection of a tumor responsible for ectopic ACTH secretion, secondary adrenal insufficiency will also be apparent. The principles of assessing recovery of the HPAA have been outlined previously in this chapter. Patient's successfully treated for ectopic ACTH secretion may also experience the withdrawal syndrome and should be advised appropriately. Patients who have not had curative resection of a primary tumor or who have undergone essentially palliative bilateral adrenalectomy, should be counselled on the extent of their disease, what they can expect from the procedures undertaken, other treatment options available including chemotherapy and radiotherapy and their long-term prognosis.

SUMMARY

For the peri-operative management of patients undergoing surgical intervention for Cushing's syndrome to be efficient a protocol for each patient should be formulated prior to surgery and all of the relevant team members advised accordingly. This should include pre-operative assessment and treatment of associated disorders e.g. diabetes mellitus, advising the patient of what they can expect in the post-operative period, investigation as to whether or not cure has occurred, assessment of activity of the HPAA and the possible occurrence of the withdrawal syndrome. Patients and all team members should be advised of the outline time-table of treatment and assessments and this should be adhered to ensuring that omissions are eliminated e.g. inadequate glucocorticoid replacement, misdiagnosis of diabetes insipidus. This protocol requires careful planning and commitment to carry it out meticulously. One clinician should supervise all aspects of management. When these principles are followed the complex peri-operative management of patients with Cushing's syndrome will be satisfactorily delivered.

REFERENCES

1. Iannacone, A., Gabrilove, J.L., Sohval A.R., Soffer L.J. (1959). The ovaries in Cushing's syndrome. New Eng J Med. 261, 775-80.
2. Luton, J.P., Thieblot P., Valcke J-C., Mahoudeau, J.A., Bricaire H. (1997). Reversible gonadotropin deficiency in male Cushing's disease. J Chin Endocrinol Metab 45, 488-495.
3. Sakakura, M., Takebe, K.V. Nakagawa S. (1975). Inhibition of lutenizing hormone secretion induced by synthetic LRH with long-term treatment with glucocorticoid s in hwnan subjects. J Clin Endocrinol Metab 40, 774-9
4. Cunninghain, S.K., McKenna, T. J. (1994). The dissociation of adrenal androgens and cortisol secretion in Cushing's syndrome. Clin Endocrinol 41, 795-850.

5. Adler, R. A., Rosen, C. J. (1994). Glucocorticoids and osteoporosis. Endocrinol Metab Clin NAmcr 23, 641-54.
6. Di Somma, C., Coloa, A., Pivonello, R., Klain M., Faggianno, A., Tripodi, F. S., Merola, B., Salvatore, M., Lombardi, G. (1998). Effectiveness of chronic treatment with alendronate in the osteoporosis of Cushmg's disease. Clin Endocrinol 48, 655-62.
7. Hermus, A. B. R., Smals, A. G., Swinkels, L. M., Huymans, D. A., Peters, G. F., Sweep, F., Corstens, F. H., Kloppenborg, P. W. (1995). Bone mineral density and bone tumover before and after cure of Cushing's syndrome. J Clin Endocrinol Metab 80, 2859-65.
8. Allen, D. B. (1 996) Growth suppression by glucocorticoid therapy. Endocrinol Metab Clin N Amer 25, 699-717.
9. Chrisoulidou, A., Kousta, E., Beschyah, S. A., Robinson, S., Johnson, D. G. (1998). How much, and by what mechanisms, does growth hormone replacement improve the quality of life in growth hormone deficient adults? Bailliere's Clin Endocrinol Metab 12, 261-79
10. Carroll, P. B., Christ, E. R., Bengtsson, B. A., Carlsson, L., Cristiansen, J. E. S., Clemmons, D., Hintz, R., Ho, K., Laron, Z., Sizonenko, P., Sonksen, P. H., Tanaka, T. Thorner, M (1998). Growth hormone deficiency in adult and the effects of growth hormone replacement: A review. J Clin Endocrinol Metab 83, 382-95.
11. Danese, R.D., Aron, D.C. (1994). Cushing's syndrome and hypertension. Endocrinol Metab Clin Amer 23, 299-324.
12. Nestler, J. E., McClahahan, M. A. (1992) Diabetes and adrenal disease. Balliere's Clin Endocrinol Metab 6, 829-47.
13. Williams, D.E., Shopra, I. J., Orgiazzi, J., et al. (1975) Acute affects of corticosteroids on thyroid activity in Graves' disease. J Clin Endocrinol Metab 41,354-61.
14. Robertson, G. L. (1 995). Diabetes insipidus. Endocrinol Metab Clin N Amer 24, 549-72.
15. Miller, J.W., Crapo, L. (1 993). The medical treatment of Cushing's syndrome. Endocr Rev 14,443-58.
16. Trainer, P.J., Lawrie, H.S.,Verheist, J. et al (1993). Transsphenoidal resection in Cushing's disease: Undectable serum cortisol as the definition of successful treatment. Clin Endocrinol 38, 73-78.
17. McCance, D.R., Besser, M., Atkinson, A. B. (1996). Assessment of cure after transsphenoidal surgery for Cushing's disease. Clin Endocrinol 44 1-6.
18. Hout, W. M., Arafah, B. M., Salazar, R., Selman, W. (1988). Evaluation of the hypothalamic-pitultary-adrenal axis immediately after adenomectomy: Is perioperative steroid therapy necessary? J Clin Endocrinol Metab *66,* 1280-12.
19. Estanean, V., Loughlin, T., Nergey, A. L., Zawadzki, J. K., Boothe, J. D., Winterer, J. C., Loriaux, D. L. (1991). Daily cortisol production rate in man determined by stable isotope dilution/mass spectrometry. J. Clin Endocrinol Metab 72, 39-45.
20. Streeten, D. H. P., Anderson, Jr., G. J. H., Bonaventura, M. M. (1996). The potential for serious consequences from misinterpreting nomal responses to the rapid adrenocorticotrophin test. J Clin Endocrino Metab 81, 285-90.
21. Mukherjee, J. J., de Castro, J., Kaltasas G., Afshar, F., Grossman, A. B., Wass, J. A. H., Besser, G. M. (1997). A comparison of the insulin tolerance/glucagon test with the short ACTH stimulation test in the assessment in the hypothalamic-pituitary-adrenal axis in the early post-operative period after hypophysectomy. Clin Endocrinol 47, 51-60.
22. Oelkers, W. (1998). The role of high-and-low dose corticotropin tests in the diagnosis of secondary adrenal insufficiency. Europ J Endocrinol 139, 567-70.
23. Streeten D. H. P. (1 999). Shortcomings in the low dose (1 μg) ACTH test for the diagnosis of ACTH deficiency states. J Clin Endocrinol Metab 84, 835-37.
24. Abdu, T. A. M., Elhaeld, T. A., Neary, E. R., Clayton, R. N. (1999). Comparison of the low dose short Synaethen test (1 μg), the conventional dose short Synacthen test (250 μg) and the insulin tolerance test for assessment for the hypothalanlic-pituitary-adrenal axis in patients with pituitary disease. J Clin Endocrinol Metab 84, 838-43.

25. van Aken, M. 0., de Herder, W. W., van der Lely, A. J., de Jong, F. H., Lainberts, S. W. (1997). Post-Terative metyrapone test in the early assessment of outcome of pituitary surgery for Cushing's disease. Clin Endocrinol 47, 149-9.
26. Auchus, H. J., Shewbridge, R. K., Sheperd, M. D. (1997). "Which patients benefit from provocative adrenal testing after transsphenoidal pituitary surgery? Clin Endocrinol 46, 217.
27. Bunoncore, C.M., Robinson, A..G., (1993). The diagnosis and management of diabetes insipidus during medical emergencies. Endocrinol Metab Clin N Amer 22, 411-23.
28. Hurel, S.J., Thompson, C.J.,Watson, M.J., Herris, M.M., Baylis, P.H., Kendall-Taylor, P. (1996). The short Synacthen and insulin stress tests in the assessment of hypothalamic pituitaiy-adrenal axis. Clin Endocrinol 44, 144-46.
29. Soule, S., Fahie-Wilson, M., Tomlinson, S. (1 996). The failure of the short ACTH test to unequivocally diagnose long-standmg symptomatic secondary hypoadrenalism. Clm Endocrinol 44, 137-40.
30. Shah A., Stanhope, R., Mathew, D. (1992). Hazards of pharmacological tests of growth hormone secretion in children. Brit Med J 304, 173-
31. Fiad, T.M., Kirby, J., Cunningham, S.K., McKenna, T.J. The overnight single-dose metyrapone test is a simple and reliable index of the hypothalamic-pituitary-adrenal axis. Clin Endocrinol. 40: 603-609, 1994.
32. Avgerinson, P. C., Chrousos, G. P., Nieman, L. K., Oldfield, E. H., Loriaux, D. L., Cutler, G. B. (1987). The corticotrophin-releasing hormone test in post-operative evaluation of patients with Cushing's syndrome. J Clin Endocrinol Metab 65, 906-13.
33. Schrell, U., Fahlbusch, R., Buchfelder, M., Liedl, S., Stalla, G. K., Muller, O.A . (1987). Corticotrophin-releasmg honuone stimulation test before and after transsphenoidal selective microadenomectomy m 30 patients with Cushing's disease. J Clin Endocrinol Metab 64,1150-9.
34. Gellner, R., Stange, M., Schiemann, U., Domschke, W., Rengst K. (1999). CRH test prior to discontinuation of long-tcrm low-dose glueocorticoid therapy. Exp E Clin Endocrinol Diabetes 107, 561-7
35. Dixon, R.B. Christy, N.P. (1980). On the various forms of glucocorticoid withdrawal syndrome Amer J Med 68, 224-301
36. Gotch, P.M. (1994). Cushing's syndrome from the patient's perspective. Endocrinol Metab Clin N Amer 23, 607-17.
37. Knappe, U. J., Ludecke, D.K. (1996). Persistent and recurrent hypercortisolism after transsphenoidal surgery for Cushing's disease. Acta Neurochir (suppl) 65, 31-4.
38. Estrada, J., Boronat, M., Mielgo, M., Magallon, R., Milian, I., Diez, S., Lucas, T., Barcelo, B. (1997). The long-term outcome of pituitary irradiation after unsuccessful transsphenoidal surgery in Cushing's disease. N Eng J Med 336, 172-7.
39. Nagesser, S. K., van Seters, A. P., Kievit, J., Hennans, J., Krans, H. M., van de Velde, C. J. (2000). Long term results of total adrenalectomy for Cushing's disease. World J Surg 24, 108-13.
40. Young, W.F. (1998). Laparoscopic adrenalectomy: An endocrinologist's perspective. Curr Opin Endocrinol Diab 6, 199-203.

SUPPORT GROUPS FOR PATIENTS AFFECTED BY CUSHING'S SYNDROME

Association for Cushing's Treatment and Help (ACTH)
54 Powney Road, Maidenhead, Berks SL6 6EQ, United Kingdom.

Cushing's Support and Research Foundation
65 E. India Row, Suite 22B, Boston, MA 02110-3389.

13
RADIATION THERAPY IN THE MANAGEMENT OF CUSHING'S DISEASE

Anthony J. Cmelak, M.D.
Department of Radiation Oncology, Vanderbilt School of Medicine, Nashville, TN

GENERAL MANAGEMENT

The clinical management of Cushing's disease is complex and should always be undertaken with a multimodality approach, utilizing specialists in endocrinology, neuroradiology, neurosurgery, neuroopthalmology, and radiation oncology (1, 2, 3). The goals of intervention are:

1. Correct hormonal abnormalities
2. Remove or control growth of tumor masses
3. Prevent or minimize normal tissue injury

The use of radiation as a therapeutic modality began in the late 19th century using gas discharge tubes for superficial skin lesions. The use of radiotherapy in the treatment of pituitary adenomas was first described in 1909 (4), shortly after the first report on the use of surgical resection (5). However, low radiation energies available at the time precluded adequate treatment of deep-seated tumors without overdosing skin and subcutaneous tissue. Not until the advent of megavoltage irradiation utilizing cobalt-60 gamma rays or linear accelerator generated x-rays in the 1950s and 1960s did the widespread and effective utilization of radiation for pituitary tumors begin. Radiotherapy has been used in a number of settings: 1) as primary treatment of unresectable tumors, 2) in patients with medical contraindications to surgery, 3) for postoperative treatment after incomplete resection or high-risk patients, or 4) as salvage treatment for tumor progression or hormonal elevation after prior surgery.

Various acceptable radiation therapy techniques are available, and the choice of technique is determined by tumor size and configuration, and also by available radiation equipment. Whatever the techniques used, the goal is:

1. Accurate delivery of a uniform dose to the tumor volume
2. Maximal sparing of dose to sensitive surrounding normal tissues, particularly optic nerves and chiasm, brainstem, and medial temporal lobes
3. Daily reproducibility of treatment setup and delivery

RADIATION TECHNIQUES AND RESULTS

Fractionated External Beam Radiotherapy

Acceptable standard radiotherapy techniques include bicoronal rotational fields, 3-field isocentric arrangement with both coronal and wedged lateral fields, and intensity modulated radiation therapy. These techniques with corresponding dose distributions are shown below. Opposed lateral fields, commonly used in the past due to its simplicity, should be avoided in order to spare the temporal lobes.

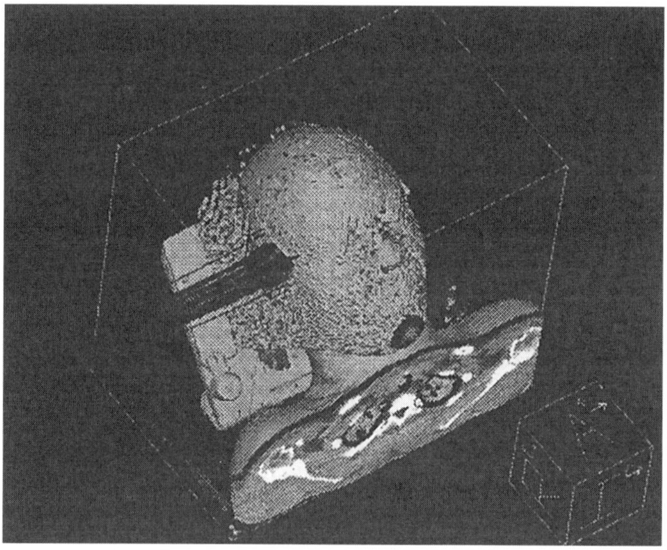

Figure 1a. Opposed lateral field arrangement. Note the patient positioning using a slanted board and Aquaplast mask.

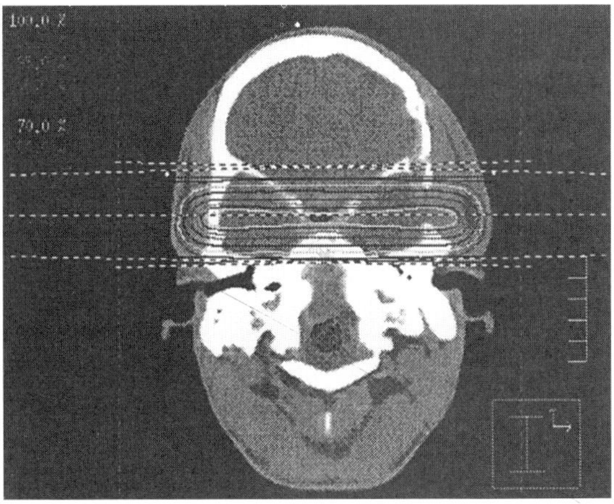

Figure 1b. Dose distribution for opposed lateral fields. Note the temporal lobes get the same 60% to 100% dose as the target region in the midline

Improved dosimetry can be obtained with a vertex beam added isocentrically to the lateral beams (Figure 2a,b). This provides excellent dose to the tumor volume and unlike a 2 field technique, less than 60% of the dose is delivered to the temporal lobes.

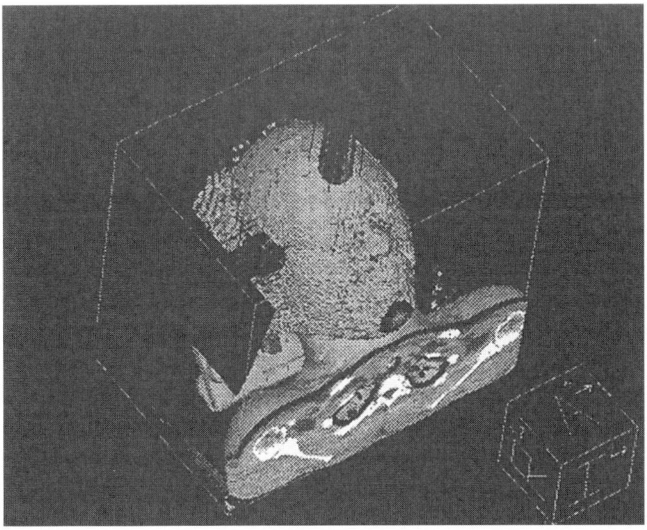

Figure 2a. Isocentric 3 filed arrangement using 2 lateral and 1 vertex field.

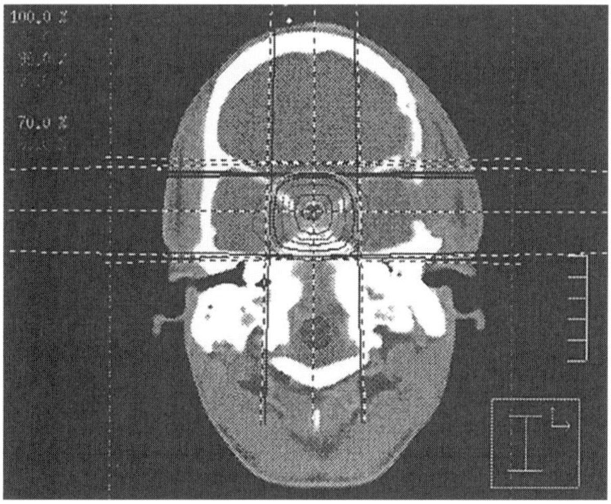

Figure 2b. Dose distribution for 3 field technique provides better protection of non- target structures.

Rotational fields using coronal arcs are also used commonly because of this technique's ability to spare non-target tissue. Wedged attenuators are usually placed with the bam while arcing over the patient's head to improve the dose distribution (Figure 3a,b).

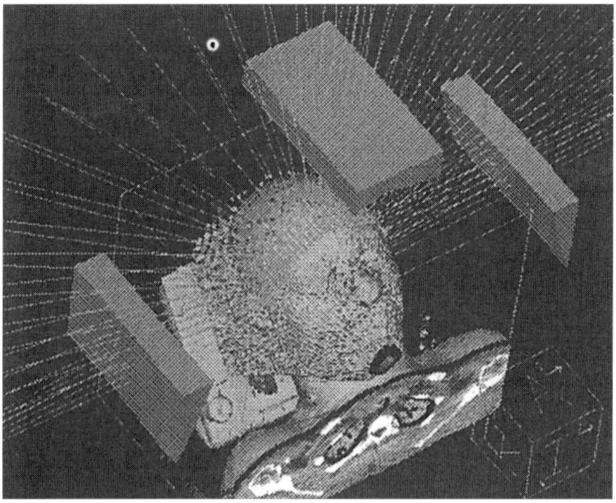

Figure 3a. Rotational field technique. The beam remains on the target while arcing over the patient's head. Note that wedges are often placed within the beam.

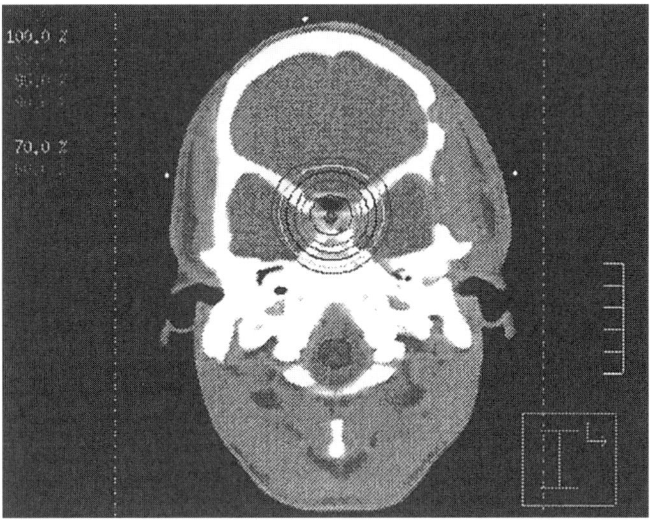

Figure 3b. Dose distribution for rotational field technique

Megavoltage energies of 6MV to 18MV via linear accelerator are preferred for better tumor dose homogeneity and minimization of dose to superficial tissues of the scalp and skull. Head immobilization with an aquaplast mask or similar device should be utilized for reproducibility of daily head positioning and dose delivery (Figure 4).

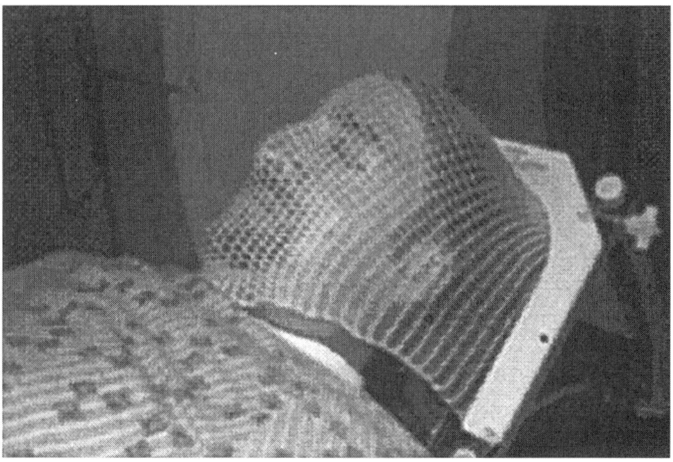

Figure 4.

Although the majority of patients can be treated with standardized techniques, each patient's skull shape and tumor size vary; therefore, the method of treatment planning should be individualized and use of CT-based treatment planning imaging is recommended. Field sizes of 4cm x 4cm to 6cm x 6cm are generally adequate to enclose the tumor volume in the 95% isodose line, with dose decreasing rapidly outside this volume. Thus, dose homogeneity within the tumor volume (as well as the optic apparatus) does not exceed 5% above that prescribed for the tumor volume. Conformality of dose to the tumor shape can be maximized using multiple intersecting beams, each individualized with wedges and beam shaping devices such as Cerrobend block blocks or multileaf collimators (Figures 5a,b).

Figure 5a. Cerrobend block mounted on plexiglass tray is used for beam shaping.

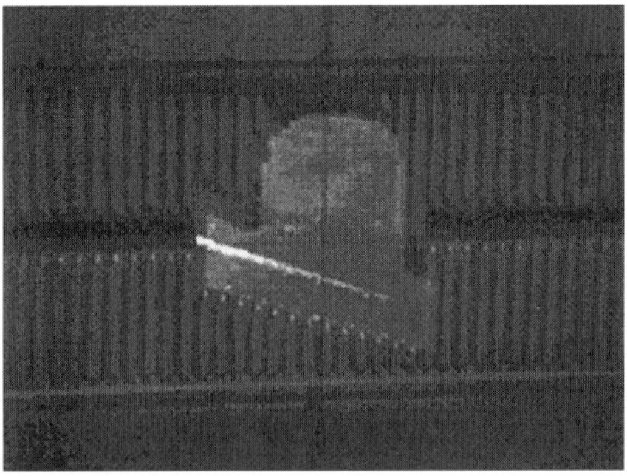

Figure 5b. Multileaf collimators of widths 3 to 10mm can be closed or opened mechanically for accurate beam shaping without the need for Cerrobend blocks.

Conventional fractionated external beam radiotherapy will control hypersecretion of ACTH in 50% to 75% of adults with Cushing's disease, and in 80% of children (6, 7). Radiation therapy appears to be equally effective for progressive tumors after transsphenoidal resection as for primary treatment of de novo tumors (8). Historically, primary radiotherapy has been utilized in patients who have medical or anatomic contraindications to surgical resection or in patients who refuse surgery. Under these circumstances, decline of ACTH levels requires three to nine months after completion of treatment. Although primary radiation is effective in relieving mass effect and associated visual loss from larger macroadenomas, the time required to observe treatment results usually make it preferable to surgically decompress the optic nerves and chiasm and then irradiate postoperatively (9). Sheline and Tyrrell reported that approximately two-thirds of patients undergoing primary radiation who presented with modest visual field deficits (not more than one quadrant) had normalization of vision in the affected eye(s). This result was comparable to surgery-alone treated patients. However, patients with more advanced tumors with greater visual deficits had better recovery of vision when receiving surgical decompression followed by postoperative irradiation. If medically inoperable, definitive irradiation can be useful in controlling adenoma progression. Chun and associates reported the results 12 patients who underwent primary radiation therapy with large and aggressive pituitary adenomas from a population of 145 other pituitary patients. Six tumors were non-functional and six were functional. With mean follow-up of 6.4 years, six (50%) were without clinical evidence of disease radiographically, and another 3 (25%) were rendered free of disease after follow-up of 5 to 8 years using salvage surgery or reirradiation (10). Salvage irradiation can result in long-term local control or patients with tumor progression after initial transsphenoidal resection. Kovalic et al. published results of salvage radiotherapy on 19 patients with recurrent adenomas with evidence of visual field deficits or decreased visual acuity, hormonal abnormalities, or radiographic evidence of extrasellar tumor extension. Recurrence had occurred 7 months to 22 years (median 3.8 years) after initial surgery, and consisted of 15 chromophobe, 3 eosinophilic, and one basophilic adenoma. With median fraction sizes of 1.8 Gy and total dose of 50.4 Gy, 10 year and 20 year progression-free survival rates were 90% and 53%. Of 4 patients experiencing a second tumor progression, only one could be salvaged by re-resection as well as reirradiation and who remains NED at 10.6 years.11 Estrada and associates reported that 48 Gy to 54 Gy in 30 patients who had previously failed transsphenoidal surgery resulted in resolution of Cushing's syndrome in 83% of patients. "Remissions" of Cushing's syndrome began 6 to 60 months after radiation therapy, but most within the first 24 months. None of the patients had a relapse of Cushing's after remission was achieved, although the majority of patients experienced hormonal deficiencies following radiotherapy (particularly growth hormone and gonadotropins) (8).

Radiosurgery

Radiosurgery is defined as high dose, single fraction stereotactic external beam irradiation. The technique was originally described by Leksell in 1949 for treating arteriovenous malformations6 and allows delivery of a focused, tumoricidal dose of radiation from multiple geometric positions (7). This technique minimizes the amount of radiation received by the non-target regions of the brain (9). Since that time, a number of reports have suggested a role of radiosurgery in the treatment of both primary and recurrent pituitary adenomas (10, 11). Ideally, tumors amenable to treatment with radiosurgery are small (<4 cm), spherically shaped with distinct margins, contrast-enhanced on imaging studies, have well-defined borders, and displace rather than infiltrate normal brain (12, 13). Different methods have been used to deliver high doses of radiation stereotactically to small volumes, including the use of linear accelerators, 201 cobalt-60 sources, and charged particle therapy (Figure 6a,b).

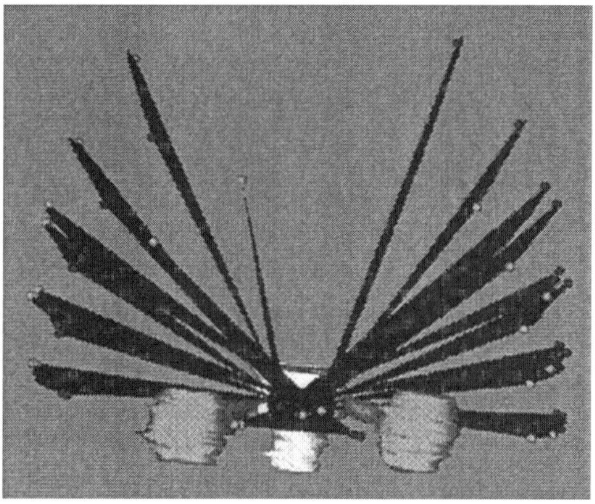

Figure 6a. Linear accelerator-based stereotactic radiosurgery plan using 2 adjacent isocenters, each using multiple non-coplaner arc rotations. Non-target critical structures such as the eyes, optic nerves, optic chiasm, and brainstem are avoided with a high degree of accuracy.

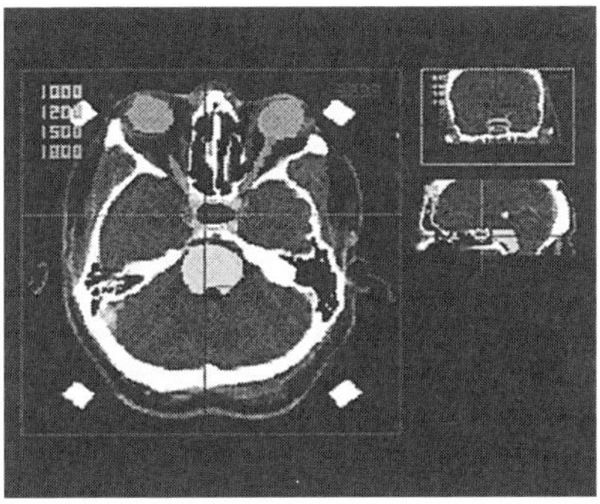

Figure 6b. Dose distribution using linear accelerator-based non-coplaner arcs and two isocenters. Note the rapid fall off of dose between 1800 cGy to 1000 cGy outside the lla region, thereby protecting critical structures.

Mokry and coworkers demonstrated that radiosurgery results in 98.3% tumor control, and an endocrinologic "cure" rate of 57%. None of the 73 patients experienced radiation induced visual damage (14). Kim and associates reported 82% of 37 patients with functioning pituitary adenomas had at least a 50% decline in hormone levels after cobalt-60 radiosurgery, with a mean latency of 7.6 months. Thirty-five percent of patients had normalization of hormone level following treatment (mean latency 22 months). Eight of 37 patients had ACTH-secreting tumors, 6 (75%) of which had hormonal response to radiosurgery, and 5 (63%) of which had hormone normalization. Early hormonal normalization was correlated with a high maximum radiation dose (>55 Gy) (15) in postoperative patients with tumor recurrence or persistent tumors. Others have reported normalization of urine free cortisol in 4 of 6 patients with Cushing's disease within 6 to 12 months following cobalt-60 radiosurgery (16). The use of radiosurgery has also been advocated for tumors invading the cavernous sinus, but has been contraindicated in the treatment of tumors abutting or invading the optic pathway (17, 18). Although some authors have concluded that single dose radiosurgery results in a more rapid decline in hormone elevation than does fractionated radiotherapy, more clinical experience is needed to define the role of radiosurgery in the treatment of Cushing's disease and other pituitary adenomas.

Brachytherapy

Brachytherapy ("close" or "contact" radiation therapy) has played a very minor role in the management of Cushing's disease and with other pituitary tumors. Like radiosurgery techniques, radioactive particles or fluids deliver very high doses to limited volumes during the process of radioactive decay. However, precise implantation of radioactive sources is difficult, and often results in significant dose inhomogeneity within the tumor volume. Areas of extremely high dose radiation cause necrosis and complications, whereas areas of underdose prevent tumor control. For this reason, only patients with intrasellar tumors without significant enroachment of the chiasm or cavernous sinuses are good candidates for this approach. However, these same patients are generally considered optimal for transsphenoidal resection. The literature is mixed on the efficacy and toxicity of brachytherapy, revealing the importance of treatment technique and patient selection when considering this technique. Fraser et al. reported a 49% complication rate using implantation of 98Y or 198Au radioactive sources (19). Others, however, have reported better results. A 77% "remission" of Cushing's disease was observed in 82 patients using either 90Y, 198Au, or a combination of the two isotopes (20). No patient had any clinical or radiographic evidence of tumor relapse with follow-up of 3 to 26 years. No visual complications occurred, although approximately 40% of patients required corticosteroid and/or thyroid replacement therapy.

COMPLICATIONS OF TREATMENT

Adverse effects from irradiation for pituitary adenoma may be mild to severe. Potential sequelae are summarized in Table X. Modern imaging techniques and radiotherapy equipment undoubtedly minimize significant complications, and it is important to keep this in mind when reviewing historical complication rates generated from treatment delivered in the 1960s and 1970s. Generally, reactions have been divided into: 1) "acute" reactions, which occur during the coarse of irradiation; 2) "early delayed" reactions, appearing from a few weeks to a few months after irradiation; and 3) "late" reactions, with onset from several months to years later (21). Reported acute and early delayed side effects include otitis media, erythema and epilation of the scalp, hypersomnolence, fatigue, scalp swelling, headache, and nausea, and transient increases in previously noted symptoms probably due to edema effects (22, 23, 24). These are usually transient and require no more therapeutic intervention than skin creams, antiemetics, and corticosteroids.

Potential late sequelae of radiation therapy is based on a number of factors:

1. Total radiation dose
2. Fraction size
3. Volume of tissue irradiated
4. Inherent susceptibility to radiation effects

Toxicities are expected to increase in frequency and severity when total radiation dose and fraction size increase, and when dose is delivered to larger volumes of tissue. Also, infant and adolescent patients may have a higher risk of late injury. Patients with collagen vascular diseases may have a higher risk of tissue necrosis, but this is controversial (25, 26).

Table 1 Complications of Radiation Therapy for Pituitary Adenomas

Acute or Subacute Sequelae	
Fatigue	Skin erythema
Nausea	Scalp swelling
Scalp epilation	
Late Sequelae	
Hypopituitarism (GH~gonadotropins>TSH>ACTH>prolactin)	
Optic nerve/chiasm injury	
Carcinogenesis (sarcoma, meningioma, glioma)	
Brain necrosis	
Accelerated atherosclerosis	

Hypopituitarism

Contrary to historical beliefs, it is now well recognized that ionizing irradiation can result in late hypopituitarism. This is most commonly manifested as GH and gonadotropin deficiency, may require years before clinically significant hormonal deficiencies develop, and is most influential in patients irradiated in pubertal or prepubertal ages (27). Sheline published results showing 21 of 36 women developed GH deficiency two to ten years after radiation followed by dopamine-agonist therapy for prolactinomas. No patient developed ACTH or TSH deficiency (28). Goldfine and associates showed that primary radiation for acromegaly in 12 patients resulted in three cases of hypopituitarism (29). After doses of 30 Gy to 85 Gy (92% (45 Gy) to the hypothalamic-pituitary axis in the treatment of nasopharyngeal carcinoma, Samaan reported 54 of 65 patients had evidence of hormonal deficiencies, and 25 had primary hypopituitarism (30). Associated sequelae in adolescent patients included growth failure and delayed bone age in 13 adolescents, but had little clinical relevance when radiated as an adult. The data on incidence of hormone deficiency after radiation is disputed, however, by Kernohan and colleagues who reported normal development of sexual characteristics, sexual function, and intellect in all 15 children treated for Cushing's disease (31). Declines in ACTH and TSH have also been reported. Eastman and colleagues reported a 32% incidence of hypoadrenalism and 10% incidence of hypothyroidism 10 years after treatment for acromegaly (32). A clear dose-response curve does not exist for irradiation and subsequent hypopituitarism. Undoubtedly, factors including patient age at treatment, prior transsphenoidal resection, and medication also play a significant role.

Vision Loss

The proximity of the optic apparatus to the pituitary makes visual loss a potential late side effect of pituitary irradiation. In addition, it has been postulated that vascular alterations occurring with some pituitary patients, particularly acromegalics, may predispose patients to visual complications, particularly if dose exceeds 45 Gy (33). In addition, daily fraction size above 1.8 Gy appear to increase risk of visual injury.

TABLE 2 Optic Apparatus Injury After Pituitary Irradiation

Author (ref)	Fraction Size	Total Dose	Incidence of Visual Injury
Aristizabel	all		4/122
	<2 Gy		0/7
	2-2.2 Gy		2/99
	>2.2 Gy		2/16
		<46 Gy	0/96
		>46 Gy	4/26
Shelin	<2 Gy	≤50 Gy	0/180
	2.25 Gy	≤50 Gy	1/1
Harris	1.75 Gy- 2.5 Gy	45 Gy-55 Gy	0/27
	2.5 Gy-3 Gy	45 Gy-55 Gy	5/28*
Atkinson	2.6 Gy-3.1 Gy	42 Gy-45 Gy	4/23**

*patients with both pituitary tumors and craniopharyngiomas included
**all patients had acromegaly

Data from UCSF on irradiation up to 50 Gy for a variety of pituitary adenomas in 181 patients showed only one patient experienced visual loss. This patient was treated with 2.25 Gy fractions (34). Aristizabal and colleagues reported 4 of 122 patients developed visual pathway injury, all 4 of which received radiation doses above 50 Gy, and 2 of these received fraction sizes of 2-2.2 Gy (99 total patients using this fractionation), and the other 2 patients received fraction sizes greater than 2.2 Gy (16 total patients had received this fraction size) (35). Harris published a report on 5 cases of optic injury; in all cases daily fraction size was at least 2.5 Gy (36). Lowering total radiation dose may not completely safeguard patients from visual complications when using large fractions sizes. Atkinson et al. reported 4 of 23 acromegalic patients suffered visual loss after receiving 42 Gy to 45 Gy delivered in 2.8 Gy to 3.1 Gy fractions (37). It appears, then, that late injury of irradiation to the optic apparatus can be maintained at less than 1% if fraction sizes of 1.8 Gy to 2 Gy are used and total doses a kept below 50 Gy.

Carcinogenesis

Secondary neoplasms after pituitary irradiation are extremely uncommon but have been reported (38, 39, 40). By definition, these tumor must occur after irradiation and within the radiation field. In general, secondary tumors begin to occur a few years after radiation and can appear many decades later. No well-defined dose-relationship exists, although most experts agree that the incidence increases with younger age at treatment. The risk of developing a radiation-induced brain tumor after conservative surgery and radiation treatment is 1.3% at 10 years and 1.9% at 20 years (41). Tumors included astrocytoma, meningioma, and sarcoma. The relative risk of developing a second brain tumor compared to the incidence in the normal population was 9.38 (3.05 to 21.89). However, no satisfactory data exists on the second tumor incidence of patients undergoing conservative surgery alone. This is necessary to determine inherent second tumor risk in this population of patients. Therefore, although ionizing radiation is an accepted promoter of secondary neoplasms, the relative contribution of radiotherapy in the appearance of these tumors is not well established.

Radiation Necrosis

Tissue necrosis, particularly of the brain, is well-established late sequelae of high-dose irradiation. It is a result of focal demyelination, microvascular injury and reduction of blood flow. Although more commonly seen for malignant neoplasms requiring high doses for local control, it is an exceedingly rare event in patients treated to 45 Gy to 50 Gy for Cushing's disease (42, 43, 44, 45, 46, 35, 47). This is particularly true when avoiding older techniques which used opposed lateral radiation fields and large fraction sizes.

TABLE 3 Reported Rates of Brain Necrosis Using Older Radiation Techniques

Author (ref)	Radiation Dose	Rate of Necrosis
Sheline and Wara	40 Gy-50 Gy	0/180
Kramer	45 Gy	0/169
Pistenmaa	57.5 Gy (average)	0/84
Emmanuel	40 Gy	0/96
Martins	66 Gy-67 Gy	2/?
Aristizabel	49 Gy	1/122
Marks (57)	55 Gy	1/66
	60 Gy	4/100
	>60 Gy	2/5

Taken in part from Perez CA, Brady LE (eds.): Principles and Practice of Radiation Oncology, pp 1108-1125. Philadelphia, JB Lippincott, 1987.

Treatment of malignant gliomas with modern treatment planning methods typically to 60 Gy carries a probability of brain necrosis of 2 to 5%. Table Z is a compilation of reports using historical treatment planning methods and fractionation schedules showing brain necrosis rates. The incidence today using modern techniques and doses would be significantly less than 1%, and is generally not considered a major determinate of contemporary treatment approach for Cushing's disease.

Cerebral Infarction

Radiation induced vascular injury is a well-documented phenomenon (48, 49). Long term studies of patients with pituitary adenoma suggest increased mortality that is independent of tumor control, with cerebrovascular accidents (CVA) as the major contributing cause. Brada and coworkers reviewed records from 331 patients treated with radiation for pituitary tumors between 1962 and 1986. The actuarial incidence of CVA after treatment was 4% at 5 years, 11% at 10 years, and 21% at 20 years. The relative risk of CVA compared to the general population was 4.1. Relative risk of CVA was higher for women than men, in patients who underwent more radical debulking surgery, and in those who received radiation does above 45 Gy.50. However in another review, Hirata and associates were unable to correlate this complication with radiation dose (51). CVA has also been reported after radiosurgery for pituitary adenomas, but the incidence and risk factors are unknown (52). Like visual injury and radiation necrosis, the overall risk of a patient developing cerebral infarction after radiation therapy using modern techniques, fractionation, and energies is expected to be a rare occurrence.

POSTTREATMENT FOLLOW-UP

Evaluation of response to radiation therapy in Cushing's disease requires evaluation using endocrine, radiographic, and visual testing. This requires a strong collaboration between the endocrinologist, neurosurgeon, radiation oncologist, and opthalmologist. It is recommended that patients have serial plasma and urine steroid levels taken and plasma ACTH levels evaluated. Response to treatment may not appear for several months to years after irradiation, and full effects are usually not evident for many years. In addition, periodic assessment of adrenal, thyroid, and gonadal function is mandatory in adults (also GH in children) to determine if radiation-induced hypopituitarism has occurred. This may need to be followed indefinitely, given that hormonal changes can occur many years after treatment. Serial MRI scans should be obtained yearly until complete tumor response or progression is seen, and formal visual field and acuity testing should be considered, particularly for patients with pretreatment deficits.

CUSHING'S IN CHILDREN

Pituitary tumors in the pediatric population are extremely rare, occurring at only 1% of the rate of tumors within the rest of the brain (53). This is in contrast to adults, where 10% of all intracranial neoplasms are of pituitary origin. Cushing's disease is the most common pituitary disorder in the pediatric age group and prolonged diagnosis of mass and hormonal effects in young patients can have devastating and lifelong sequelae. In addition, treatment approaches must be similarly determined relatively more by potential late toxicities than is needed in an adult population. Irradiation of the developing brain can potentially cause intellectual decline, bone growth delay, stroke, seizures, hearing loss or permanent hypopituitarism. Histopathologically, diffuse microhemorrhages, dystrophic calcifications, and leukoencephalopathy are recognized late sequelae of overirradiation, even when attempting to avoid late toxicity using hyperfractionation (54). Therefore, radiation must be used with discretion and assurance used to minimize irradiation of non-tumor structures.

Jennings and associates reported 80% of 15 children with Cushing's had normalization of both mean plasma cortisol and 24-hour urine corticosteroid excretion within 18 months of external beam irradiation. Sexual development proceeded normally in all 15, and no deficits in intellectual function were observed (55). Grigsy and colleagues at Washington University retrospectively reviewed results of 11 children (less than 19 years of age) treated with subtotal resection and radiation (S+R) or radiation alone (R) for a variety of pituitary adenomas. Median follow-up was 15.6 years. Of four patients undergoing S+R, one developed recurrence. The other 7 all had radiation alone, with one failure in this group. No patient suffered a visual complication, but 8 of 11 (including all 4 who received S+R) required long term medication for hypopituitarism (56). Others have reported that pituitary adenomas in adolescents may be biologically more aggressive, with younger patients having larger tumors and shorter time to tumor progression after treatment (57). The authors suggested that very young patients require both surgery and postoperative radiotherapy optimal tumor control.

REIRRADIATION FOR PERSISTENCE OR RECURRENCE

Patients who fail after initial radiation therapy, with or without surgery are generally referred for surgical salvage. Reirradiation has been used sparingly due to concerns over optic chiasm, brainstem, and temporal lobe radiotolerance. However, a course of retreatment has been attempted in clinical scenarios where no treatment alternatives exist and a long interval from initial radiation has elapsed. Flickinger et al. reported the results of 6 pituitary patients and 4 others with suprasellar tumors retreated from 35 Gy to 49.6 Gy after initial treatment doses of 36 Gy to 53.65 Gy. Eight remain free of tumor progression 1.6 to 20.5 years after retreatment, with the other 2 dying of intercurrent disease. Only one patient developed optic neuropathy

1.3 years after 40 Gy reirradiation (total dose 86 Gy) (58). Similar results were reported in 15 patients by the group at the University of California, San Francisco (59). Stereotactic radiosurgery using both linear accelerator and cobalt-60 techniques have been used in the setting of tumor recurrence after prior fractionated radiation therapy. However, published results on these patients are usually included in combination with patients treated without prior irradiation, and therefore, no definitive conclusions can be drawn (60). It is currently an area of active clinical investigation, particularly in patients with no medical or surgical alternative.

CONCLUSION

The management of Cushing's disease is complex and requires a multidisciplinary approach to effectively consider all treatment options. The goals of treatment include not only tumor control, but also control of hypersection of ACTH, correct other hormonal abnormalities, and minimization of injury to the hypothalamic-pituitary axis, optic apparatus, brain parenchyma, and cerebral vasculature. Great strides have been made in the last few decades in the fields of medical management, imaging, microsurgical techniques, and radiation delivery technology. The specific role of radiotherapy in the management of Cushing's disease has become more clearly defined with the advent of high-resolution MRI, ultra-sensitive hormone assays, and understanding both the benefits and limitations of transsphenoidal resection. Surgery is generally considered first in order to obtain rapid improvement in hormonal abnormalities and mass effect. However, overly aggressive attempts to remove all tumor is not warranted given the efficacy of postoperative irradiation in controlling residual disease or for tumor recurrence. In addition, primary radiation is an excellent alternative for patients wishing to avoid surgery, in the very elderly, or in those who are medically inoperable.

REFERENCE

1. Ciric I, Mikhall M, Stafford T, Lawson L, Garces R. Transsphenoidal microsurgery of pituitary macroadenomas with long term follow-up results. J Neurosurg 59:395-401, 1983.
2. Bloom HTG. Radiotherapy of pituitary tumors. In Pituitary Tumors, J.S. Jenkins (Ed.), London, Butterworth, 1973, pp. 165-197.
3. Hayes TP, Davis RA, Raventos A. The treatment of pituitary chromophobe adenomas. Radiology 98:149-153, 1971.
4. Gramegna A. Un cas d'acromegalie traite' par la radiotherapie. Rev Neurol 17:15, 1909.
5. Schloffer J. Zur Frage der operationen an der Hypophyse. Beitr Klin Chir 50:767, 1906.
6. Orth DN, Liddle GW. Results of treatment in 108 patients with Cushing's syndrome. N Engl J Med 285:243, 1971.

7. Jennings AS, Liddle GW, Orth DN. Results of treating childhood Cushing's disease with pituitary irradiation. N Engl J Med 297:957, 1977.
8. Estrada J, Boroat M, Mielgo M, et al. The long-term outcome of pituitary irradiation after unsuccessful transsphenoidal surgery in Cushing's disease. NEJM 336(3):172-77, 1997.
9. Sheline GE, Tyrell JB. Pituitary Tumors. In: Principles and Practice of Radiation Oncology. Perez CA and Brady LW (Eds.), J.B. Lippincott, Philadelphia, pp 1108-1125, 1987.
10. Chun M, Masko GB, Hetelekidis S. Radiotherapy in the treatment of pituitary adenomas. Int J Radiat Oncol Biol Phys 15(2):305-309, 1988.
11. Kovalic JJ, Grigsby PW, Fineberg BB. Recurrent pituitary adenomas after surgical resection: the role of radiation therapy. Radiol 177:273-275, 1990.
12. Leksell L. A stereotaxic apparatus for intercerebral surgery. Acta Chir Scand 99:229-33, 1949.
13. Leksell L. Stereotactic radiosurgery. J Neurol Neurosurg Psychiatry 46:797-803, 1983.
14. Mokry M, Ramschak-Schwarzer S, Simbrunner J, Ganz JC, Pendl G. A six-year experience with the postoperative radiosurgical management of pituitary adenomas. Stereotact Funct Neurosurg 72 Suppl 1:88-100, 1999.
15. Kim SH, Huh R, Chang JW, Park YG, Chung SS. Gamma Knife radiosurgery for functioning pituitary adenomas. Stereotact Funct Neurosurg 72 Suppl 1:101-110, 1999.
16. Morange-Ramos I, Regis J, Dufour H, et al. Sort-term endocrinologic results after Gamma Knife surgery of pituitary adenomas. Stereotact Funct Neurosurg 70(sppl 1):127-38, 1998.
17. Jackson IM, Noren G. Role of gamma knife therapy in the management of pituitary tumors. Endocrinol Metab Clin North Am 28(1):133-42, 1999.
18. Inoue HK, Kohga H, Hirato M, Sasaki T, Ishihara J, Shibazaki T, Ohye C, Andou Y. Pituitary adenomas treated by microsurgery with or without Gamma Knife surgery: experience in 122 cases. Stereotact Funct Neurosurg 72 Suppl 1:125-131, 1999.
19. Fraser R, Doyle GF, Joplin CW, et al. The assessment of endocrine effect and the effectiveness of ablative pituitary treatment by 98Y and 198Au implantation. In Kohler PO, Ross GT (Eds.), In: Pituitary tumors. Elsevier, New York, p 35, 1973.
20. Sandler LM, Richards NT, Carr DH, et al. Long-term follow-up of patients with Cushing's disease treated by interstitial irradiation. J Clin Endocrinol Metab 65:441, 1989.
21. Sheline GE, Wara WM, Smith V. Therapeutic irradiation and brain injury. Int J Radiat Oncol Biol Phys 6(9):1215-1228, 1980.
22. Noell KT. Prolactin and other hormone producing pituitary tumors; radiation therapy. Clin Obstet Gynecol 23(2): 441-452, 1980.
23. Eastman RC, Roth J. Conventional supervoltage irradiation is an effective treatment for acromegaly. J Clin Endocrinol Metab 48:931-940, 1979.
24. Baglan R, Marks J. Soft-tissue reactions following irradiation of primary brain and pituitary tumors. Int J Rad Oncol Biol Phys 7:455, 1981.
25. Morris MM, Powell SN. Irradiation in the setting of collagen vascular disease: acute and late complications. J Clin Oncol 15(7):2728-2735, 1997.
26. Ross JG, Hussey DH, Mayr NA, Davis CS. Acute and late reactions to radiation therapy in patients with collagen vascular diseases. Cancer 71(11):3744-3752, 1993.
27. Richards GE, Wara WM, Grumbach MM, et al. Delayed onset of hypopituitarism: Sequelae of therapeutic radiation of central nervous system, eye, and middle ear tumors. J Pediatr 989:553, 1976.
28. Sheline GE, Grossman A, Jones AE, et al. Radiation therapy for prolactinomas. In Black PM, Zervas NT, Ridgway ED, et al (Eds.), Secretory Tumors of the Pituitary Gland, pp 93-108, New York, Raven, 1984.
29. Goldfine ID, Lawrence AM. Hypopituitarism in acromegaly. ArchInt Med 130:720, 1972.

30. Samaan N, Maor M, Sampiere VA, et al. Hypopituitarism after external irradiation of nasopharyngeal cancer. In Recent Advances in the Diagnosis and Treatment of Pituitary Tumors, Linfoot JA (Ed.), pp 315-330, New York, Raven, 1979.
31. Kernohan JW, Sayre GP. Tumors of the Pituitary Gland and infundibulum. Section X, fascicle 26, p. 7, Washington D.C., Armed Forces Institute of Pathology, 1956.
32. Eastman RC, Gorden P, Roth J. Conventional supervoltage irradiation is an effective treatment for acromegaly. J Clin Endocrinol Metab 48:931, 1979.
33. Kramer S. Personal communication, as reported in Sheline GE, Tyrrell B. Pituitary Adenomas. Radiation Oncology Annual, Phillips TL and Pistenmaa (Eds.), Raven Press, pp 1-34, 1984.
34. Sheline GE. The role of conventional radiation therapy in the treatment of functional pituitary tumors. In: Recent Advances in the Diagnosis and Treatment of Pituitary Tumors. Linfoot JA (ed), pp289-315, New York, Raven Press, 1979.
35. Aristizabal S, Caldwell WL, Avila J. The relationship of time-dose fractionation factors to complications in the treatment of pituitary tumors by irradiation. Int J Radiat Oncol Biol Phys 2:667-673, 1977.
36. Harris JR, Levene MB. Visual complications following irradiation for pituitary adenomas and craniopharyngiomas. Radiology 120:167-171, 1976.
37. Atkinson AB, Allen IV, Gordon DS, Hadden DR, Maguire CJF, Trimble ER, et al. Progressive visula failure in acromegaly following external pituitary irradiation. Clin Endocrinol (oxf), 10:469-479, 1979.
38. Waltz TA, Brownell B. Sarcomas: A possible late result of effective radiation therapy for pituitary adenoma: report of two cases. J Neurosurg 24:901, 1966.
39. Rubenstein IJ, Garg UK. Secondary tumors of the central nervous system. In Tumors of the Central Nervous System, pp 313-314, Bethesda MD, Armed Forces Institute of Pathology, 1972.
40. Powell HC, Marshall LF, Igneizi RJ. Post-irradiation pituitary sarcoma. Acta Neuropathol 39:165,1977
41. Brada M, Ford, Ashley s, Bliss JM, Crowley S, Mason B, Traish D. Risk of second brain tumor after conservative surgery and radiotherapy for pituitary adenoma. Br J Med 304:1343-6, 1992.
42. Sheline GE, Wara WM. Radiation therapy of acromegaly and nonsecretory chromophobe adenomas of the pituitary. In Seydel HG (ed.): Tumors of the Central Nervous System, pp 199-131. New York, Wiley, 1975.
43. Kramer S. Personal communication, 1978. In Sheline GE, Tyrrell JB. Pituitary Tumors. In Perez CA, Brady LE (eds.): Principles and Practice of Radiation Oncology, pp 1108-1125. Philadelphia, JB Lippincott, 1987.
44. Pistenmaa DA, Goffinett DR, Bagshaw MA, et al. Treatment of chromophobe adenomas with megavoltage irradiation. Cancer 35:1574, 1975.
45. Emmanuel IG. Symposium on pituitary tumors, IV. Historical aspects of radiotherapy, present treatment technique and results. Clin Radiol 17:154, 1966.
46. Martins AN, Hayes GH, Kempe LG. Invasive pituitary adenomas. J Neurosurg 22:268, 1965.
47. Pistenmaa DA, Goffinet DR, Bagshaw MA, et al. Treatment of acromegaly with megavoltage radiation therapy. Int J Radiat Oncol Biol Phys 1:885, 1976.
48. Cheng SW, Wu LL, Ting AC, Lau H, Lam LK, Wei WI. Irradiation-induced extracranial carotid stenosis in patients with head and ncck malignancies. Am J Surg 178(4):323-28, 1999.
49. Carmody BJ, Arora S, Avena R, Curry KM, Simpkins J, Cosby K, et. al Accelerated carotid artery disease after high-dose head and neck radiotherapy: is there a role for routine carotid Doppler surveillance? J Vasc Surg 30(6):1045-51, 1999.
50. Brada M, Burchell L, Ashley S, Traish D. The incidence of cerebrovascular accidents in patients with pituitary adenoma. Int J Radiat Oncol Biol Phys 45(3):693-8, 1999.

51. Hirata Y, Matsukado Y, Mihara Y, et al. Occlusion of the internal carotid artery after radiation therapy for the chiasmal lesion. Acta Neurochir (wien) 74:141, 1985.
52. Lim YJ, Leem W, Park JT, Kim TS, Rhee BA, Kim GK. Serebral infarction with ICA after Gamma Knife radiosurgery for pituitary adenoma: a case report. Stereotact Funct Neurosurg 72 suppl 1:132-39, 1999.
53. Hoffman HJ. Pituitary adenomas. In: American Association of Neurological Surgeons: Surgery of the Developing Nervous System, pp 493-499, Grune and Stratton, New York, 1982.
54. Freeman CR, Bourgouin PM, Sanford RA, Cohen ME, Friedman HS, Kun LE. Long term survivors of childhood brain stem gliomas treated with hyperfractionated radiotherapy. Clinical characteristics and treatment related toxicities. The Pediatric Oncology Group. Cancer 77(3):555-62, 1996.
55. Jennings AS, Liddle GW, Orth DN. Results of treating childhood Cushing's disease with pituitary irradiation. N Eng J Med 297(18):957-62, 1977.
56. Grigsby PW, Thomas PR, Simpson JR, Fineberg BB. Long-term results of radiotherapy in the treatment of pituitary adenomas in children and adolescents. Am J Clin Oncol 11(6):607-611, 1988.
57. Fisher BJ, Gaspar LE, Stitt LW, Noone BE. Pituitary adenoma in adolescents: a biologically more aggressive disease? Radiology 192(3):869-72, 1994.
58. Flickinger JC, Deutsch M, Lunsford LD. Repeat megavoltage irradiation of pituitary and suprasellar tumors. Int J Radiat Oncol Biol Phys 17(1):171-75, 1989.
59. Schoenthaler R, Albright NW, Wara WM, Phillips TL, Wilson CB, Larson DA. Re-irradiation of pituitary adenoma. Int J Radiat Oncol Biol Phys 24(2):307-14, 1992.
60. Mitsumori M, Shrieve DC, Alexander E 3rd, Kaiser UB, Richardson GE, Black PM, Loeffler JS. Initial clinical results of LINAC-based stereotactic radiosurgery and stereotactic radiotherapy for pituitary adenomas. Int J Radiat Oncol Biol Phys 42(3):573-80, 1998.
61. Marks LB, Spencer DP. The influence of volume on the tolerance of the brain to radiosurgery. In t J Radiat Oncol Biol Phys 75:177-80, 1991.

14
ADRENAL SURGERY FOR CUSHING'S

Richard E. Goldstein, M.D., Ph.D.
Division of Surgical Oncology, Department of Surgery, Vanderbilt University School of Medicine, Nashville, TN

ANATOMY

In order to safely resect the adrenal glands in the setting of Cushing's disease or syndrome, it is imperative that the surgeon has a good understanding of the anatomy of these glands. The two adrenal glands are located in the retroperitoneum above and slightly medial to the upper poles of the kidney. The weight of a normal adrenal gland is usually in the range of 3-5 grams. On sectioning, adrenal glands are seen to grossly consist of an outer cortical and an inner medullary layer. The blood supply to each adrenal gland is supplied by branches of the inferior phrenic artery, the aorta and the renal artery. Practically, these tend to manifest as inferior, middle, and superior vascular bundles approaching the adrenal gland from the medial side. On the right, venous drainage is primarily through the very short adrenal vein originating from the medial aspect of the adrenal gland and draining directly into the vena cava. The surgeon needs to be especially wary of an occasional vein originating from the superior medial aspect of the adrenal gland and draining into the right hepatic vein. The primary venous drainage from the left adrenal gland is via a long left adrenal vein draining into the superior aspect of the left renal vein. There can also be significant venous drainage into diaphragmatic branches of the phrenic vein running along the medial aspect of the left adrenal gland and additional venous branches draining into the renal vein can occur on both the right and left sides. Significant superior lateral veins are also frequently encountered on both the right and left sides.

SURGICAL APPROACHES TO THE ADRENAL GLAND

The surgical approach to the adrenal gland is largely credited with having started with publications by Charles Mayo (1) and Cesar Roux (2), although neither surgeon started their respective operations with the intent of resecting the adrenal gland. In the United States, Charles Mayo operated on a 30-year-old music teacher with severe hypertensive attacks (1). The initial intent was to interrupt the sympathetic chain in the abdomen. An unexpected 4x6 cm right suprarenal mass was encountered and resected, resulting in resolution of the hypertensive spells. In retrospect, this was the first resection of a pheochromocytoma. The entity of Cushing's Disease was not described until five years later with the landmark publication of Harvey Cushing in which he described the syndrome now associated with his name (3). In the publication he reported on the correlation between the clinical findings associated with the syndrome and basophylic adenomas of the pituitary gland. He also speculated that some patients might have a syndrome due a primary adrenal tumor, an insight that would prove correct.

Prior to resecting adrenal glands in patients with Cushing's syndrome or disease, steroid support should be initiated as soon as the patient enters the operating room, regardless of the surgical technique employed. It is the current practice at Vanderbilt University to administer one hundred mg of hydrocortisone intravenously and the dose is repeated six and 12 hours later. Subsequent doses may be less as a slow taper is begun. All patients are discharged from the hospital with exogenous oral steroid support. Weaning from this support as the patient regains a properly functioning pituitary adrenal axis can be lengthy (4) and should be well supervised. It is preferable to maintain close communication between the surgeon and the endocrinologist in order to decrease the risk of severe adrenal crisis.

OPEN ADRENALECTOMY

Initial surgical approaches of the adrenal glands were often through an anterior transabdominal approach (5). Scott et al. reported on the surgical treatment of fifteen patients with Cushings syndrome due to adrenal cortical tumor (6). In the first seven patients, a transabdominal approach was used for removal of the tumor. In most of these patients, a long transverse abdominal or bucket handle incision was utilized. Prior to the development of sensitive imaging techniques, this transverse abdominal approach allowed examination of both adrenal glands. With large tumors, extension of the incision into the eighth or ninth intercostal space greatly improved exposure (7).

For a tumor involving the right adrenal gland, exposure of the right adrenal gland begins by incising the peritoneum lateral to the duodenum and a Kocher maneuver is performed to retract the first, second and third portions of the duodenum away from the adrenal gland (5). The peritoneum of the hepatocolic and the gastrocolic ligaments is divided. If needed, the triangular ligaments of the right hepatic lobe can be incised to facilitate mobilization of the right hepatic lobe off the right adrenal tumor. Planes are then developed between the medial aspect of the adrenal tumor and the vena cava identifying and ligating the adrenal vein while being cognizant of potential venous drainage into the right hepatic vein. If the adrenal tumor appears clinically to be malignant and the kidney is involved by tumor, an en bloc resection should be performed. In the current era, preoperative imaging should be performed to confirm function and size of the opposite kidney if adrenal resection is contemplated. If sufficient exposure cannot be obtained with a large transverse or bucket handle incision, further exposure can be gained by either "teeing up" the incision along the midline to the base of the sternum or through a thoracoabdominal extension to the incision. If the abdominal incision needs to be extended to a thoracoabdominal approach, the incision is extended to the costal margin and the thoracic musculature is divided in the eighth or ninth intercostal space to the mid axillary line. The cartilage of the eighth or ninth rib is transected. The thoracotomy incision is carried to the interspace and the diaphragm is divided in the direction of the intercostal incision. If the kidney is to be removed as part of the en bloc resection, the renal vein, renal artery and ureter should be dissected and the controlling vessel-loops should be placed around the artery and vein. The renal artery is double suture ligated and divided followed by double suture ligation of the renal vein and ligation of the ureter.

On the left side, the same principles apply. If the lesion is likely to be benign and the tissue planes are well defined, the left adrenal gland is exposed by mobilizing the spleen and the tail of the pancreas by means of a lateral peritoneal incision (5). One incises the avascular peritoneal reflection along the convexity of the spleen down to the splenic flexure of the colon. The lienocolic and lienorenal ligaments are divided. Gerota's fascia is incised vertically to expose the upper portion of the kidney. Retracting the kidney inferiorly and the spleen and tail of pancreas medially, the left adrenal gland is exposed. Dissection should expose the long left adrenal vein that is ligated and divided. Occasionally, a lumbar vein may arise from the posterior surface of the adrenal vein and this will need to be controlled.

If malignancy is suspected and the tissue planes are indistinct, radical adrenalectomy involving en bloc resection of a large left adrenal tumor with the kidney, the tail of the pancreas, and the spleen may be required (5). The omentum below the spleen is dissected away from the splenic flexure and the left colon allowing the colon to be retracted downward. The gastrosplenic omentum along the greater curvature of the gastric fundus is divided and the omental incision is carried down to the transverse colon. The spleen is mobilized off of the diaphragm after making a lateral incision in the lienophrenic peritoneal reflection around the convexity of the spleen. If the pancreatic tail is adherent to the larger adrenal tumor, the body and tail of the pancreas should be mobilized. The splenic artery and vein should be isolated, double suture ligated, and divided. The body of the pancreas can be transected by first controlling the vascular supply to the pancreatic tail by ligating the superior and inferior edges of the pancreas with 3-0 Surgilon sutures followed by firing the TIA stapler across the entire width of the pancreas at the point of transection. An attempt should be made to identify the transected duct of Wirsung and securely closing it with a figure-of-eight of 3-0 Surgilon suture. Whenever pancreatic resection is performed or there is concern about pancreatic duct injury, a closed section drain should be placed in the left upper quadrant.

While the anterior transabdominal approach remains appropriate for very large tumors responsible for Cushing's syndrome and malignant adrenal lesions producing cortisol, the open posterior approach became the preferred open approach for benign adenomas during much of the past 30 yrs (8). The procedure is based on an approach described by Hugh Young in 1936 which allowed for the simultaneous exposure of both adrenal glands by dissecting though the muscles of the back (9). Prior to preoperative localization techniques, this technique allowed two teams to operate simultaneously, identify both adrenal glands, and then resect the side found to contain the adenoma. While Young originally described this approach involving the sub pleural resection of the twelfth rib, most surgeons today using a similar posterior approach will resect the eleventh rib.

In the original description of the posterior approach by Young, an incision is made posteriorly 5cm from the median line and about 3cm above the twelfth rib (9). From there it is directed outward and downward to a point 8cm from the median line to the crest of the iliac. While it is possible with two surgical teams to expose both adrenals simultaneously, working conditions are quite tight. The incision is then carried down through the fat and fascia exposing the latissimus dorsi muscle. This exposes the serratus magnus muscle that is also divided and an incision is made through the conjoined portion of the lumbodorsal fascia just lateral to the outer border of the sacrospinalis muscle. Tissues are freed from the undersurface of the costovertebral ligament with care taken to avoid tearing the pleura. If the pleural cavity is breached, this can usually be repaired by inserting a red rubber catheter into the pleural space and re-approximating the pleural defect using interrupted 3-0 Vicryl suture. The red rubber catheter is removed while applying suction to evacuate air that may have entered the pleural space and quickly closing the final defect with a 3-0 Vicryl suture placed in a figure-of-eight manner. Either the 12th rib can be resected or it may be free enough to be moved outward and forward, allowing exposure of the underlying structures. Gerota's fascia is identified and incised. After incising the fascia, the kidney is retracted inferiorly with a sponge stick to expose the adrenal gland. It is important to maintain the attachment between the kidney and the adrenal initially until the superior aspect of the adrenal gland is dissected free from its upper attachments.

However, the exposure created using this posterior can be quite limiting, particularly in the clinical setting of Cushing's syndrome in which it is common for significant accumulation of retroperitoneal fat to have occurred. In addition, as the incision is directed vertically, it is difficult to obtain a significant increase in exposure if needed (e.g. bleeding from the vena cava). If significant bleeding is encountered, it may become necessary to flip the patient back to a supine position and make a midline laparotomy incision to adequately control a tear in the inferior vena cava. Hence, many surgeons have adopted a modified approach to the posterior approach utilizing a "hockey stick" incision (10-14).

In general, more recent open posterior approaches utilize this "hockey stick" shaped incision extending out along the eleventh or twelfth rib. Utilizing this incision, the patient can be either in a prone (14) or lateral (12) position. To approach the adrenal gland using such a modified posterior approach, the patient is turned to the prone position with hips and shoulders flexed. Supports need to be placed under the hips and chest. An incision is marked along the course of the 11th rib and is extended superiorly for 5-6 cms, 5 cms lateral to the center of the spine. The latissimus and serratus posterior muscles are divided with the electrocautery over the length of the 11th rib. After the periosteum is stripped from the rib, a subperiosteal resection or the entire rib is performed. As before, the pleura is identified and dissected up to avoid injury, Gerota's fascia is incised exposing the upper pole of the kidney and the adrenal gland is identified and resected. In order to facilitate more exposure should it be needed, some authors advocate using the same "hockey stick" incision" but placing the patient in a lateral position and then rolling the patient ventrally 15-30 degrees (12). This position has the advantage of easily allowing an anterior flank extension if needed (13). Using this approach, operative times averaging approximately 160 min and tumors up to 10 cms in diameter have been successfully resected (12).

Comparisons have been made between open anterior approaches (either midline or bucket handle incisions) and open posterior approaches for Cushing's syndrome or disease. Occasional perioperative deaths are reported in several series but are unrelated to the surgical approach (6, 12). In general, the posterior approach is associated with less morbidity and mortality than anterior open approaches (15). In a large though retrospective series directly comparing anterior and posterior approaches involving 315 patients, patients operated on via a posterior approach required fewer blood transfusions and had shorter hospital stays (16).

Scott et al. reported on over 300 patients with Cushing's syndrome and disease treated at Vanderbilt University form 1953 to 1983 (8). In this series, the preferential surgical approach was the posterior approach using a "hockey stick incision". Again, the posterior approach was associated with decreased morbidity when compared with the anterior approach. However, the anterior approach was, and remains, the preferred approach when adrenal carcinoma is suspected. The differential diagnosis for tumors demonstrating large size, indistinct margins, or a heterogeneous texture on CT or MRI imaging should include adrenal carcinoma and an anterior approach should be strongly considered in such cases (17). It is also important to remember that most reports comparing different surgical approaches are retrospective in nature and considerable selection bias is likely to be present. Larger tumors and those lesions highly likely to be malignant likely would have been selected to be resected using an anterior approach and an increased morbidity would be expected in this patient population.

LAPAROSCOPIC ADRENALECTOMY

In 1992 and 1993, Michel Gagner reported on the successful application of laparoscopic techniques to the resection of adrenal tumors (18, 19). The laparoscopic technique has now become widely accepted as the preferred technique to resect benign adrenal adenomas or adrenal hyperplasia associated with Cushing's syndrome and disease. There are two broad categories into which the technique of laparoscopic adrenalectomy can be grouped: the transperitoneal and the retroperitoneal approach.

The transperitoneal laparoscopic approach has been the standard approach used at Vanderbilt University since 1995 (20). In performing a laparoscopic adrenalectomy by the transabdominal approach, the patient is positioned in the lateral decubitus position. The table is broken and the kidney rest extended and the patient's position adjusted such that the kidney rest contacts the patient midway between the umbilicus and the lower portion of the sternum. This position facilitates the widening of the space between the iliac crest and the inferior aspect of the rib cage. Initially, a 12 mm Hassan trocar is introduced into the abdominal cavity in the midclavicular line approximately 4 cms below the rib cage using an open technique. The abdominal cavity is insufflated to 15 mm Hg with carbon dioxide. Following initial inspection of the upper abdominal cavity using a 30° angled laparoscope, three additional trocars are placed. The 30° angled laparoscope is used for both right and left laparoscopic adrenalectomies.

If the patient is to undergo a right adrenalectomy, a 5 mm and two 10 mm trocars are placed in the right midaxillary, right anterior axillary, and right subcostal line, respectively. If needed, the right hepatic flexure can be mobilized to facilitate placement of the more lateral trocars. The triangular ligament of the right lobe of the liver is divided up toward the suprahepatic vena cava. The peritoneum is further separated inferiorly and the liver retracted using a fan retractor. This allows adequate exposure to the adrenal gland and the inferior vena cava. The majority of the dissection is done with the hook electrocautery or the monopolar scissors. Some surgeons have preferred the harmonic scalpel but we have not noted any significant advantage to this device. As the adrenals have a very rich blood supply and the adrenal capsule tends to be fragile, it is very important not to directly grasp the adrenal gland. Once the right hepatic lobe is dissected off the adrenal gland, the space between the gland and the inferior vena cava is carefully explored in an inferior to superior direction. The short right adrenal vein is encountered early and should be dissected with great respect. It should be ligated after applying two sets of clips to the caval side of the vein and one or two clips to the adrenal side.

Next, the superior and medial aspect of the adrenal should be dissected from with care taken to be cognizant of potential venous drainage into the right hepatic vein. Once these veins are ligated, the adrenal gland will separate from the vena cava allowing the arterial vessels to be identified and ligated. The inner edge of the right adrenal will tend to lie more medial and inferior than one might expect and there is almost always a vascular pedicle at the inferior medial tip that needs to be controlled. The dissection should proceed posteriorly enough to expose muscle. The inferior aspect of the adrenal is dissected free from the upper pole of the kidney and the lateral and superior aspects of the adrenal are freed last. A useful technique to employ during these latter steps is to elevate the adrenal out of the retroperitoneum using either a straight suction device or grasper to expose the underlying fatty tissue that can then be coagulated. The fatty tissue surrounding the adrenal gland will contain numerous small vessels. It is important for these to be controlled with coagulation, as blunt dissection alone will tend to cause bleeding; even minor bleeding can make visualization through the laparoscope extremely difficult. After the adrenal is freed completely, it is placed in a bag and retrieved out of the abdomen. The fascia of the abdominal wall should be re-approximated at all sites in which a 10 or 12 mm trocar was used.

The left adrenal gland is approximated using mirror image trocar placement with the patient placed in a left flank position. Again, the hips are rotated back an extra ten degrees. The spleen is mobilized by dividing the peritoneum posterior to the spleen and by dividing the phrenolienal ligament. This allows the spleen to fall medially away from the adrenal and kidney, facilitating the exposure. The dissection continues by developing the space between the left adrenal gland and the pancreatic tail with the goal of identifying the underside of the splenic vein. Once the splenic vein is identified, the dissection continues in an inferior manner. This will allow the identification of the diaphragmatic vein, which will lead into the left adrenal vein. The long left adrenal vein should be cleared for a distance of several cms allowing the identification of its confluence with the left renal vein. The left adrenal vein should be doubly clipped on the renal vein side and generally singly clipped on the adrenal side. The diaphragmatic branch is ligated and the dissection proceeds back up the medial side of the adrenal gland. The middle and superior arterial branches are controlled and the dissection should expose muscle posteriorly. The inferior, superior, and lateral aspects of the adrenal gland are then dissected free and the gland removed. As on the right side, division of the adrenal vein is the key step and allows the gland to rotate laterally and anteriorly, facilitating further dissection.

A report from our institutions documented 19 laparoscopic adrenalectomies (20). In four patients, the indication for adrenalectomy was Cushing's syndrome. Average operative time and blood loss was 164 min and 109 ml, respectively. The average size of the resected lesion was 2.5 cms with the largest lesion measuring 5 cms in diameter. No patients had to be converted to an open procedure and average length of stay was 2.3 days. Gagner, et. al. reported on 97 adrenalectomies and biopsies, 21 of which were performed for Cushing's adenomas or disease (21). The average operative time and blood loss was 123 min and 70 ml, respectively. The average size of the resected lesion was 4.95 cms in diameter with the largest lesion being 14 cms. Only three patients required conversion to an open laparotomy.

RETROPERITONEAL ENDOSCOPIC APPROACHES

In addition to the transperitoneal laparoscopic approach, one can also approach the adrenal gland through a lateral retroperitoneal endoscopic approach (22). The patient is positioned similarly as for a laparoscopic transperitoneal approach, except that hips are not angled back the extra 10 degrees. After making a small incision just at the caudal tip of the 11th rib and then creating a small space with finger dissection, a dissection balloon is introduced and aimed posteriorly. After inflating the balloon and identifying the quadratus laborum muscle, the balloon is removed and replaced with a trocar and the space insufflated to a maximum pressure of 12 mmHg. A 5 mm trocar placed at the caudal tip of the 12th rib and the peritoneal sac is dissected off of the anterior abdominal muscles. Next, a 10 mm trocar is placed near the inferior aspect of the 10th rib and a 5 mm trocar slightly more anteriorly. Most of the dissection can be done using a 0° scope, but a 30° scope can be used if needed later in the dissection. The kidney is identified after incising Gerota's fascia. Dissecting superiorly and medially, the adrenal gland is identified and resected.

Using this technique, Bonjer et al. reported their results of resecting adrenal lesions in 10 patients, including several with Cushing's adenomas (22). The procedure was deemed safe with an average operative time that was felt to be quite reasonable at 75 min. The average diameter of the resected lesion was 3.0 cms. Conversion to an open approach was necessary in one patient. That patient had a pheochromocytoma with dense vascular adhesions necessitating the conversion.

A more posterior endoscopic approach to the adrenal gland has also been described and is based on the open posterior approach (23). To use this technique, the patient is positioned prone and a 1.5 cm transverse incision is made just below the tip of the 12th rib. As in the lateral retroperitoneal approach, after a small space is created with finger dissection, the retroperitoneal space is opened using a balloon. After the balloon is removed and replaced with a 10 mm trocar, the retroperitoneal space is maintained by CO_2 insufflation to 12 to 20 mmHg. Two additional 5 mm trocars are placed, the kidney is identified, and the adrenal gland is located and removed.

In a report of 30 posterior retroperitoneoscopic adrenalectomies, a number of which had Cushing's syndrome and disease, 25 of 30 were successfully performed endoscopically (23). Five required conversion to an open approach, generally due to technical problems and difficulties with the exposure. Again, the procedure appeared safe, no blood transfusions were required, and the average operative time was 124 minutes. The largest adrenal lesion resected had a diameter of 7 cm.

For all laparoscopic or endoscopic techniques other than the posterior retroperitoneoscopic approach, bilateral adrenalectomies performed for bilateral cortical hyperplasia involve completing the procedure on one side, then repositioning the patient in order to perform the procedure on the remaining adrenal gland. However, with the operative team in place, the time necessary to complete the second side has been reported to be less than the expected time for a unilateral procedure (21).

COMPARISON OF TECHNIQUES

There have been several reports comparing the technique of laparoscopic and endoscopic adrenalectomy with open adrenalectomy. Brunt et al. compared open transabdominal, open posterior, and laparoscopic transabdominal approaches (24). Although the study was a retrospective analysis, the results demonstrated shorter hospital stays and decreased postoperative narcotic requirements in the group who received laparoscopic adrenalectomy. In a report from Vanderbilt University, laparoscopic transabdominal adrenalectomy was compared to open adrenalectomy utilizing either a transabdominal or posterior approach (20). Again, this was a retrospective study but the results demonstrated similar operative times with fewer complications and decreased length of stay in the laparoscopic group. Operative times were longest in the laparoscopic subgroup where one endocrine surgeon was operating with a surgical resident as opposed to the procedure being performed by two attending surgeons. However, over time methods to efficiently instruct housestaff have been developed and operative times have significantly decreased.

Bonjer *et al* compared the open posterior approach to the laparoscopic transperitoneal and endoscopic retroperitoneal approaches (22). Operative times were slightly longer for the transperitoneal approach and postoperative analgesic requirements were least after the retroperitoneal approach. Several patients had persistent burning flank pain after an open posterior approach, a finding that has been observed at our institution as well. Multiple studies have also reported shorted hospital stays after laparoscopic or endoscopic adrenalectomies when compared to open approaches (22, 25-27).

In directly comparing transperitoneal to retroperitoneal video assisted approaches, there are few measurable differences (22, 28-30). While some studies reported small differences in operative times, essentially, the choice of procedure is largely dictated by individual surgeon preference, training, and comfort. However, as most studies are retrospective, it is difficult to factor in additional factors that may affect hospital length of stay such as pressure from medical insurance carriers, physician and patient cultures, and time to adjust oral steroid replacement therapy after adrenalectomy. Nevertheless, most patients who undergo a videoscopic adrenalectomy are ready for discharge on the second postoperative from the standpoint of recovery after the surgical procedure.

CURRENT RECOMMENDATIONS FOR ADRENALECTOMY IN CUSHING'S SYNDROME AND DISEASE

The clear conclusion from an analysis of recent reports strongly support laparoscopic or endoscopic adrenalectomy as the procedure of choice for most patients with Cushing's Syndome, manifested by a cortisol producing adenoma, or for most patients with Cushing's Disease whose pituitary disease cannot be effectively managed without bilateral adrenalectomy. Whereas the open posterior approach largely supplanted the open transperitoneal approach prior to 1990, the laparoscopic and endoscopic approaches have now supplanted all open approaches for benign disease with a few exceptions.

Exceptions to a transabdominal laparoscopic adrenalectomy include patients who have a history of significant upper abdominal surgery. Due to extensive adhesions, a laparoscopic transabdominal adrenalectomy may not be a viable option. The decision to attempt a laparoscopic adrenalectomy may depend on the individual surgeon's comfort in entering an abdominal cavity with extensive adhesions. For these patients, adrenalectomy performed using the open posterior approach may be the best option.

An open transperitoneal approach has been, and remains, the recommended approach for patients in whom Cushing's syndrome is caused by a lesion highly likely to be malignant. Lesions that are large, heterogeneous, or which may have indistinct borders suggesting local invasion should be approached through an open transabdominal technique (17). As primary resection for cure is the best hope for these patients, the open approaches allow the greatest possibility for a wide en-bloc resection to be performed. In general, the transverse or "bucket-handle" incision is recommended. As discussed earlier in this chapter, this incision can be extended either through "teeing" the incision up toward the sternum or by extending the incision to create a thoracoabdominal.

A large (greater than 6 cm) primary adrenal lesion, though assessed to be likely benign preoperatively, is a relative contraindication to a laparoscopic or endoscopic approach. Clearly, in general, tumors up to 5-6 cms in diameter can be safely resected using endoscopic or laparoscopic approaches (20, 22, 24). There is, however, no firm size criteria above which laparoscopic or endoscopic techniques cannot be considered. Benign tumors larger than 6 cm have been successfully resected using laparoscopic techniques. Duh et al reported the resection of tumors up to 13 cm in diameter using the laparoscopic transperitoneal approach (30). Similarly, Gagner et al. reported the resection of tumors up to 14 cm using a similar approach. Whether or not the endoscopic retroperitoneal approach can be used routinely for the resection of benign adrenal tumors greater than 6 cms is more questionable. Duh et al. reported more difficulty in resecting adrenal tumors larger than 6 cms using the posterior endoscopic approach (30). Benign adrenal tumors producing cortisol can be resected using these newer approaches. However, whether or not the tumors should be resected using these techniques will depend on number of factors, including the judgment and experience of the specific surgeon.

In summary, Cushing's Disease and Syndrome involve excess cortisol production from the adrenal glands. With the advances in the management of pituitary Cushing's Disease, the need to perform bilateral adrenalectomy for hyperplasia is relatively infrequent. In general, the most frequent indication for adrenalectomy in the setting of Cushing's Syndrome will relate to a single cortisol producing adenoma. The latter half of the twentieth century has seen a significant evolution in the surgical techniques of adrenalectomy. While open techniques still have a place, they are largely relegated to patients likely to have adrenal carcinoma or patients with prior significant abdominal surgery. Laparoscopic and endoscopic techniques have become the standard approaches used to resect adrenal glands in the setting of excess cortisol production secondary to single adrenal adenomas and bilateral hyperplasia.

REFERENCES

1. Mayo, Charles 1927. Paroxysmal hypertension with tumor of retroperitoneal nerve. *J. A. M. A.* 89:1047-1950.
2. Roux, Cesar 1958. Le pheochromocytome bilateral. *Union Medical du Canada* 87:165-178.
3. Cushing, Harvey 1932. The basophil adenomas of the pituitary body and their clinical manifestations. *Bulletin of the Johns Hopkins Hospital* 50:137-195.
4. Doherty Gerard M., Lynnette K. Nieman, Gordon B. Jr. Cutler, George P. Chrousos, and Jeffrey A. Norton. 1990. Time to recovery of the hypthalamic-pituitary-adrenal axis after curative resection of adrenal tumors in patients with Cushing's syndrome. *Surgery* 108:1085-1090.
5. Scott, H. William Jr., Naji N. Abumrad, and W. Scott McDougal. 1990. Surgical approaches for bilateral adrenalectomy and resection of benign and malignant tumors. In Surgery of the Adrenal Glands. H. William Jr. Scott, editor. J.B. Lippincott Company, Philadelphia. 297-310.
6. Scott, H. William, John H. Foster, Grant Liddle, and Eugene T. Davidson. 1965. Cushing's syndrome due to adrenocortical Tumor: 11-year review of 15 patients. *Ann Surg* 162:505-516.
7. Scott, H. William,Jr., John A. Oates, A. S. Nies, Henry Burko, D. L. Page, and R. K. Rhamy. 1976. Pheochromocytoma: Present diagnosis and management. *Ann Surg* 183:587-593.

8. Scott, H. William,Jr., Naji N. Abumrad, and David N. Orth. 1985. Tumors of the adrenal cortex and Cushing's sydrome. *Ann Surg* 201:586-594.
9. Young, Hugh H. 1936. A technique for the simultaneous exposure and operation on the adrenals. *Surg Gynecol Obstet* 54:179-188.
10. Vaughan, E. Darracott,Jr. and Harriet Phillips. 1987. Modified posterior approach for right adrenalectomy. *Surg Gynecol Obstet* 165:453-455.
11. Brunicardi, F. Charles, Paul M. Rosman, Kathy Lee Lesser, and Dana K. Andersen. 1985. Current status of adenalectomy for Cushing's disease. *Surgery* 98:1127-1134.
12. Proye, C. A. G., J. Y. Huart, X. D. Cuvillier, N. M. L. Assez, B. Gambardella, and B. M. L. Carnaille. 1993. Safety of the posterior approach in adrenal surgery: Experience in 105 cases. *Surgery* 114:1126-1131.
13. Raynor, B. Winfield and Louis R. M. DelGuercio. 1986. The eleventh rib transcostal incision: An extrapleural, transperitoneal approach to the upper abdomen. *Surgery* 99:95 101.
14. Scott, H. William, Grant W. Liddle, J. L. Jr. Mulherin, T. J. McKenna, S. L. Stroup, and R. K. Rhamy. 1977. Surgical Experience with Cushing's Disease. *Ann Surg* 185:524-534.
15. Bruining, Hajo A., Steven W. J. Lamberts, Eveline G. L. Ong, and Andries J. van Seyen. 1984. Results of adrenalectomy with various surgical approaches in the treatment of different diseases of the adrenal glands. *Surg Gynecol Obstet* 158:367-369.
16. Russell, Colin F., Bertil Hamberger, Jonathan A. van Heerden, Anthony J. Edis, and Duane M. Ilstrup. 1982. Adrenalectomy: Anterior or Posterior Approach? *Am J Surg* 144:322- 324.
17. Barnett, C. C.,Jr., D. G. Varma, A. K. El-Naggar, A. P. Dackiw, G. A. Porter, A. S. Pearson, A. P. Kudelka, R. F. Gagel, D. B. Evans, and J. E. Lee. 2001. Limitations of size as a criterion in the evolution of adrenal tumors. *Surgery* 128:973-983.
18. Gagner, Michel, Andre Lacroix, and Edouard Bolte. 1992. Laparoscopic adrenalectomy in Cushing's syndrome and pheochromocytoma. *New England Journal of Medicine* 327:1033.
19. Gagner, Michel, Andre Lacroix, Richard A. Prinz, Edouard Bolte, David Albala, Claude Potvin, Pavel Hamet, Otto Kuchel, Serge Querin, and Alfons Pomp. 1993. Early experience with laparoscopic approach for adrenalectomy. *Surgery* 114:1120-1125.
20. Jacobs, J. Kenneth, Richard E. Goldstein, and Richard J. Geer. 1997. Laparoscopic adrenalectomy: A new standard of care. *Ann Surg* 225:495-502.
21. Gagner, Michel, Alfons Pomp, B. Todd Heniford, Daniel Pharand, and Andre Lacroix. 1997. Laparoscopic adrenalectomy: Lessons learned from 100 consecutive procedures. *Ann Surg* 226:238-247.
22. Bonjer, H. J., J. F. Lange, G. Kazemier, W. W. De Herder, E. W. Steyerberg, and H. A. Bruining. 1997. Comparison of three techniques for adrenalectomy. *Br J Surg* 84:679-682.
23. Walz, Martin K., Klaus Peitgen, Rudolf Hoermann, Reiner M. Giebler, Klaus Mann, and Friedrich W. Eigler. 1996. Posterior retroperitoneoscopy as a new minimally invasive approach for adrenalectomy: Results of 30 adrenalectomies in 27 patients. *World J Surg* 20:769-774.
24. Brunt, L. Michael, Gerard M. Doherty, Jeffrey A. Norton, Nathaniel J. Soper, Mary A. Quasebarth, and Jeffrey F. Moley. 1996. Laparoscopic adrenalectomy compared to open adrenalectomy for benign adrenal neoplasms. *Journal of the American College of Surgeons* 183:1-10.
25. Prinz, Richard A. 1995. A comparison of laparoscopic and open adrenalectomies. *Archives of Surgery* 130:489-494.
26. Korman, Jeremy E., Ho Taylor, Jonathan R. Hiatt, and Edward H. Phillips. 1997. Comparison of laparoscopic and open adrenalectomy. *American Surgeon* 63:908-912.

27. Linos, Dimitrious, Nikolaos Stylopoulos, Michalis Boukis, Athanasios Souvatzoglou, Sotirios Raptis, and John Papadimitriou. 1997. Anterior, posterior, or laparoscopic approach for the management of adrenal diseases? *Am J Surg* 173:120-125.
28. Fernandez-Cruz, Laureano, Alejandro Saenz, Guerson Benarroch, Emiliano Astudillo, Pilar Taura, and Luis Sabater. 1996. Laparoscopic unilateral and bilateral adrenalectomy for Cushing's syndrome: Transperitoneal and retroperitoneal approaches. *Ann Surg* 224:727-736.
29. Chee, Christopher, Tharmalingham Ravinthiran, and Christopher Cheng. 1998. Laparoscopic adrenalectomy: Experience with transabdominal and retroperitoneal approaches. *Urology* 51:29-32.
30. Duh, Quan-Yang, Allan E. Siperstein, Orlo H. Clark, William P. Schecter, Jan K. Horn, Michael R. Harrison, Thomas K. Hunt, and Lawrence W. Way. 1996. Laparoscopic adrenalectomy: Comparison of the lateral and posterior approaches. *Archives of Surgery* 131:870-875.

15
THE MEDICAL MANAGEMENT OF CUSHING'S SYNDROME

Tom Kason, M.D. and Lewis S. Blevins, Jr., M.D.
Division of Endocrinology, Diabetes and Metabolism, Department of Medicine, Vanderbilt School of Medicine, Nashville, TN, USA

Physicians treating patients with Cushing's syndrome often need to rely on the use of pharmacologic agents in order to normalize cortisol secretion. The most effective agents interrupt the biosynthesis of cortisol via inhibition of specific adrenal cytochrome P450 enzymes. In patients with ACTH-dependent forms of hypercortisolism, the efficacy of these agents is often additive and improved by the concomitant administration of irradiation or other antitumor therapies. Occasional patients "escape" from therapy and rises in cortisol excretion rates may follow apparently successful therapy. In this chapter, we will briefly discuss specific drugs that have been employed successfully in the management of patients with Cushing's syndrome and then offer a practical approach to the selection and use of these agents.

KETOCONAZOLE

Ketoconazole, an imidazole derivative with antimycotic properties, inhibits several of the cytochrome P450 enzymes involved in cortisol biosynthesis including 11-beta hydroxylase, side-chain cleavage enzyme, and 17,20-lyase/17-alpha hydroxylase. It may also act as an antagonist of the glucocorticoid receptor and impair ACTH release from pituitary adenoma cells.(1-3) Reports of the efficacy of ketoconazole in patients with Cushing's syndrome first began to appear in the mid-1980's. (1) Since then, numerous reports have documented that ketoconazole is effective in the long-term management of patients with hypercortisolism. (1, 4-10) Many now consider ketoconazole to be the first-line pharmacologic agent in patients with refractory hypercortisolism.

In a review of the literature to 1993, Engelhardt and Weber found that ketoconazole had normalized urine cortisol levels in 57 of 82 patients (70%) with Cushing's disease. (9) Further, they reported normalization of urine cortisol levels in four of 9 patients with ectopic ACTH hypersecretion, all seven patients with adrenal adenoma, all five patients with adrenal hyperplasia, and only one of five patients with adrenocortical carcinoma. Subsequent reports have documented that ketoconazole normalizes urine cortisol levels in only one-half of patients with the syndrome of ectopic ACTH hypersecretion. (10) In most patients who respond, urinary cortisol levels decrease to normal or near normal within the first week of therapy. Clinical features attributed to hypercortisolism usually improve during prolonged therapy. Effective doses of ketoconazole range from 400 to 1200 mg/d. Liver toxicity, ranging from an asymptomatic elevation in serum transaminases to overt clinical hepatitis, occurs in 15% of patients. Other side-effects include skin rash, nausea, dyspepsia, gynecomastia, hypogonadism, and hypocalcemia. (7,9,10)

AMINOGLUTETHIMIDE

The anti-seizure drug aminoglutethimide has been used since the mid-1960's to treat Cushing's syndrome. (11,12) Aminoglutethimide inhibits the side chain cleavage, 21-hydroxylase, 17-alpha hydroxylase/ C17-20 lyase, 11-beta hydroxylase, aromatase, and 18-hydroxylase enzymes. Nearly one-half of treated patients enjoy normalization of cortisol levels in response to treatment.(11-13) In their review of the literature of 66 patients treated with aminoglutethimide, Misbin et al noted clinical and biochemical improvement in 13 of 21 patients with metastatic adrenocortical carcinoma, four of 6 patients with ectopic ACTH production, all six patients with adrenocortical adenomas, and 14 of 33 patients with Cushing's disease. (12) Effective doses typically range from 250 mg/d to 2000 mg/d. Common side-effects include sedation, anorexia, nausea, and rash. Sedation, observed in 30% of patients, typically improves with continued therapy. (12)

METYRAPONE

Metyrapone, a pyridine derivative, inhibits 11-hydroxylase at low concentrations and, in high concentrations, inhibits side chain cleavage enzyme. First introduced in the late-1950's, this agent has been employed in the medical management of patients with Cushing's syndrome for more than 30 years. In a study by Verhelst et al, three-quarters of patients treated for 1-16 weeks normalize plasma cortisol levels in response to treatment.(14) Twenty-four patients with Cushing's disease received metyrapone and pituitary irradiation. The median duration of treatment was 27 months. Successful control of hypercortisolism was documented in 20 of 24 patients (83%). The investigators noted control of cortisol levels in 13 of 18 patients

(70%) with ectopic ACTH hypersecretion and in 13 of 16 patients with adrenocortical tumors. (14) Typical doses range from 750 mg/d to 2000 mg/d given in two to three divided doses. Side effects include hirsutism, dizziness, lethargy, ataxia, edema, hypokalemia, nausea, and rash. The hirsutism, edema and hypokalemia are due to 11-hydroxylase inhibition and subsequent accumulations of adrenal androgens and mineralocorticoids, respectively.

MITOTANE

Mitotane, o,p'-DDD (a congener of the insecticide DDT) has been used since the 1960's to treat Cushing's syndrome. (15) This drug inhibits the side chain cleavage and 11-hydroxylase enzymes. Its metabolites disrupt mitochondrial function in the zonae fasciculata and reticularis ultimately leading to mitochondrial destruction and adrenocortical cell death. (13) The usefulness of this adrenolytic agent in the management of adrenocortical cancer is discussed in the chapter on adrenal Cushing's syndrome. More than four-fifths of patients with Cushing's disease achieve biochemical remission of their hypercortisolism during therapy. Long-term remissions are, however, evident in only 30% of patients following discontinuation of therapy. Higher response rates are seen in patients with Cushing's disease who receive concomitant pituitary irradiation.(15) Early reports employed doses of 6-12 grams/d for 6-8 months. Administration of doses ranging from 2 to 4 g/d may be just as effective and also limit toxicity. Side-effects include gastrointestinal discomfort, fatigue, weakness, dizziness, gynecomastia, rash, elevation in serum transaminases and alkaline phosphatase levels, hypercholesterolemia, and hypouricemia. Adrenal insufficiency may occur and usually necessitates replacement therapy. Hydrocortisone is recommended for replacement as mitotane accelerates the metabolism of halogenated steroids such as dexamethasone. Mineralocorticoid therapy is not usually required as mitotane does not affect the zona glomerulosa.

ETOMIDATE

Etomidate, also an imidazole derivative, is perhaps the most potent inhibitor of 11-beta hydroxylase Its use is restricted to hospitalized patients. The usual dose is 0.3 mg/kg/h. Patients must be monitored for excessive sedation and adrenal insufficiency. (16)

MIFEPRISTONE

Mifepristone (RU-486) is a progesterone and glucocorticoid antagonist that is FDA-approved to terminate pregnancy. It may be used to reverse psychosis, major vegetative depression, and suicidal ideation in patients with hypercortisolism. Effective doses range form 5-20 mg/kg/d. (13,17) ACTH and cortisol levels increase in patients with tumors that have functional glucocorticoid receptors due to loss of negative feedback. Side-effects include relative adrenal insufficiency, hypoglycemia, and eosinophillia. Treatment of symptoms of adrenal insufficiency is difficult as it is impossible to determine the degree of receptor blockade and the dose of glucocorticoid replacement needed to compete at the receptor level with mifepristone.

OCTREOTIDE

Octreotide, a long-acting somatostatin analog, inhibits ACTH in patients with adrenocortical insufficiency. (18) This observation has prompted a number of investigators to employ octreotide in the management of patients with ACTH-dependent forms of Cushing's syndrome. (19-21) Normalization of cortisol levels can be expected in only a minority of those with Cushing's disease. Approximately one-third of patients with ectopic ACTH hypersecretion respond to treatment.

OTHER AGENTS

The use of neuromodulatory agents, including bromocriptine, cyproeptadine, and valproic acid in the management of patients with ACTH-dependent forms of hypercortisolism has been reviewed elsewhere. (13) In our experience, beneficial responses to treatment with these agents are short-lived. We do not routinely employ these drugs in the management of our patients.

A RATIONAL APPROACH

The use of pharmacologic agents in the management of patients with hypercortisolism is largely a matter of personal preference. We employ these agents in patients in whom other measures to control hypercortisolism have failed. These tend to be patients with Cushing's disease who failed transsphenoidal surgery and are undergoing radiotherapy, patients with severe hypercortisolism who are in need of medical and physical rehabilitation prior to initial surgery, patients with metastatic or residual adrenocortical carcinoma, and patients with the syndrome of ectopic ACTH hypersecretion due to metastatic or untreatable disease.

We usually first employ ketoconazole to inhibit cortisol biosynthesis. Patients are started on 200 mg twice daily. A 24h urine free cortisol one week after the initiation of therapy is used to determine if any dose adjustments are needed. If patients remain hypercortisolemic, the dose of ketoconazole is increased to 400 mg twice daily and increased in a stepwise fashion to a maximum of 400 mg four times daily as needed in attempt to normalize the urine cortisol excretion rate. We check serum transaminase levels at monthly intervals for three months and then every three months thereafter. Ketoconazole is discontinued whenever the serum transaminases exceed a value three times the upper limit of normal.

Aminoglutethimide is our preferred drug when ketoconazole is ineffective or must be discontinued. We also select this drug to use in combination wth ketoconazole in those instances when a partial response is obtained. We initiate therapy with 125 mg twice daily and increase the dose at weekly intervals based on urine cortisol levels. We have had great success with aminoglutethimide in doses ranging 250 mg/d to 750 mg/d in divided doses. Slowly ramping up the dose of the drug seems to limit the sedation experienced by patients.

Occasionally, patients fail to respond to either ketoconazole or aminoglutethimide alone or in combination. Under these circumstances, we employ metyrapone, usually starting with a dose of 250 mg three times daily. We have used this drug successfully alone, in combination with aminoglutethimide, and also in combination with ketoconazole. One of our patients with metastatic medullary thyroid carcinoma required 750 mg of metyrapone four times daily to normalize her urine free cortisol excretion rate. Again, we typically asses urine cortisol levels at weekly to bi-weekly intervals and alter drug dosages as needed.

We reserve the use of mitotane for patients with adrenocortical carcinoma and those with ACTH-dependent forms of Cushing's disease who prefer a medical adrenalectomy to surgical excision of their adrenal glands. We usually initiate treatment with 500 mg at bedtime. The dose is escalated as tolerated at weekly intervals by 500 mg/week until a daily dose of 4g is reached. After three months, and especially when patients begin to note more side effects, we lower the dose to 2 g daily. We note better tolerance of the medication when 50% of the daily dose is administered at bedtime and the remainder of the daily dose is divided and administered at breakfast and either lunch or the evening meal. Provided that normalization of the urine cortisol excretion rate had been achieved, in patients with ACTH dependent forms of hypercortisolism we withdraw therapy after 6 to 9 months of treatment to reasses need for continued therapy.

We employ octreotide in all patients with ectopic ACTH hypersecretion when measures directed against the primary tumor have been unsuccessful. Most of these patients ultimately proceed to bilateral adrenalectomy. We initiate therapy with 50 mcg three times daily and increase the dose to a maximum of 500 mcg three times daily if needed. Responders are ultimately treated with the LAR preparation of octreotide and the dose is adjusted based on need.

We usually withhold adrenal replacement therapy until successfully treated patients manifest urine free cortisol excretion rates in the lower half of the normal range, serum cortisol levels < 5 mcg/dL, or develop symptoms and signs of steroid withdrawal or cortisol deficiency. We prefer adrenal replacement with dexamethasone 0.5 mg at bedtime for our patients unless they are treated with mitotane in which case we instruct the patients to take hydrocortisone 20 mg each morning and 5 to 10 mg in the late afternoon.

Metryapone and aminogluthimide are probably safe to use in patients with Cushing's syndrome during pregnancy. Ketoconazole is teratogenic and must be avoided. Mitotane will result in destruction of the fetal adrenal glands and should also be avoided during pregnancy. Mifepristone is an abortifacient and should not be used during the pregnancy if a successful outcome is desired.

REFERENCES

1. Loli P, Berselli ME, Tagliaferri M. Use of ketoconazole in the treatment of Cushing's syndrome. J clin Endocrinol Metab 63:1365-1371,1986.
2. Loose DS, Kan PB, Hirst MA, Marcus RA, Feldman D. Ketoconazole blocks adrenal steroidogenesis by inhibiting cytochrome P450-dependent enzymes. J Clin Invest 71:1495, 1983.
3. Berwaerts JJ, Verhelst JA, Verhaert GC, Verhaegen AA, Abs RE. Corticotropin-dependent Cushing's syndrome in older people: Presentation of five cases and therapeutical use of ketoconazole. J Am Geriatr Soc 46:881-885, 1998.
4. Angeli A, Frairia R. Ketoconazole therapy in Cushing's disease. Lancet 1:821, 1985.
5. Sonino N, Boscaro M, Merola G, Mantero F. Prolonged treatment of Cushing's disease by ketoconazole. J Clin endocrinol Metab 61:718 1985.
6. McCance DR, Hadden DR, Kennedy L, Sheridan B, Atkinson AB. Clinical experience with ketoconazole as a therapy for patients with Cushing's syndrome. Clin Endocrinol 27:593-599, 1987.
7. Sonino N, Boscaro M, Paoletta A, Mantero F, Ziliotto D. Ketoconazole treatment in Cushing's syndrome: Experience in 34 patients. Clin Endocrinol 35:347-352, 1991.

8. Tabarin A, Navarranne A, Guerin J, Corcuff JB, Parneix M, Roger P. Use of ketoconazole in the treatment of Cushing's disease and ectopic ACTH syndrome. Clin Endocrinol 34:63-69, 1991.
9. Engelhardt D, Weber MM. Therapy of Cushing's syndrome with steroid biosynthesis inhibitors. J Steroid Biochem 49:261-267, 1994.
10. Winquist EW, Laskey J, Crump M, Khamsi F, Shepherd FA. Ketoconazole in the management of paraneoplastic Cushing's syndrome secondary to ectopic adrenocorticotropin production. J Clin Oncol 13:157-164, 1995.
11. Fishman LM, Liddle GW, Island DP, Fleischer N, Kuchel O. Effects of aminoglutethimide on adrenal function in man. J Clin Endocrinol Metab 27:481-490,1967.
12. Misbin RI, Canary J, Willard D. Aminoglutethimide in the treatment of Cushing's syndrome. J Clin Pharmacol 16:645-651,1976.
13. Miller JW, Crapo L. The medical treatment of Cushing's syndrome. Endocr Rev 14:443-458,193.
14. Verhelst JA, Trainer PJ, Howlett TA, Perry L, Rees LH, Grossman AB, Wass AH, Besser GM. Short and long-term responses to metyrapone in the medical management of 91 patients with Cushing's syndrome. Clin Endocrinol 35:169-178,1991.
15. Luton JP, Mahoudeau JA, Bouchard P, Thieblot P, Hautecouverture M, Simon D, Laudat MH, Tourtou Y, Bricare H. Treatment of Cushing's disease by o,p'DDD. Survey of 62 cases. N Engl J Med 300:459-464,1979.
16. Schulte HM, Benker G, Reinwein D, et al. Infusion of low-dose etomidate: Correction of hypercortisolemia in patients with Cushing's syndrome and dose-response relationship in normal subjects. J Clin endocrinol Metab 70:1426- ,1990.
17. Bertagna X, Bertagna C, Laudat MH, et al. Pituitary-adrenal response to the antiglucocorticoid action of RU 586 in Cushing's syndrome. J Clin endocrinol Metab 63:639- ,1986.
18. Fehm HL, Voigt KH, Lang R, Beinert KE, Raptis S, Pfeiffer EF. Somatostatin: A potent inhibitor of ACTH-hypersecretion in adrenal insufficiency. Klin Wochenschr 54:173-175,1976.
19. Invitti C, De Martin M, Brunani A, Piloni M, Cavagnini F. Treatment of Cushing's syndrome with the long-acting somatostatin analogue SMS 201-995 (Sandostatin). Clin endocrinol 32:275-281,1990.
20. Woodhouse NJY, Dagogo-Jack S, Ahmed M, Judzewitsch R. Acute and long-term effects of octrteotide in patients with ACTH-dependent Cushing's syndrome. Am J Med 95:305-38,1993.
21. Vignati F, Loli P. Additive effect of ketoconazole and octreotide in the treatment of severe adrenocorticotropin-dependent hypercortisolism. J Clin Endocrinol Metab 81:2885-2890,1996.